VOLUME NINETY EIGHT

ADVANCES IN
BOTANICAL RESEARCH

Plant Metabolomics in Full Swing

ADVANCES IN BOTANICAL RESEARCH

Series Editor

Jean-Pierre Jacquot
Membre de l'Institut Universitaire de France, Unité Mixte de Recherche INRA, UHP 1136 "Interaction Arbres Microorganismes", Université de Lorraine, Faculté des Sciences, Vandoeuvre, France

VOLUME NINETY EIGHT

ADVANCES IN BOTANICAL RESEARCH

Plant Metabolomics in Full Swing

Volume Editors

PIERRE PÉTRIACQ

UMR 1332 Biologie du Fruit et Pathologie, Université de Bordeaux, INRAE, Centre INRAE Bordeaux Nouvelle Aquitaine, Plateforme Bordeaux Metabolome, MetaboHUB, PHENOME, Villenave d'Ornon, France

ALAIN BOUCHEREAU

Institut de Génétique, Environnement et Protection des Plantes (IGEPP); Plateforme de Profilage Métabolique et de Métabolomique (P2M2); UMR 1349, INRAE-Institut Agro-Univ. Rennes 1; Rennes-Le Rheu, France

ELSEVIER

ACADEMIC PRESS

An imprint of Elsevier

Academic Press is an imprint of Elsevier
125 London Wall, London, EC2Y 5AS, United Kingdom
The Boulevard, Langford Lane, Kidlington, Oxford OX5 1GB, United Kingdom
50 Hampshire Street, 5th Floor, Cambridge, MA 02139, United States
525 B Street, Suite 1650, San Diego, CA 92101, United States

First edition 2021

Notices

Knowledge and best practice in this field are constantly changing. As new research and experience broaden our understanding, changes in research methods, professional practices, or medical treatment may become necessary.

Practitioners and researchers must always rely on their own experience and knowledge in evaluating and using any information, methods, compounds, or experiments described herein. In using such information or methods they should be mindful of their own safety and the safety of others, including parties for whom they have a professional responsibility.

To the fullest extent of the law, neither the Publisher nor the authors, contributors, or editors, assume any liability for any injury and/or damage to persons or property as a matter of products liability, negligence or otherwise, or from any use or operation of any methods, products, instructions, or ideas contained in the material herein.

ISBN: 978-0-12-821688-0
ISSN: 0065-2296

For information on all Academic Press publications
visit our website at https://www.elsevier.com/books-and-journals

Publisher: Zoe Kruze
Acquisitions Editor: Sam Mahfoudh
Editorial Project Manager: Leticia M. Lima
Production Project Manager: James Selvam
Cover Designer: Victoria Pearson

Typeset by SPi Global, India

Contents

Section 2
Metabolomics and plant-environment interactions

Contributors

J. William Allwood
Environmental and Biochemical Sciences, The James Hutton Institute, Dundee, Scotland, United Kingdom

Wagner L. Araújo
Departamento de Biologia Vegetal, Universidade Federal de Viçosa, Viçosa, Minas Gerais, Brazil

Caroline Baroukh
LIPM, Université de Toulouse, INRAE, CNRS, Castanet-Tolosan, France

Rawen Ben Malek
Institut Jean-Pierre Bourgin, Université Paris-Saclay, INRAE, AgroParisTech, Versailles, France

Cédric Bertrand
PSL Research University: EPHE-UPVD-CNRS, USR 3278. CRIOBE, Université de Perpignan; Société Akinao, Perpignan Cedex, France

Li Chen
Institute for Plant Sciences, Cluster of Excellence on Plant Sciences (CEPLAS), University of Cologne, Cologne, Germany

Massimiliano Corso
Institut Jean-Pierre Bourgin, Université Paris-Saclay, INRAE, AgroParisTech, Versailles, France

Olivier Fernandez
EA4707 Laboratoire Résistance Induite et Bioprotection des Plantes, Univ. Reims Champagne-Ardenne, Reims Cedex 2, France

Clément Frainay
Toxalim (Research Center in Food Toxicology), Université de Toulouse, INRAE, ENVT, INP-Purpan, UPS, Toulouse, France

Jordi Gamir
Metabolic Integration and Cell Signaling Laboratory, Plant Physiology Section, Dept Ciencias Agrarias y del Medio Natural, Universitat Jaume I, Castellón, Spain

Léo Gerlin
LIPM, Université de Toulouse, INRAE, CNRS, Castanet-Tolosan, France

Yves Gibon
UMR1332, Université de Bordeaux, INRAE, Bordeaux Metabolome, MetaboHUB, PHENOME, Villenave d'Ornon, France

Azucena Gonzalez-Coloma
Instituto de Ciencias Agrarias, CSIC, Madrid, Spain

Antoine Gravot
IGEPP, AGROCAMPUS OUEST, INRAE, Univ Rennes, Le Rheu, France

Maxime R. Hervé
IGEPP, INRAE, Institut Agro, Univ Rennes, Rennes, France

Richard P. Jacoby
Institute for Plant Sciences, Cluster of Excellence on Plant Sciences (CEPLAS), University of Cologne, Cologne, Germany

Fabien Jourdan
Toxalim (Research Center in Food Toxicology), Université de Toulouse, INRAE, ENVT, INP-Purpan, UPS, Toulouse, France

Stanislav Kopriva
Institute for Plant Sciences, Cluster of Excellence on Plant Sciences (CEPLAS), University of Cologne, Cologne, Germany

Jenna Krumbach
Institute for Plant Sciences, Cluster of Excellence on Plant Sciences (CEPLAS), University of Cologne, Cologne, Germany

Loïc Lepiniec
Institut Jean-Pierre Bourgin, Université Paris-Saclay, INRAE, AgroParisTech, Versailles, France

Grégoire Loupit
EGFV, Univ. Bordeaux, Bordeaux Sciences Agro, INRAE, ISVV, Villenave d'Ornon, France

Emilie J. Millet
Department of Plant Sciences, Mathematical and Statistical Methods—Biometris, Wageningen, The Netherlands

Annick Moing
UMR1332, Université de Bordeaux, INRAE, Bordeaux Metabolome, MetaboHUB, PHENOME, Villenave d'Ornon, France

Grégory Mouille
Institut Jean-Pierre Bourgin, Université Paris-Saclay, INRAE, AgroParisTech, Versailles, France

Sonia Osorio
Department of Molecular Biology and Biochemistry, Instituto de Hortofruticultura Subtropical y Mediterránea "La Mayora", University of Málaga–Consejo Superior de Investigaciones Científicas (IHSM-UMA-CSIC), Málaga, Spain

François Perreau
Institut Jean-Pierre Bourgin, Université Paris-Saclay, INRAE, AgroParisTech, Versailles, France

Kristian Peters
Bioinformatics and Scientific Data, Leibniz Institute of Plant Biochemistry, Halle (Saale), Germany

Pierre Pétriacq
UMR1332, Université de Bordeaux, INRAE, Bordeaux Metabolome, MetaboHUB, PHENOME, Villenave d'Ornon, France

Sylvain Prigent
UMR1332, Université de Bordeaux, INRAE, Bordeaux Metabolome, MetaboHUB, PHENOME, Villenave d'Ornon, France

Claire Prigent-Combaret
Université Lyon, Université Claude Bernard Lyon 1, CNRS, INRA, VetAgro Sup, UMR5557 Ecologie Microbienne, Villeurbanne, France

Loïc Rajjou
Institut Jean-Pierre Bourgin, Université Paris-Saclay, INRAE, AgroParisTech, Versailles, France

Tristan Richard
Bordeaux Metabolome Facility, Unité de recherche Oenologie, Axe Molécules d'Intérêt Biologique, EA 4577, USC 1366 INRAE, Université de Bordeaux, ISVV, Villenave d'Ornon, France

Renaud Rincent
UMR 1095 Génétique, Diversité et Ecophysiologie des Céréales, INRAE, Clermont-Ferrand, France

Melina Schwier
Institute for Plant Sciences, Cluster of Excellence on Plant Sciences (CEPLAS), University of Cologne, Cologne, Germany

Maria Sorokina
Institute for Inorganic and Analytical Chemistry, Friedrich Schiller University, Jena, Germany

Henriette Uthe
German Centre for Integrative Biodiversity Research (iDiv) Halle-Jena-Leipzig; Institute of Biodiversity, Friedrich Schiller University, Jena, Germany

José G. Vallarino
Max-Planck-Institut für Molekulare Pflanzenphysiologie, Potsdam-Golm, Germany

Josep Valls Fonayet
Bordeaux Metabolome Facility, Unité de recherche Oenologie, Axe Molécules d'Intérêt Biologique, EA 4577, USC 1366 INRAE, Université de Bordeaux, ISVV, Villenave d'Ornon, France

Nicole M. van Dam
German Centre for Integrative Biodiversity Research (iDiv) Halle-Jena-Leipzig; Institute of Biodiversity, Friedrich Schiller University, Jena, Germany

Heather J. Walker
biOMICS Facility, Department of Animal and Plant Sciences, University of Sheffield, Sheffield, United Kingdom

Alexander Weinhold
German Centre for Integrative Biodiversity Research (iDiv) Halle-Jena-Leipzig; Institute of Biodiversity, Friedrich Schiller University, Jena, Germany

Alex Williams
School of Earth and Environmental Sciences, The University of Manchester, Manchester, United Kingdom

Preface

In the last two decades, the advent of metabolomics as a postgenomic and a systems biology strategy has greatly contributed to rapidly enhancing our knowledge of complex biological and metabolic machinery such as plants. As an integrative, descriptive, and functional methodology, metabolomics gathers insight from the different omics scales including the genome, epigenome, transcriptome, proteome, and enzyme activome. In this sense, the metabolome (i.e., the complete set of low-molecular-weight compounds) embodies not only the eventual phenotype influenced by gene expression and the alteration of protein function from genetic and environmental origins, but also the feedback and response that these cellular processes constantly undergo.

Metabolomics refers nowadays to both untargeted and targeted approaches that are used to describe metabolomes through global pictures and precise concentrations of metabolites, respectively, thereby characterizing organismal chemistries. As a follow-up of volume 67 *Metabolomics coming of age with its technological diversity* published in *Advances in Botanical Research*, the present volume aims at reviewing some of the current major contributions of metabolomics to plant research. Through 10 chapters contributed by leading international researchers, this volume of *Advances in Botanical Research* showcases novel biological outputs that are divided into three sections: *Metabolomics and plant development*, *Metabolomics and plant-environment interactions*, and *Metabolomics applications in plant science*. Insofar as the emerging issues of climate change, ecological, food, and energy transition will have a major impact on our vision of plant systems and their metabolic properties, metabolomics applied to plants has a very bright future.

Finally, we would like to acknowledge the contributors of these book chapters for their enthusiastic work despite the COVID19 pandemic, as well as the reviewers for their time to ensure high scientific quality of this volume. We hope that the readers find something within the diverse chapters that entices their curiosity and feeds their interest. We would also like to thank our colleagues at Elsevier for handling this volume.

Alain Bouchereau
Pierre Pétriacq

SECTION 1

Metabolomics and plant development

CHAPTER ONE

Developmental metabolomics to decipher and improve fleshy fruit quality

J. William Allwood[a], Yves Gibon[b], Sonia Osorio[c], Wagner L. Araújo[d], José G. Vallarino[e], Pierre Pétriacq[b], and Annick Moing[b,*]
[a]Environmental and Biochemical Sciences, The James Hutton Institute, Dundee, Scotland, United Kingdom
[b]UMR1332, Université de Bordeaux, INRAE, Bordeaux Metabolome, MetaboHUB, PHENOME, Villenave d'Ornon, France
[c]Department of Molecular Biology and Biochemistry, Instituto de Hortofruticultura Subtropical y Mediterránea "La Mayora", University of Málaga–Consejo Superior de Investigaciones Científicas (IHSM-UMA-CSIC), Málaga, Spain
[d]Departamento de Biologia Vegetal, Universidade Federal de Viçosa, Viçosa, Minas Gerais, Brazil
[e]Max-Planck-Institut für Molekulare Pflanzenphysiologie, Potsdam-Golm, Germany
*Corresponding author: e-mail address: annick.moing@inrae.fr

Contents

Advances in Botanical Research, Volume 98
ISSN 0065-2296
https://doi.org/10.1016/bs.abr.2020.09.016

Abstract

Fruit quality, that remains crucial to human nutrition and health, is closely related to biochemical composition, mostly, although not only, due to low molecular-weight metabolites. Final fruit quality results from coordinated physiological processes during development from anthesis to growth and ripe stages. Fruit development and quality may be modulated in response to (a)biotic stress, as well as through compositional evolution postharvest. Biotic stress effects are not addressed here.

Comprehensive biochemical analyses of the metabolome have been used to describe and generate knowledge concerning fruit biochemical composition and metabolism, and their developmental and spatial changes before and after harvest. They have involved NMR- or MS-based fingerprinting, profiling or imaging strategies, possibly combined with targeted analyses (e.g., antioxidant metabolites). In this chapter, we first stress analytical aspects crucial for fruit metabolomics. Then, recent examples concerning metabolomics of temperate and tropical fleshy fruits (fresh, not processed), whether they are models or species of agronomical interest, are discussed in the context of possible impacts on fruit agricultural practices or breeding. These cases concern metabolism reprogramming during fruit set, development and postharvest, environmental effects, gene functional analysis, genetics and systems biology. Major current challenges of specific interest for fruit metabolomics are also discussed as examples of the problems and possibilities for this exciting research field.

1. Introduction

Fleshy fruits have high economic importance with ~1,200 Mt produced in 2017 (fao.org). Tomatoes, watermelons and bananas have a yearly production higher than 100 Mt. Fresh fleshy fruits are also crucial for human nutrition and health (Nile & Park, 2014; Pott, Osorio, & Vallarino, 2019; Rodriguez-Casado, 2016). Fruit quality includes organoleptic and nutritional aspects tightly related to their biochemical composition, especially low-molecular-weight organic compounds (metabolites, including sugars, organic acids, amino acids, phenolic compounds, isoprenoids, alkaloids), besides minerals, starch and cell walls. Such biochemical compositions have been tuned firstly through domestication, then by breeding and cultural practices, and are furthermore influenced by environmental parameters. The biochemical composition of ripe fruit results from complex metabolic changes during fruit development (Gillaspy, Ben-David, & Gruissem, 1993). Additional modifications also occur after harvest, during final ripening, storage and transport (Fortes, Granell, Pezzotti, & Bouzayen, 2017). Detailed measurements of metabolites during fruit development have been performed for decades using analyses targeted to given biochemical families. Recently, untargeted metabolomic approaches based on proton nuclear magnetic resonance (^{1}H NMR) or mass spectrometry (MS) are commonly used and combined with targeted approaches (de Vos, Hall, & Moing, 2011; D'Urso, Piacente, Pizza, & Montoro, 2018; Sobolev et al., 2015). Model fruit metabolomics, e.g., for tomato (de Vos et al., 2011), is increasingly used for a range of fruit species of agronomical interest (Stewart, Shepherd, Hall, & Fraser, 2018).

Fleshy fruits are complex organs of variable color, size and shape (van der Knaap & Østergaard, 2018), and composed of several tissues. Although the development of these tissues is coordinated, their composition may significantly differ. Therefore, tissular biochemical studies may be of interest using dissection or imaging. Ten years ago, Hanhineva and Aharoni (2010) reviewed the application of metabolomics to follow metabolism in transgenic plants, mutants or populations, and the corresponding outputs for the identification of genomic regions associated with metabolite traits. The authors stated that most metabolomic assays focused on tomato and strawberry, they also successfully predicted that metabolomics would later encompass numerous other species. A later review on tomato metabolomics at different developmental stages, environmental conditions, or following genetic perturbation (Tohge & Fernie, 2015), described how the approach contributed to improved understanding of the influence of primary metabolism on fruit growth and development, as well as structure and evolution of specialized metabolic pathways.

Focusing on recent studies analyzing several stages of fruit development, the present chapter addresses metabolism reprogramming during fruit set, development and postharvest, and compositional changes induced by abiotic effects (Fig. 1). We provide examples of gene functional analyses, genetic studies and systems biology approaches, based on or including metabolomic techniques. Finally, we discuss major current points of attention and challenges, for the analytical or scientific domains, of relevance to fruit metabolomics, and outline strategies to improve both qualitative and quantitative traits of fruit crops.

2. Analytical strategies for fruit metabolomics

NMR and MS (Emwas, 2015) have predominated within plant metabolomics. ^{1}H NMR spectroscopy is appropriate to analyze bulk metabolites, GC-MS to determinate volatile organic compounds (VOC's) and nonvolatile metabolites after derivatization, while LC-MS is applied to polar and nonpolar compounds including specialized metabolites of

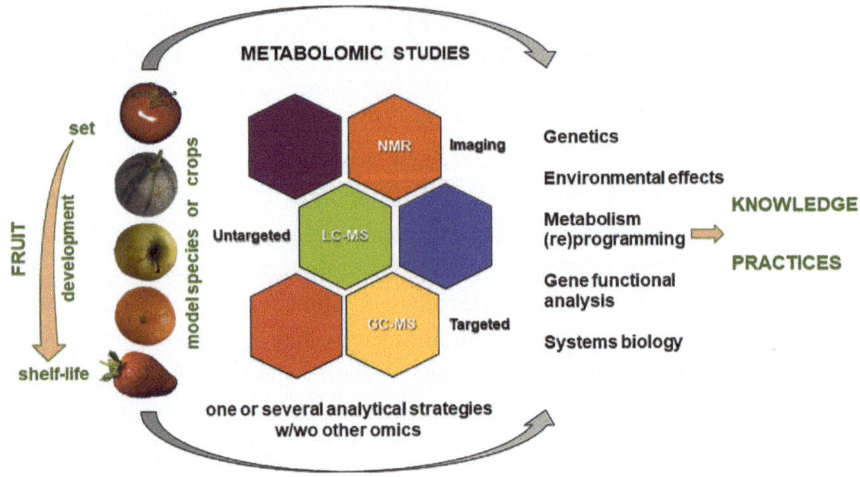

Fig. 1 Overview of the major outputs of fruit metabolome studies. These studies are performed at different stages of fruit development or after harvest, with one or several analytical strategies.

importance to fruit (Fig. 2). Therefore, combinations of multiple analytical techniques are commonly applied to fruit metabolism.

2.1 NMR-based strategies

NMR-based metabolomics has been applied across diverse scientific domains including plant science (Deborde et al., 2017; Markley et al., 2017; Nagana Gowda & Raftery, 2017). Although NMR is less sensitive than MS, it offers several major advantages as it is reproducible, nondestructive, nonspecific and quantitative. For fruits (Sobolev et al., 2015), 1D liquid-state ^{1}H NMR remains more commonly used than High Resolution Magic Angle Spinning NMR (HR-MAS). 2D homonuclear or heteronuclear NMR experiments are usually dedicated to structural identification of compounds of interest (Sobolev et al., 2015), often in combination with MS. For fruit samples collected throughout development or postharvest, 1D liquid-state NMR fingerprinting or profiling has been used to characterize soluble compounds of polar extracts of tomato (Lemaire-Chamley et al., 2019), melon (Moing et al., 2011) and kiwifruit (Capitani et al., 2010), semipolar extracts of grape berry (Ali et al., 2011) and banana (Yuan et al., 2017), and apolar extracts of juniper berries (Falasca et al., 2014). Several solvent systems can be applied, as performed in black raspberry (Kim et al., 2011), persimmon (Maulidiani et al., 2018) and juniper

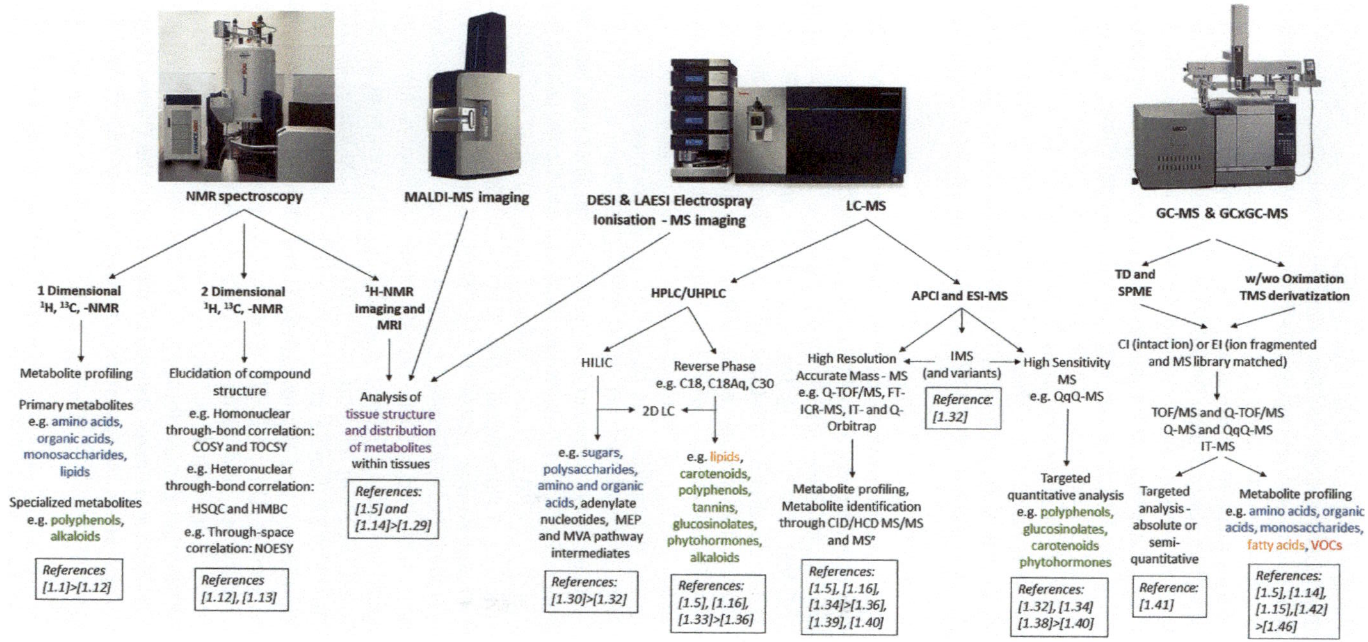

Fig. 2 See legend on next page.

Fig. 2 Analytical platforms applied in developmental metabolomics studies of fruit. A summary of the major analytical technologies and approaches. Compounds are colored according to their biochemical families: primary metabolites, blue; specialized metabolites, green; lipids, orange; volatiles, red. Nuclear Magnetic Resonance (NMR) spectroscopy: Correlation spectroscopy (COSY); Total correlation spectroscopy (TOCSY); Heteronuclear single-quantum correlation spectroscopy (HSQC); Heteronuclear multiple-bond correlation spectroscopy (HMBC); Nuclear Overhauser effect spectroscopy (NOESY); Magnetic resonance imaging (MRI). Liquid Chromatography (LC): High Performance (HP) and Ultra High Performance (UHP) Liquid Chromatography (LC); C18 Reverse Phase LC (C18); C18 aqueous end-capped Reverse Phase LC (C18Aq); C30 Reverse Phase LC (C30); Hydrophobic Interaction Liquid Chromatography (HILIC); Two Dimensional LC (2D-LC) utilizes two columns, most commonly reversed-phase C18 and HILIC; Gas Chromatography (GC): Thermal Desorption (TD); Solid Phase Micro Extraction (SPME); Chemical Ionization (CI); Electron Impact (EI); Single dimension Gas Chromatography (GC), Two-Dimensional Gas Chromatography (GCxGC); Volatile Organic Compounds (VOCs); Mass Spectrometry (MS): Atmospheric Pressure Chemical Ionization (APCI); Electro Spray Ionization (ESI); Ion mobility spectrometry (IMS); Fourier-transform ion cyclotron resonance mass spectrometry (FT-ICR-MS); Quadrupole Time of Flight Mass Spectrometry (Q-TOF/MS); Time of Flight Mass Spectrometry (TOF/MS); Single Quadrupole Mass Spectrometry (Q-MS); Triple Quadrupole Mass Spectrometry (QqQ-MS); Ion Trap (IT); Ion Trap (IT)-Orbitrap, e.g., LTQ-Orbitrap and Orbitrap-Fusion series; Quadrupole (Q)-Orbitrap, e.g., Q-Exactive series. References: [1.1] Deborde et al. (2017); [1.2] Markley et al. (2017); [1.3] Nagana Gowda and Raftery (2017); [1.4] Lemaire-Chamley et al. (2019); [1.5] Moing et al. (2011); [1.6] Capitani et al. (2010); [1.7] Ali et al. (2011); [1.8] Yuan et al. (2017); [1.9] Falasca et al. (2014); [1.10] Kim et al. (2011); [1.11] Maulidiani et al. (2018); [1.12] Sobolev et al. (2015); [1.13] Jézéquel et al. (2015); [1.14] Biais et al. (2009); [1.15] Biais et al. (2010); [1.16] Hanhineva et al. (2008); [1.17] Bhandari et al. (2015); [1.18] Enomoto, Sato, Miyamoto, Ohtsuka, and Yamane (2018); [1.19] Wiseman, Ifa, Song, and Cooks (2006); [1.20] Etalo, De Vos, Joosten, and Hall (2015); [1.21] Nakamura et al. (2017); [1.22] Veličković et al. (2014); [1.23] Srivastava, Talluri, Beebi, and Kumar (2018); [1.24] Banerjee, George, Bharathwaj, and Chandrakumar (2009); [1.25] Cheng, Wang, Chen, and Lin (2011); [1.26] Musse et al. (2009); [1.27] Ramadan, Lin, and Stanwell (2013); [1.28] Dona, Pagès, and Kuchel (2011); [1.29] Podda, Delli Castelli, Digilio, Gullino, & Aime, 2015; [1.30] Arapitsas and Mattivi (2018); [1.31] Yang et al. (2019); [1.32] Venter et al. (2018); [1.33] Okazaki and Saito (2018); [1.34] Allwood et al. (2019); [1.35] De Vos et al. (2007); [1.36] Redestig et al. (2018); [1.37] Šimura et al. (2018); [1.38] Carvalho et al. (2013); [1.39] Hövelmann, Hahn, Hübner, and Humpf (2019); [1.40] Allwood and Goodacre (2010); [1.41] Bassi and Spencer (1985); [1.42] Roessner-Tunali et al. (2003); [1.43] Lisec, Schauer, Kopka, Willmitzer, and Fernie (2006); [1.44] Allwood et al. (2014); [1.45] Uekane et al. (2017); [1.46] Aith et al. (2020). Image permissions: The image of the Bruker Ascend 800 NMR system and Bruker MALDI-TIMS-TOF flex are used with the permission of Bruker BioSpin and Bruker Daltonics. The image of the Thermo Dionex U3000 UHPLC-Orbitrap Fusion series MS system is used with the permission of ThermoFisher Scientific. The image of the LECO Pegasus BT GC-MS system is used with permission of LECO Corporation.

berry (Falasca et al., 2014). The control and adjustment of pH is important in fruit polar extracts, to limit uncontrolled chemical shifts (Deborde et al., 2017). With certain precautions, absolute quantification is possible with 1D (Deborde et al., 2017) or 2D (Jézéquel et al., 2015) NMR experiments.

2.2 MS-based strategies

GC-MS is commonly used to determine fruit polar metabolites (after derivatization, following early work on tomato (Roessner-Tunali et al., 2003)), or VOCs, for instance in melon (Allwood et al., 2014). Two ionization sources are applied, EI (Electron Impact) or CI (Chemical Ionization) (Allwood & Goodacre, 2010). Whilst Q-MS is more commonly applied with GC, TOF-MS is often favored since it provides higher sensitivity and scan speeds (Lisec et al., 2006). In addition, comprehensive two-dimensional GCxGC-MS is used to increase resolution, with rapid separations on the second column requiring high scan rates, thus TOF-MS is routinely applied. GCxGC-MS is extremely powerful for flavor analysis in fruit (Uekane et al., 2017) and fruit products (Aith et al., 2020).

For LC-MS analysis, reversed-phase (RP) C18 HPLC (Allwood & Goodacre, 2010) is commonly employed, being suited to a range of nonpolar lipid species (Okazaki & Saito, 2018) and fruit metabolites such as flavonoids (e.g., soft fruits) and alkaloids (e.g., tomato) (Allwood et al., 2019; De Vos et al., 2007). RP C30 HPLC is suited to the analysis of fruit tetraterpenoid carotenoids (Amorim-Carrilho, Cepeda, Fente, & Regal, 2014). Hydrophobic interaction liquid chromatography (HILIC) is suited to polar metabolite separations and has been applied to grapes/wines (Arapitsas et al., 2016) and blackcurrants (Yang et al., 2019). HILIC and RP methods can be serially coupled, thus increasing resolution (Venter et al., 2018). Several manufacturers have coupled ion mobility to MS, with promising applications on flavonoids, tannins and phenolic acids (Venter et al., 2018).

Recently, MS detectors (e.g., TOF, QqQ, IT and FT) have gone through an impressive series of developments. The ability to collect accurate mass MS^n information on intact metabolites and substructures provides a powerful tool for fruit metabolite identification (Allwood et al., 2019). Although significant progress has been achieved in computational techniques and data processing, proper usage and development of MS-based spectral databases for metabolite identification is still a hot topic (Tsugawa, 2018).

2.3 Targeted analyses

Although a range of analytical strategies can be used for targeted approaches in fruit, we focus here on those applying GC- or LC-MS for absolute quantification (with labeled standards), that are required for low concentration metabolites such as phytohormones. Quantitative UHPLC-QqQ-MS methods for cytokinins, auxins, brassinosteroids, gibberellins, jasmonates, salicylates and abscisates, are available (Šimura et al., 2018). Ethylene, of key importance in climacteric fruit ripening, requires targeted GC (Bassi & Spencer, 1985) or photo-acoustic laser spectrophotometry (Lloyd et al., 2011).

Fruits, especially citrus and berry fruits, contain a plethora of antioxidant metabolites like terpenoids and polyphenols, providing fruit color (e.g., anthocyanins and cyanidins, carotenoids) and conferring health benefits through scavenging or processing free radicals (Decros et al., 2019). Carotenoids demand targeted extraction (Saini & Keum, 2018) and RP C30 HPLC-MS (Amorim-Carrilho et al., 2014). The fruit polyphenols and alkaloids, although amenable to untargeted LC-MS, can be more accurately quantified by LC-QqQ-MS (Carvalho et al., 2013; Hövelmann et al., 2019).

2.4 Combined strategies

Greater depths of information can be gained by combining multiple analytical strategies covering wider ranges of metabolite classes. A multiplatform metabolomics approach was employed in tomato, with LC-MS providing coverage of lipids, phenolics and alkaloids, CE-MS of charged molecules (e.g., amino, organic and nucleic acids, water-soluble vitamins, cofactors) and GC-MS of polar metabolites (amino and organic acids, sugars) (Kusano et al., 2011; Redestig et al., 2018). In several studies of spatial metabolite changes in developing melon fruits, an unprecedented range of platforms were applied, including ^{1}H NMR and GC-MS of polar metabolites (Moing et al., 2011; Mori et al., 2019), SPME-GC-MS of volatiles, LC-MS and Inductively Coupled Plasma (ICP)-MS of mineral elements (Moing et al., 2011), as well as determination of enzyme activities (Mori et al., 2019).

For a mechanistic insight, combining multiple "omics" strategies is crucial. Systems level approaches have applied a combination of gene expression data (microarray or RNAseq) with primary metabolite data (concentration and flux), commonly obtained via ^{1}H NMR or GC-MS (Osorio et al., 2013; Vallarino et al., 2018), as well as enzyme activity data (Biais et al., 2014), to confirm links between gene expression and changes in metabolite levels and/or fluxes. Such approaches have increased our systematic understanding of tomato fruit development (Mounet et al., 2009; Pascual et al., 2013), identifying key regulatory genes (Mounet et al., 2009; Zanor et al., 2009) throughout tomato ripening (Osorio et al., 2013), and inheritance of traits such as fruit size (Pascual et al., 2013). Although many studies have focused on tomato, more recent studies have also involved strawberry (Vallarino et al., 2015, 2018) and tropical fruits (Wang et al., 2019).

2.5 Fruit spatial metabolomics

Fruits represent a wide diversity of tissue structures and therefore metabolism. Tissue sectioning, fixation and embedding with resinous materials causes undesirable biochemical changes, limiting its use to histology. Such limitations have favored spatial analyses of fruits and flowers applying classic dissection and acquisition of metabolite profiles on a section-by-section basis, allowing reconstruction of low-resolution metabolic maps (e.g., Biais et al., 2009, 2010; Moing et al., 2011) or tissue-specific metabolite profiles (e.g., Hanhineva et al., 2008; Lemaire-Chamley et al., 2019). Recently, laser capture microdissection (LCM) has permitted isolation of pure cell populations or tissues (Bhandari et al., 2015; Dong et al., 2016; Sumner et al., 2018).

MS imaging approaches are commonly based upon MALDI (Matrix Assisted Laser Desorption Ionization, (Enomoto et al., 2018)), DESI (Desorption Electrospray Ionization, (Wiseman et al., 2006)), Laser Ablation (LA, (Etalo et al., 2015)), and to a lesser extent, Direct Analysis in Real-Time (DART)- and Secondary Ion (SI)- MS. Since MS imaging approaches are direct sampling techniques without chromatography, they cannot differentiate isomeric compounds unless differential fragmentation spectra are obtained through MS/MS. Examples of MS imaging applications in fruits have included distribution analyses of anthocyanins, sugars and organic acids in strawberry (Enomoto et al., 2018; Enomoto, Takahashi, Takeda, & Hatta, 2019), imaging across developmental stages (Nakamura et al., 2017) and of cutin and suberin in tomato (Veličković et al., 2014).

Magnetic resonance can be used for localized spectroscopy or imaging based upon spectral attributes such as intensity, chemical shift, relaxation time and molecule movement, that can be measured in a spatially resolved manner in vivo, its use in fruit has been reviewed recently (Srivastava et al., 2018). Volume-localized spectroscopy has been applied to major soluble sugars and organic acids during citrus fruit ripening (Banerjee et al., 2009), and sugars and lycopene during tomato ripening (Cheng et al., 2011). Magnetic resonance imaging (MRI, based on proton signal intensity or relaxation times of water molecules) has been applied to follow fruit structural changes during development and ripening in tomato (Musse et al., 2009). Diffusion MRI, was used to study internal microstructures of grape tissues during development (Dean et al., 2014). Magnetic resonance spectroscopic imaging (MRSI) or 3D-MRSI, based on the collection of one NMR spectrum per pixel or voxel, as performed in vivo for human organs (Ramadan et al., 2013), has potential for fruit tissues also. MRI using the Chemical Exchange Saturation Transfer contrast enhancement technique (CEST-MRI) (Dona et al., 2011) is also promising for spectral imaging of fruit, as applied to asparagine distribution in plum (Podda et al., 2015).

3. Fruit metabolomics for fruit set, development, ripening and postharvest studies

3.1 Fruit set and early fruit growth

Metabolomic studies of fruit set or early development (cell division stage) remain rare compared to later stages of development and especially ripening. A study applying tomato transformants revealed that ovaries are more sensitive to metabolic intervention than fruitlets (Ru et al., 2017). A study of stenospermocarpy in grape showed regulation of sugar- and hormone-mediated pathways and repression of specialized metabolism (Domingos, Fino, Paulo, Oliveira, & Goulao, 2016), expanding knowledge of the molecular events occurring during inflorescence fruit set. During early fruit growth, a study on tomato highlighted regulatory genes (Mounet et al., 2009), and a recent review addressed Citrus fruit growth (Tadeo, Terol, Rodrigo, Licciardello, & Sadka, 2020) with classical and metabolomic approaches.

3.2 Fruit ripening

For many years breeding programs have focused on improving agronomic characteristics such as fruit size and sweetness, yield or resistance to stress. More recently, breeding has aimed to maintain or improve nutraceutical content and flavor. For this aim, comparative metabolomics has been applied to explore the natural/genetic variability of various wild, landrace/traditional and cultivated accessions, to identify new genes that contribute to flavor and neutraceutical quality. In strawberry, investigating variation in the primary and specialized metabolism of domesticated and wild accessions showed that *Fragaria* subspecies can be identified solely on their metabolic profiles, with these differences in metabolite associations serving to rank metabolites whose regulation has been modified during domestication (Vallarino et al., 2018).

Metabolomics has contributed much to our increased understanding of the molecular basis of fruit coloration involving isoprenoids (Fraser & Bramley, 2006) and anthocyanins (He et al., 2010). For instance, combined metabolomic and transcriptomic analyses in fig (*Ficus carica* L.) revealed that cyanidin 3-*O*-malonylhexoside increased in the purple-peeled cultivar compared to the green peeled (Wang, Cui, Vainstein, Chen, & Ma, 2017), with relevant transcription factors such as R2R3-MYB regulated at the mature stage (Wang et al., 2017). Expression of three UDP-glucose flavonoid 3-*O*-glucosyltransferases (UFGT) were positively associated with the accumulation of two anthocyanins (malvidin 3-*O*-glucose, delphinidin 3-*O*-glucose) in jujube (Zhang et al., 2020). Red color fading of "Red Bartlett" pear (*Pyrus communis*) was shown to result from reduced anthocyanin biosynthesis, increased anthocyanin degradation and suppression of transport (Wang, Du, et al., 2017). Seven flavonoid metabolites and six genes were identified as candidates involved in the pigmentation of all-red-fleshed and all-green-fleshed *Actinidia arguta* fruit (Li et al., 2018). In *Prunus mira*, a study focusing on ripening identified five anthocyanins associated with flesh coloration and provided information on how pigmentation development in flesh is controlled at the transcriptional level (Ying et al., 2019).

3.3 Postharvest

Long shelf-life is crucial for commercial fruits. Postharvest, fruit respiration affects primary metabolism, directly influencing sugar, organic acid, protein and fat levels, that act as major respiratory substrates during shelf-life. The main sugars (i.e., glucose, fructose and sucrose) display species-dependent behavior postharvest. While these sugars tend to decrease in tomato or blackberries (Kim et al., 2019; Osorio et al., 2020), in other fruits (kiwi, banana) they increase after starch hydrolysis (Salzano et al., 2019; Yuan et al., 2017). Previous studies suggested that sucrose is imported from photosynthetic tissue and transiently stored as starch. Recently, metabolite analysis on tomato fruit ripening on (naturally) or off (commercially) the vine, suggested that considerable sucrose delivery occurs beyond the time at which the abscission zone forms (Osorio et al., 2020). Minor sugar and sugar-alcohols (i.e., galactinol, trehalose, myoinositol, sorbitol, raffinose) seemingly play a remarkable role in alleviating negative effects of postharvest abiotic stress (Uarrota et al., 2019).

Amino acids are also affected by postharvest storage, since their catabolism can stabilize reductions in electron supply from the TCA cycle (Araújo et al., 2011). Levels of malate, one of the most abundant organic acids in fruit, influence shelf life (Centeno et al., 2011), as corroborated by metabolic analysis of recombinant inbred lines originating from the cultivated tomato *Solanum lycopersicum* and the wild-type *Solanum pimpinellifolium* (López et al., 2015). In a recent study (Osorio et al., 2020), several tomato shelf-life mutants revealed an association between malate levels and postharvest storage. Interestingly, a link was found between, decreases in malate, citrate, aconitate and fumarate, and loss of fruit quality, in postharvest pomelo (Sun et al., 2013).

Alongside postharvest changes in the fruit volatilome, targeted approaches have led to the characterization of several genes involved in branched-chain amino acid catabolism, fatty acid cleavage and ethanol fermentation, that are proposed to be responsible for increases in branched-chain esters ("fruity," "overripe" aroma), fatty acid-derived volatiles ("musty" notes) and ethanol (Tietel, Feldmesser, Lewinsohn, Fallik, & Porat, 2011). Further, the activation of anaerobic fermentative metabolism is important in the induction of "off-aroma" compounds, mainly ethanol, acetaldehyde and ethyl acetate (Wu et al., 2016). These compounds are described as key metabolites in fruit quality decline (Almenar, Hernández-Muñoz, Lagarón, Catalá, & Gavara, 2006).

4. Characterization of environmental effects

4.1 Seasonal variations and abiotic stresses

^{1}H NMR- and/or MS-based metabolomic profiling can be used to describe fruit ripening changes, across the season, as for tomato (Deborde et al., 2009), in different environments, as for melon (Bernillon et al., 2013), or different cultural practices, as for strawberry (Akhatou, González-Domínguez, & Fernández-Recamales, 2016). Studying the effects of abiotic stress across fruit development is rarer. Following proteins and metabolites at different ripening stages provided information on tomato fruit quality improvement when grown in saline substrates (Tang, Zhang, Gong, Yan, & Shi, 2020), where NaCl induced expression of beta-fructofuranosidase and lipoxygenase, increased the accumulation of alanine and the expression of branched-chain-amino-acid aminotransferase (a key enzyme involved in ester catalyzing synthesis), as well as inhibiting the expression of beta-galactosidase and endoglucanase, resulting in firmer fruits. In white grapes, combined transcriptome and metabolite profiling revealed that prolonged drought modulated the phenylpropanoid and terpenoid pathways (Savoi et al., 2016). Grape acclimation to excessive solar irradiance lead to repartitioning between major flavonoids (Reshef, Agam, & Fait, 2018), where increased solar exposure was associated with lower levels of anthocyanins and flavan-3-ols, and higher levels of flavonols. In blackcurrant, 18 metabolites that were responsive to different day length conditions were identified, including naringin glucoside (Allwood et al., 2019).

4.2 Global warming

Fruiting plants strictly regulate the production of photoassimilates and the source-sink response to changing environments (Osorio, Ruan, & Fernie, 2014; Ruan, 2014). Combining proteomic and metabolomic profiling revealed that high temperatures strongly impaired carbohydrate and energy metabolism, as well as altered key proteins driving the development and ripening of grapes (Lecourieux et al., 2020). In addition to primary metabolism, plants also modify specialized metabolites in response to high temperatures. Allwood et al. (2019) applied LC-MS metabolomics to study the effect of elevated growth temperature on blackcurrant fruits. The accumulation of the two most abundant anthocyanins in blackcurrant, delphinidin-3-*O*-rutinoside and cyanidin-3-*O*-rutinoside, was first elevated between 12 and 18°C, before being strongly reduced at the highest ripening temperature of 24°C. This result indicates that total anthocyanins will be reduced strongly under future climate scenarios. A range of organic acids including ascorbate and sugars including sucrose, were also observed to significantly decrease under elevated temperatures (Allwood et al., 2019).

5. Fruit metabolomics for gene functional analyses

Combined analyses of fruit metabolomes and transcriptomes have been largely applied to propose candidate genes or pathways involved in the regulation of metabolite content or key fruit traits, especially for "models" such as tomato (de Vos et al., 2011; Rothan, Diouf, & Causse, 2019). Recent examples over the past 2 years now include a range of fruit species of agronomical interest including temperate and tropical fruits:

e.g., peach (Wang, Zhang, Yang, & Yuan, 2018), actinidia (Commisso et al., 2019; Li et al., 2018), pumpkin (Huang et al., 2019), acerola (dos Santos et al., 2019), banana (Li et al., 2019) and litchi (Wang et al., 2019). Among major metabolites, the regulation of soluble sugars has been studied in tomato, with overexpression of the SlFgr gene, encoding a plasma membrane-localized glucose efflux transporter of the SWEET family, being shown to reduce glucose levels, which permits increased fructose (Shammai et al., 2018).

Redox metabolism is central to many physiological roles in fruit (reviewed in Decros, Baldet, et al., 2019). As such, functional analyses of redox biology have benefited from metabolomics. In tomato, some studies illustrate a critical role for the mitochondrial malate valve in transmitting redox status to the plastids, thereby impacting plastidial metabolic pathways (Centeno et al., 2011; Osorio et al., 2013). Analyses of pyridine nucleotides, integrated to RNAseq analyses and protein profiling, revealed regulation clusters of nucleotide synthesis and recycling during fruit development with specific links to mitochondrial contributions (Decros et al., 2019). Further, ascorbate metabolism is detrimental to maintain the redox poise during fruit development. A combination of omics data in tomato fruit, using RNAi lines for several ascorbate metabolic enzymes, pointed toward the influence of ascorbate redox state on the expression of genes involved in ribosomal function, cellular protein synthesis and stability (Stevens et al., 2018). In addition to major redox buffers, terpenoids and polyphenols contribute to antioxidant properties (Muñoz & Munné-Bosch, 2018). QTL dissection has been used to identify candidate genes underlying antioxidant content in tomato fruit (Calafiore et al., 2016).

Concerning specialized metabolites, a recent review on strawberry metabolomics (Haugeneder et al., 2018) shows examples of MYB transcription factors or genes involved in the biosynthesis of polyphenols. Regarding metabolites involved in color, aroma or taste, several multiomics-based analyses have allowed identification and functional characterization of genes expressed in fruit other than strawberry. For fruit color, two pink-fruit tomato mutants were characterized by combining metabolomics, gene expression analysis and in silico searches to identify potential target genes (Fernandez-Moreno et al., 2016), revealing that mutants failed to accumulate flavonol glycosides in their skin. Transcriptome analysis provided insights into the role of SlMYB12 in metabolism homeostasis, through the identification of potential candidate target genes of SlMYB12 activity. Concerning isoprenoids, integrated proteomic and metabolomic analyses in banana pulp on two varieties differing for carotenoid accumulation, suggested that besides carotenogenesis-associated genes, the glycolytic pathway supports the higher accumulation of carotenoids (Heng et al., 2019). DE-ETIOLATED1 downregulation in tomato induced widespread effects on diverse metabolic pathways (impacting carotenoids, tocopherol, phenylpropanoids, flavonoids, and anthocyanidins), although the main effect of DET1 is a loss of function in the coordinated activation of core metabolic processes in cell types amenable to plastid biogenesis. The TOMATO AGAMOUS-LIKE 1 gene implicated in the development and ripening of tomato was shown to regulate not only ripening, but also to affect the synthesis and levels of amino acids, organic acids, carotenoids and α-tomatine (Zhao, Yuan, Chen, Meng, & Fu, 2018).

For volatiles, a combination of omic approaches allowed identification of the NON-SMOKY GLYCOSYLTRANSFERASE1 gene responsible for the key tomato fruit (*Solanum lycopersicum*) aroma attribute, termed "smoky," related to phenylpropanoid volatiles released from their glycosylated precursors (Tikunov et al., 2013). For fruit taste, bitterness in Solanaceae is related to steroidal glycoalkaloids (SGA) (Cárdenas et al., 2019). Fine QTL mapping and functional characterization showed that 2-OXOGLUTARATE-DEPENDENT-DIOXYGENASE enzymes catalyze SGA-hydroxylation across various *Solanum* species and this enzyme prevents bitterness in ripe tomato fruit consumed today. Next-generation toolsets for functional genomics in plant sciences will certainly also allow derivation of improved metabolic models for fruit (Rai, Yamazaki, & Saito, 2019).

6. Fruit metabolomics for genetics and breeding

Although ripening and its regulation have been extensively studied in fruits with different characters, the application of metabolite profiling during crop breeding remains challenging (Fernandez et al., 2021). However, both the metabolic characterization and the understanding of its genetic basis in fruits, present themselves as an adequate tool allowing the potential manipulation of metabolite levels during fruit development and providing significant insights into their interaction with the environment (Obata et al., 2015; Omena-Garcia et al., 2019).

Several studies have identified genetic factors related to metabolite levels in fruits (Fridman, Carrari, Liu, Fernie, & Zamir, 2004), enabling the characterization and enhanced comprehension of different pathways in fruit metabolism. An example describing the potential for the combination of metabolic and genetic analyses was the identification and characterization of the QTL associated with total soluble solid content in tomato fruits, which contain the wild allele of Invertase5 (LIN5) (Fridman et al., 2004; Zanor et al., 2009). Overall, these studies have substantially enhanced our understanding, not only of primary metabolism in tomato fruits, but also its impact on fruit yield and quality with clear applications in tomato breeding programs.

Although fruits are our major source of health-related compounds such as antioxidants, domestication has seemingly reduced their concentrations, suggesting that there is a trade-off with growth and thus productivity (Gest, Gautier, & Stevens, 2012). For example, in wild relatives of cultivated tomato and kiwifruit, the ascorbate content is approximately 5 and 20 times higher than values observed in commercial cultivars (reviewed in Gest et al., 2012). Although the reasons behind such trade-offs remain rather unclear, domestication-related traits often arose with loss-of-function mutations in single loci (Doebley, Gaut, & Smith, 2006). Accordingly, it seems reasonable to assume that the introduction of domestication-related mutations into wild species is likely easier than the introduction of adaptation-related alleles into crops. In strong agreement, it has been recently demonstrated that, in tomato, loci targeted for selection of important traits such as fruit size and disease resistance, in some cases have culminated with direct effects on metabolic characteristics, as well as indirect consequences through the action of linked loci (Zhu et al., 2018). This multiomics approach of several hundred accessions of tomato revealed that such unintended and sometimes undesirable consequences likely affected not only tomato but also other important crops. It seems reasonable to suggest that these identified metabolic loci will provide useful targets for further metabolic engineering for fruit metabolite composition using state-of-the-art molecular breeding or gene editing technologies (e.g., as for genetic addition used to alter both architecture and flowering times in tomatoes (Kwon et al., 2020)).

The identification of key metabolites and pathways in specific fruits will likely accelerate the design and improvement of breeding programs. We anticipate that once the genomic bases are known for metabolites of interest, both crop breeding and biotechnological programs will be able to properly use this information to incorporate metabolic traits of interest and further improve fruit yield and/or quality, especially of plants grown under field conditions. Likely, this knowledge will also enable researchers to pursue novel avenues to increase our understanding of the complex networks governing metabolite function in different crop species. Furthermore, the high level of biological insight captured within such plant metabolite profiling studies clearly demonstrates the utility of this approach, particularly when combined with other "omics" (i.e., transcripts, proteins, enzymatic determinations of metabolite concentrations, and kinetic modeling). Genome-wide association studies (GWASs, e.g., for melon fruit quality traits (Zhao et al., 2019)) assisted by metabolomics techniques (mGWAS, e.g., for date sugars and organic acids (Hazzouri et al., 2019)) and metabolic quantitative trait loci (mQTLs, as for tomato flavor (Garbowicz et al., 2018) or strawberry vitamin C (Vallarino et al., 2019)) remain powerful tools to detect genetic variations linked with metabolic traits in fruit species. Epialleles may also have to be taken into account, as shown for vitamin E in tomato fruit (Quadrana et al., 2014). In summary, we provide evidence for the potential of metabolomics-assisted breeding for fruit crop improvement and its future applications in accelerating breeding, as previously discussed (Omena-Garcia et al., 2019; Razzaq, Sadia, Raza, Khalid Hameed, & Saleem, 2019).

7. Contribution to fruit systems biology

Systems Biology, studies the function of organisms in a holistic and no longer reductionist fashion (Edwards, 2017), with iterations between modeling and experimentation (Auffray, Imbeaud, Roux-Rouquié, & Hood, 2003), by which models are constructed to link environmental and/or biological variables to phenotypic traits of interest. Such models make it possible to identify biomarkers, better understand the underlying mechanisms and/or predict phenotypes. Thus, linking metabolism and yield (Riedelsheimer et al., 2012) will make it possible to identify biomarkers useful for the selection or design of enhanced-performance crops (Fernandez et al., 2016).

Certain approaches are based on correlation searches between potentially explanatory variables at level x and target variables at level y (e.g., D'Angelo et al., 2019). Several levels of organization can be used, as in tomato, for which the integration for several hundred accessions of genotypic, transcriptomic and metabolomic data, revealed how domestication and selection influenced metabolism, in particular the reduction of antinutritional properties involving five major loci (Zhu et al., 2018). Such modeling lends itself to nontargeted metabolomics and does not require absolute metabolite concentrations. Widening the analysis to lipid or even to cell-wall composition seems promising as they were recently found to be strongly linked to fruit relative growth rate (Roch et al., 2020). Data, even if semiquantitative, need to be interoperable, through the use of multiple internal standards (Sysi-Aho, Katajamaa, Yetukuri, & Orešič, 2007), "nonchanging" metabolites (De Livera et al., 2012) or "quality control" (QC) samples (Sangster, Major, Plumb, Wilson, & Wilson, 2006). Solutions providing interpretable data from experiments carried out at different times, on different machines or even different laboratories, will facilitate data mining (Buendia et al., 2019; Peng, Chen, Chen, & Li, 2014).

In other approaches that assemble reductionist knowledge, to extract interpretable information from complex systems, mathematical models are applied to describe, for example, properties of metabolic networks (structure, nature of reactions) or their components (enzymes capacities and properties). Knowing and understanding the distribution of fluxes in metabolic networks remains a major challenge, however, the use of isotope-labeled substrates has made it possible to obtain flux maps in various species and organs (e.g., Ratcliffe & Shachar-Hill, 2006), although it is very difficult to label fruit homogeneously. An alternative is the use of stoichiometric models, which by using simple linear equations, can describe metabolic networks up to the genome level and do not require detailed metabolomic data (Shi & Schwender, 2016). Their resolution (calculation of flux maps) requires estimations of the output fluxes from the modeled network and the use of so-called "objective" functions (e.g., minimum sum of fluxes, maximum growth) to limit the solution space (Orth, Thiele, & Palsson, 2010). Thanks to a relatively detailed and quantitative description of the biomass throughout tomato fruit development (Colombié et al., 2015), a better understanding of the climacteric crisis was obtained (Colombié et al., 2017). Although stoichiometric modeling has much to gain from quantitative metabolomics data regarding output fluxes (e.g., accumulated compounds, intra-tissue and intracellular localization), it can be combined with semiquantitative metabolomics in order to better understand flux behavior (Sajitz-Hermstein, Töpfer, Kleessen, Fernie, & Nikoloski, 2016), an approach that should be applied to fruit. Predicting concentrations in addition to fluxes represents another challenge for understanding metabolic networks. Kinetic models, capable of predicting both, but also of studying their control, are constructed using ordinary differential equations that describe the reactions involved (Rohwer, 2012). These models, even when describing relatively small metabolic networks, have many parameters that should be measured rather than fitted using dedicated algorithms (Tummler, Lubitz, Schelker, & Klipp, 2014). This involves measuring enzyme capacities, which vary along fruit development (Biais et al., 2014; Desnoues et al., 2014), or estimating them from quantitative proteomic data (Stitt & Gibon, 2014), while other parameters supposed more stable such as K_m can be found in literature (Beauvoit et al., 2018). When models are compartmentalized, estimates of subcellular volumes and carrier capacities are also required, although to date the latter is impossible to measure (Beauvoit et al., 2018). Quantitative measurements of metabolic intermediates are also required, in particular, to allow the fitting of parameters that cannot be measured. A kinetic model describing sucrose metabolism in tomato fruit (Beauvoit et al., 2014), later adapted to peach (Desnoues, Génard, Quilot-Turion, & Baldazzi, 2018), identified steps whose manipulation can influence sugar content, such as the retro-inhibition of vacuolar invertase and glucokinase or vacuolar transport of sugars. Strikingly, the tedious construction of kinetic models is already envisaged at the genome-scale (Fröhlich, Kaltenbacher, Theis, & Hasenauer, 2017), a wager which will undoubtedly require the integration of quantitative multiomic data (Jensen, Gudmundsson, & Herrgård, 2018). Modeling fruit metabolism will also benefit from considering environmental factors, known to influence fruit yield and quality, by integrating environmental variables with metabolomic data or process-based models with kinetic or stoichiometric models (Beauvoit et al., 2018).

8. Challenges and promises for fruitful fruity metabolomics

Advances in plant metabolomics have been particularly fruitful in recent times and have made it possible to profile a plethora of metabolites of interest and changes in their concentrations, which can provide mechanistic evidence for biochemical processes involved in fruit biology, as exemplified above. Current metabolomics tends to combine a range of techniques for greater coverage of the metabolome (Fig. 2), from single metabolite quantification to high-throughput methods producing thousands of metabolic profiles. This strategy will be systemically implemented in future studies to address more ambitious questions related to fruit biology. As stressed above, most works have focused on model crops, most particularly tomato, with respect to development and abiotic stress responses (e.g., recently (Lemaire-Chamley et al., 2019; Zhang et al., 2019; Zhao, Yuan, Chen, Fu, & Jiang, 2019)). Nonetheless, future research on various species like tropical fruits should inevitably provide greater insight into fruit metabolism

(Kaneria, Rakholiya, Marsonia, Dave, & Golakiya, 2018; Khakimov et al., 2016; Roch et al., 2019). Importantly, the improvement of high-throughput methods shall be key to reach this goal, either for untargeted profiling or targeted analyses of specific metabolites (Zampieri, Sekar, Zamboni, & Sauer, 2017; Zhu et al., 2018).

One challenge from metabolomics applications in the context of fruit studies is the difficulty to work with heterogeneous tissues (Lemaire-Chamley et al., 2019). Most studies focus on certain tissue types, such as the pericarp or skin and cuticles, or bulked tissue samples, which undoubtedly biases the precise depiction of fruit metabolites. A possible solution is given through spatial metabolomics although this approach possesses specific disadvantages (e.g., quantitation, spatial resolution, (Jorge, Mata, & António, 2016; Nakamura et al., 2017)).

Another major challenge for fruit metabolomics lies in metabolite annotation, more specifically for MS-based strategies (Viant, Kurland, Jones, & Dunn, 2017). As for other plant organs, fruit metabolomes are chemically diverse due to numerous specialized compounds including colored, flavored and odorous metabolites, gained throughout the plants evolution (Francini, Pintado, Manganaris, & Ferrante, 2020; Pott et al., 2019). Metabolite annotation remains central for all untargeted metabolomic studies (Allard, Genta-Jouve, & Wolfender, 2017; Chaleckis, Meister, Zhang, & Wheelock, 2019), most significantly for LC-MS-based profiling. Collective initiatives have emerged in plant sciences, which in turn provide promising tools that will help compute and annotate large MS datasets (Peters et al., 2018; Ruttkies, Neumann, & Posch, 2019; Stanstrup et al., 2019). It is reasonable to assume that such developments will benefit fruit metabolomics in the short term.

Acknowledgments

We thank Dr. Catherine Deborde and Dr. Guilhem Pagès for critical reading of the NMR sections, MetaboHUB (ANR-11-INBS-0010, France) project, RTI2018-099797-B-100 (Ministerio de Ciencia, Innovación y Universidades, Spain), the European Union H2020 GoodBerry Programme (grant number 679303), and Scottish government RESAS program, for partial financing.

References

Aith, B. J., Primieri, N. K., Souza-Silva, É. A., Camarão Telles Biasoto, A., Welke, J. E., & Alcaraz Zini, C. (2020). Volatile profile and aroma potential of tropical Syrah wines elaborated in different maturation and maceration times using comprehensive two-dimensional gas chromatography and olfactometry. *Food Chemistry*, *308*, 125552. https://doi.org/10.1016/j.foodchem.2019.125552.

Akhatou, I., González-Domínguez, R., & Fernández-Recamales, Á. (2016). Investigation of the effect of genotype and agronomic conditions on metabolomic profiles of selected strawberry cultivars with different sensitivity to environmental stress. *Plant Physiology and Biochemistry*, *101*, 14–22. https://doi.org/10.1016/j.plaphy.2016.01.016.

Ali, K., Maltese, F., Fortes, A. M., Pais, M. S., Choi, Y. H., & Verpoorte, R. (2011). Monitoring biochemical changes during grape berry development in Portuguese cultivars by NMR spectroscopy. *Food Chemistry*, *124*(4), 1760–1769. https://doi.org/10.1016/j.foodchem.2010.08.015.

Allard, P.-M., Genta-Jouve, G., & Wolfender, J.-L. (2017). Deep metabolome annotation in natural products research: Towards a virtuous cycle in metabolite identification. *Current Opinion in Chemical Biology*, *36*, 40–49. https://doi.org/10.1016/j.cbpa.2016.12.022.

Allwood, J. W., Cheung, W., Xu, Y., Mumm, R., de Vos, R. C. H., Biais, B., et al. (2014). Metabolomics in melon: A new opportunity for aroma analysis. *Phytochemistry*, *99*, 61–72. https://doi.org/10.1016/j.phytochem.2013.12.010.

Allwood, J. W., & Goodacre, R. (2010). An introduction to liquid chromatography–mass spectrometry instrumentation applied in plant metabolomic analyses. *Phytochemical Analysis*, *21*(1), 33–47. https://doi.org/10.1002/pca.1187.

Allwood, J. W., Woznicki, T. L., Xu, Y., Foito, A., Aaby, K., Sungurtas, J., et al. (2019). Application of HPLC-PDA-MS metabolite profiling to investigate the effect of growth temperature and day length on blackcurrant fruit. *Metabolomics*, *15*(1), 12. https://doi.org/10.1007/s11306-018-1462-5.

Almenar, E., Hernández-Muñoz, P., Lagarón, J. M., Catalá, R., & Gavara, R. (2006). Controlled atmosphere storage of wild strawberry fruit (*Fragaria vesca* L.). *Journal of Agricultural and Food Chemistry*, *54*(1), 86–91. https://doi.org/10.1021/jf0517492.

Amorim-Carrilho, K. T., Cepeda, A., Fente, C., & Regal, P. (2014). Review of methods for analysis of carotenoids. *TrAC Trends in Analytical Chemistry*, *56*, 49–73. https://doi.org/10.1016/j.trac.2013.12.011.

Arapitsas, P., Corte, A. D., Gika, H., Narduzzi, L., Mattivi, F., & Theodoridis, G. (2016). Studying the effect of storage conditions on the metabolite content of red wine using HILIC LC-MS based metabolomics. *Food Chemistry*, *197*, 1331–1340. https://doi.org/10.1016/j.foodchem.2015.09.084.

Arapitsas, P., & Mattivi, F. (2018). LC-MS untargeted protocol for the analysis of wine. In G. A. Theodoridis, H. G. Gika, & I. D. Wilson (Eds.), *Metabolic profiling: Methods and protocols* (pp. 225–235). New York, NY: Springer.

Araújo, W. L., Nunes-Nesi, A., Osorio, S., Usadel, B., Fuentes, D., Nagy, R., et al. (2011). Antisense inhibition of the iron-sulphur subunit of succinate dehydrogenase enhances photosynthesis and growth in tomato via an organic acid–mediated effect on stomatal aperture. *The Plant Cell*, *23*(2), 600–627. https://doi.org/10.1105/tpc.110.081224.

Auffray, C., Imbeaud, S., Roux-Rouquié, M., & Hood, L. (2003). From functional genomics to systems biology: Concepts and practices. *Comptes Rendus Biologies*, *326*(10-11), 879–892. https://doi.org/10.1016/j.crvi.2003.09.033.

Banerjee, A., George, C., Bharathwaj, S., & Chandrakumar, N. (2009). Postharvest ripening study of sweet lime (*Citrus limettioides*) in situ by volume-localized NMR spectroscopy. *Journal of Agricultural and Food Chemistry*, *57*(4), 1183–1187. https://doi.org/10.1021/jf803159u.

Bassi, P. K., & Spencer, M. S. (1985). Comparative evaluation of photoionization and flame ionization detectors for ethylene analysis. *Plant, Cell & Environment*, *8*(2), 161–165. https://doi.org/10.1111/j.1365-3040.1985.tb01224.x.

Beauvoit, B., Belouah, I., Bertin, N., Cakpo, C. B., Colombie, S., Dai, Z. W., et al. (2018). Putting primary metabolism into perspective to obtain better fruits. *Annals of Botany*, *122*(1), 1–21. https://doi.org/10.1093/aob/mcy057.

Beauvoit, B. P., Colombié, S., Monier, A., Andrieu, M.-H., Biais, B., Bénard, C., et al. (2014). Model-assisted analysis of sugar metabolism throughout tomato fruit development reveals enzyme and carrier properties in relation to vacuole expansion. *The Plant Cell*, *26*(8), 3224–3242. https://doi.org/10.1105/tpc.114.127761.

Bernillon, S., Biais, B., Deborde, C., Maucourt, M., Cabasson, C., Gibon, Y., et al. (2013). Metabolomic and elemental profiling of melon fruit quality as affected by genotype and environment. *Metabolomics*, *9*(1), 57–77. https://doi.org/10.1007/s11306-012-0429-1.

Bhandari, D. R., Wang, Q., Friedt, W., Spengler, B., Gottwald, S., & Römpp, A. (2015). High resolution mass spectrometry imaging of plant tissues: Towards a plant metabolite atlas. *Analyst*, *140*(22), 7696–7709. https://doi.org/10.1039/C5AN01065A.

Biais, B., Allwood, J. W., Deborde, C., Xu, Y., Maucourt, M., Beauvoit, B., et al. (2009). 1H-NMR, GC-EI-TOFMS, and data set correlation for fruit metabolomics: Application to spatial metabolite analysis in melon. *Analytical Chemistry*, *81*(8), 2884–2894. https://doi.org/10.1021/ac9001996.

Biais, B., Beauvoit, B., Allwood, J. W., Deborde, C., Maucourt, M., Goodacre, R., et al. (2010). Metabolic acclimation to hypoxia revealed by metabolite gradients in melon fruit. *Journal of Plant Physiology*, *167*(3), 242–245. https://doi.org/10.1016/j.jplph.2009.08.010.

Biais, B., Bénard, C., Beauvoit, B., Colombié, S., Prodhomme, D., Ménard, G., et al. (2014). Remarkable reproducibility of enzyme activity profiles in tomato fruits grown under contrasting environments provides a roadmap for studies of fruit metabolism. *Plant Physiology*, *164*(3), 1204–1221. https://doi.org/10.1104/pp.113.231241.

Buendia, P., Bradley, R. M., Taylor, T. J., Schymanski, E. L., Patti, G. J., & Kabuka, M. R. (2019). Ontology-based metabolomics data integration with quality control. *Bioanalysis*, *11*(12), 1139–1154. https://doi.org/10.4155/bio-2018-0303.

Calafiore, R., Ruggieri, V., Raiola, A., Rigano, M. M., Sacco, A., Hassan, M. I., et al. (2016). Exploiting genomics resources to identify candidate genes underlying antioxidants content in tomato fruit. *Frontiers in Plant Science*, 7. https://doi.org/10.3389/fpls.2016.00397.

Capitani, D., Mannina, L., Proietti, N., Sobolev, A. P., Tomassini, A., Miccheli, A., et al. (2010). Monitoring of metabolic profiling and water status of Hayward kiwifruits by nuclear magnetic resonance. *Talanta*, *82*(5), 1826–1838. https://doi.org/10.1016/j.talanta.2010.07.080.

Cárdenas, P. D., Sonawane, P. D., Heinig, U., Jozwiak, A., Panda, S., Abebie, B., et al. (2019). Pathways to defense metabolites and evading fruit bitterness in genus Solanum evolved through 2-oxoglutarate-dependent dioxygenases. *Nature Communications*, *10*(1), 5169. https://doi.org/10.1038/s41467-019-13211-4.

Carvalho, E., Franceschi, P., Feller, A., Palmieri, L., Wehrens, R., & Martens, S. (2013). A targeted metabolomics approach to understand differences in flavonoid biosynthesis in red and yellow raspberries. *Plant Physiology and Biochemistry*, *72*, 79–86. https://doi.org/10.1016/j.plaphy.2013.04.001.

Centeno, D. C., Osorio, S., Nunes-Nesi, A., Bertolo, A. L. F., Carneiro, R. T., Araujo, W. L., et al. (2011). Malate plays a crucial role in starch metabolism, ripening, and soluble solid content of tomato fruit and affects postharvest softening. *Plant Cell*, *23*(1), 162–184. https://doi.org/10.1105/tpc.109.072231.

Chaleckis, R., Meister, I., Zhang, P., & Wheelock, C. E. (2019). Challenges, progress and promises of metabolite annotation for LC–MS-based metabolomics. *Current Opinion in Biotechnology*, *55*, 44–50. https://doi.org/10.1016/j.copbio.2018.07.010.

Cheng, Y.-C., Wang, T.-T., Chen, J.-H., & Lin, T.-T. (2011). Spatial–temporal analyses of lycopene and sugar contents in tomatoes during ripening using chemical shift imaging. *Postharvest Biology and Technology*, *62*(1), 17–25. https://doi.org/10.1016/j.postharvbio.2011.04.006.

Colombié, S., Beauvoit, B., Nazaret, C., Bénard, C., Vercambre, G., Le Gall, S., et al. (2017). Respiration climacteric in tomato fruits elucidated by constraint-based modelling. *New Phytologist*, *213*(4), 1726–1739. https://doi.org/10.1111/nph.14301.

Colombié, S., Nazaret, C., Bénard, C., Biais, B., Mengin, V., Solé, M., et al. (2015). Modelling central metabolic fluxes by constraint-based optimization reveals metabolic reprogramming of developing *Solanum lycopersicum* (tomato) fruit. *The Plant Journal*, *81*(1), 24–39. https://doi.org/10.1111/tpj.12685.

Commisso, M., Negri, S., Bianconi, M., Gambini, S., Avesani, S., Ceoldo, S., et al. (2019). Untargeted and targeted metabolomics and tryptophan decarboxylase in vivo characterization provide novel insight on the development of kiwifruits (*Actinidia deliciosa*). *International Journal of Molecular Sciences*, *20*(4), 897. https://doi.org/10.3390/ijms20040897.

D'Angelo, M., Zanor, M. I., Burgos, E., Asprelli, P. D., Boggio, S. B., Carrari, F., et al. (2019). Fruit metabolic and transcriptional programs differentiate among Andean tomato (*Solanum lycopersicum* L.) accessions. *Planta*, *250*(6), 1927–1940. https://doi.org/10.1007/s00425-019-03274-4.

De Livera, A. M., Dias, D. A., De Souza, D., Rupasinghe, T., Pyke, J., Tull, D., et al. (2012). Normalizing and integrating metabolomics data. *Analytical Chemistry*, *84*(24), 10768–10776. https://doi.org/10.1021/ac302748b.

de Vos, R. C. H., Hall, R., & Moing, A. (2011). Metabolomics of a model fruit: Tomato. In R. Hall (Ed.), *Vol. 43. Biology of plant metabolomics* (pp. 109–155). Oxford: Wiley-Blackwell Ltd.

De Vos, R. C. H., Moco, S., Lommen, A., Keurentjes, J. J. B., Bino, R. J., & Hall, R. D. (2007). Untargeted large-scale plant metabolomics using liquid chromatography coupled to mass spectrometry. *Nature Protocols*, *2*(4), 778–791. https://doi.org/10.1038/nprot.2007.95.

Dean, R. J., Stait-Gardner, T., Clarke, S. J., Rogiers, S. Y., Bobek, G., & Price, W. S. (2014). Use of diffusion magnetic resonance imaging to correlate the developmental changes in grape berry tissue structure with water diffusion patterns. *Plant Methods*, *10*(1), 35. https://doi.org/10.1186/1746-4811-10-35.

Deborde, C., Maucourt, M., Baldet, P., Bernillon, S., Biais, B., Talon, G., et al. (2009). Proton NMR quantitative profiling for quality assessment of greenhouse-grown tomato fruit. *Metabolomics*, *5*(2), 183–198. https://doi.org/10.1007/s11306-008-0134-2.

Deborde, C., Moing, A., Roch, L., Jacob, D., Rolin, D., & Giraudeau, P. (2017). Plant metabolism as studied by NMR spectroscopy. *Progress in Nuclear Magnetic Resonance Spectroscopy*, *102-103*, 61–97. https://doi.org/10.1016/j.pnmrs.2017.05.001.

Decros, G., Baldet, P., Beauvoit, B., Stevens, R., Flandin, A., Colombié, S., et al. (2019). Get the balance right: ROS homeostasis and redox signalling in fruit. *Frontiers in Plant Science*, *10*(1091). https://doi.org/10.3389/fpls.2019.01091.

Decros, G., Beauvoit, B., Colombié, S., Cabasson, C., Bernillon, S., Arrivault, S., et al. (2019). Regulation of pyridine nucleotide metabolism during tomato fruit development through transcript and protein profiling. *Frontiers in Plant Science*, *10*(1201). https://doi.org/10.3389/fpls.2019.01201.

Desnoues, E., Génard, M., Quilot-Turion, B., & Baldazzi, V. (2018). A kinetic model of sugar metabolism in peach fruit reveals a functional hypothesis of a markedly low fructose-to-glucose ratio phenotype. *The Plant Journal*, *94*(4), 685–698. https://doi.org/10.1111/tpj.13890.

Desnoues, E., Gibon, Y., Baldazzi, V., Signoret, V., Génard, M., & Quilot-Turion, B. (2014). Profiling sugar metabolism during fruit development in a peach progeny with different fructose-to-glucose ratios. *BMC Plant Biology*, *14*(1), 336. https://doi.org/10.1186/s12870-014-0336-x.

Doebley, J. F., Gaut, B. S., & Smith, B. D. (2006). The molecular genetics of crop domestication. *Cell*, *127*(7), 1309–1321. https://doi.org/10.1016/j.cell.2006.12.006.

Domingos, S., Fino, J., Paulo, O. S., Oliveira, C. M., & Goulao, L. F. (2016). Molecular candidates for early-stage flower-to-fruit transition in stenospermocarpic table grape (*Vitis vinifera* L.) inflorescences ascribed by differential transcriptome and metabolome profiles. *Plant Science*, *244*, 40–56. https://doi.org/10.1016/j.plantsci.2015.12.009.

Dona, A. C., Pagès, G., & Kuchel, P. W. (2011). Kinetics of starch hydrolysis and glucose mutarotation studied by NMR chemical exchange saturation transfer (CEST). *Carbohydrate Polymers*, *86*(4), 1525–1532. https://doi.org/10.1016/j.carbpol.2011.06.056.

Dong, Y., Li, B., Malitsky, S., Rogachev, I., Aharoni, A., Kaftan, F., et al. (2016). Sample preparation for mass spectrometry imaging of plant tissues: A review. *Frontiers in Plant Science*, 7(60). https://doi.org/10.3389/fpls.2016.00060.

dos Santos, C. P., Batista, M. C., da Cruz Saraiva, K. D., Roque, A. L. M., de Souza Miranda, R., Alexandre e Silva, L. M., et al. (2019). Transcriptome analysis of acerola fruit ripening: Insights into ascorbate, ethylene, respiration, and softening metabolisms. *Plant Molecular Biology*, *101*(3), 269–296. https://doi.org/10.1007/s11103-019-00903-0.

D'Urso, G., Piacente, S., Pizza, C., & Montoro, P. (2018). Metabolomics of healthy berry fruits. *Current Medicinal Chemistry*, *25*(37), 4888–4902. https://doi.org/10.2174/0929867323666161206100006.

Edwards, L. M. (2017). Metabolic systems biology: A brief primer. *The Journal of Physiology*, *595*(9), 2849–2855. https://doi.org/10.1113/jp272275.

Emwas, A.-H. M. (2015). The strengths and weaknesses of NMR spectroscopy and mass spectrometry with particular focus on metabolomics research. In J. T. Bjerrum (Ed.), *Metabonomics: Methods and protocols* (pp. 161–193). New York, NY: Springer New York.

Enomoto, H., Sato, K., Miyamoto, K., Ohtsuka, A., & Yamane, H. (2018). Distribution analysis of anthocyanins, sugars, and organic acids in strawberry fruits using matrix-assisted laser desorption/ionization-imaging mass spectrometry. *Journal of Agricultural and Food Chemistry*, *66*(19), 4958–4965. https://doi.org/10.1021/acs.jafc.8b00853.

Enomoto, H., Takahashi, S., Takeda, S., & Hatta, H. (2019). Distribution of flavan-3-ol species in ripe strawberry fruit revealed by matrix-assisted laser desorption/ionization-mass spectrometry imaging. *Molecules*, *25*(1), 103. https://doi.org/10.3390/molecules25010103.

Etalo, D. W., De Vos, R. C. H., Joosten, M. H. A. J., & Hall, R. D. (2015). Spatially resolved plant metabolomics: Some potentials and limitations of laser-ablation electrospray ionization mass spectrometry metabolite imaging. *Plant Physiology*, *169*(3), 1424–1435. https://doi.org/10.1104/pp.15.01176.

Falasca, A., Melck, D., Paris, D., Saviano, G., Motta, A., & Iorizzi, M. (2014). Seasonal changes in the metabolic fingerprint of *Juniperus communis* L. berry extracts by ^{1}H NMR-based metabolomics. *Metabolomics*, *10*(1), 165–174. https://doi.org/10.1007/s11306-013-0566-1.

Fernandez, O., Millet, E., Rincent, R., Prigent, S., Pétriacq, P., & Gibon, Y. (2021). Plant metabolomics and breeding. In P. Pétriacq, & A. Bouchereau (Eds.), *Vol. 98. Plant metabolomics in full swing*. Burlington, UK: Academic Press. In this issue.

Fernandez, O., Urrutia, M., Bernillon, S., Giauffret, C., Tardieu, F., Le Gouis, J., et al. (2016). Fortune telling: Metabolic markers of plant performance. *Metabolomics*, *12*(10), 158. https://doi.org/10.1007/s11306-016-1099-1.

Fernandez-Moreno, J. P., Tzfadia, O., Forment, J., Presa, S., Rogachev, I., Meir, S., et al. (2016). Characterization of a new pink-fruited tomato mutant results in the identification of a null allele of the SlMYB12 transcription factor. *Plant Physiology*, *171*(3), 1821–1836. https://doi.org/10.1104/pp.16.00282.

Fortes, A. M., Granell, A., Pezzotti, M., & Bouzayen, M. (2017). Editorial: Molecular and metabolic mechanisms associated with fleshy fruit quality. *Frontiers in Plant Science*, *8*(1236). https://doi.org/10.3389/fpls.2017.01236.

Francini, A., Pintado, M., Manganaris, G. A., & Ferrante, A. (2020). Editorial: Bioactive compounds biosynthesis and metabolism in fruit and vegetables. *Frontiers in Plant Science, 11*, 129. https://doi.org/10.3389/fpls.2020.00129.

Fraser, P. D., & Bramley, P. M. (2006). Metabolic profiling and quantification of carotenoids and related isoprenoids in crop plants. In K. Saito, R. A. Dixon, & L. Willmitzer (Eds.), *Plant Metabolomics* (pp. 229–242). Berlin, Heidelberg: Springer Berlin Heidelberg.

Fridman, E., Carrari, F., Liu, Y.-S., Fernie, A. R., & Zamir, D. (2004). Zooming in on a quantitative trait for tomato yield using interspecific introgressions. *Science, 305*(5691), 1786–1789. https://doi.org/10.1126/science.1101666.

Fröhlich, F., Kaltenbacher, B., Theis, F. J., & Hasenauer, J. (2017). Scalable parameter estimation for genome-scale biochemical reaction networks. *PLoS Computational Biology, 13*(1), e1005331. https://doi.org/10.1371/journal.pcbi.1005331.

Garbowicz, K., Liu, Z., Alseekh, S., Tieman, D., Taylor, M., Kuhalskaya, A., et al. (2018). Quantitative trait loci analysis identifies a prominent gene involved in the production of fatty acid-derived flavor volatiles in tomato. *Molecular Plant, 11*(9), 1147–1165. https://doi.org/10.1016/j.molp.2018.06.003.

Gest, N., Gautier, H., & Stevens, R. (2012). Ascorbate as seen through plant evolution: The rise of a successful molecule? *Journal of Experimental Botany, 64*(1), 33–53. https://doi.org/10.1093/jxb/ers297.

Gillaspy, G., Ben-David, H., & Gruissem, W. (1993). Fruits: A developmental perspective. *The Plant Cell, 5*(10), 1439–1451. https://doi.org/10.1105/tpc.5.10.1439.

Hanhineva, K., & Aharoni, A. (2010). Metabolomics in fruit development. In S. M. Jain, & D. S. Brar (Eds.), *Molecular techniques in crop improvement* (2nd ed., pp. 675–693). Heidelberg, Germany: Springer.

Hanhineva, K., Rogachev, I., Kokko, H., Mintz-Oron, S., Venger, I., Kärenlampi, S., et al. (2008). Non-targeted analysis of spatial metabolite composition in strawberry (*Fragaria* × *ananassa*) flowers. *Phytochemistry, 69*(13), 2463–2481. https://doi.org/10.1016/j.phytochem.2008.07.009.

Haugeneder, A., Trinkl, J., Härtl, K., Hoffmann, T., Allwood, J. W., & Schwab, W. (2018). Answering biological questions by analysis of the strawberry metabolome. *Metabolomics, 14*(11), 145. https://doi.org/10.1007/s11306-018-1441-x.

Hazzouri, K. M., Gros-Balthazard, M., Flowers, J. M., Copetti, D., Lemansour, A., Lebrun, M., et al. (2019). Genome-wide association mapping of date palm fruit traits. *Nature Communications, 10*(1), 4680. https://doi.org/10.1038/s41467-019-12604-9.

He, F., Mu, L., Yan, G.-L., Liang, N.-N., Pan, Q.-H., Wang, J., et al. (2010). Biosynthesis of anthocyanins and their regulation in colored grapes. *Molecules, 15*(12), 9057–9091. https://doi.org/10.3390/molecules15129057.

Heng, Z., Sheng, O., Huang, W., Zhang, S., Fernie, A. R., Motorykin, I., et al. (2019). Integrated proteomic and metabolomic analysis suggests high rates of glycolysis are likely required to support high carotenoid accumulation in banana pulp. *Food Chemistry, 297*, 125016. https://doi.org/10.1016/j.foodchem.2019.125016.

Hövelmann, Y., Hahn, M., Hübner, F., & Humpf, H.-U. (2019). Detection of novel cytotoxic imidazole alkaloids in tomato products by LC-MS/MS. *Journal of Agricultural and Food Chemistry, 67*(13), 3670–3678. https://doi.org/10.1021/acs.jafc.9b00461.

Huang, H.-X., Yu, T., Li, J.-X., Qu, S.-P., Wang, M.-M., Wu, T.-Q., et al. (2019). Characterization of *Cucurbita maxima* fruit metabolomic profiling and transcriptome to reveal fruit quality and ripening gene expression patterns. *Journal of Plant Biology, 62*(3), 203–216.

Jensen, K., Gudmundsson, S., & Herrgård, M. J. (2018). Enhancing metabolic models with genome-scale experimental data. In N. Rajewsky, S. Jurga, & J. Barciszewski (Eds.), *Systems biology* (pp. 337–350). Cham: Springer International Publishing.

Jézéquel, T., Deborde, C., Maucourt, M., Zhendre, V., Moing, A., & Giraudeau, P. (2015). Absolute quantification of metabolites in tomato fruit extracts by fast 2D NMR. *Metabolomics*, *11*(5), 1231–1242.

Jorge, T. F., Mata, A. T., & António, C. (2016). Mass spectrometry as a quantitative tool in plant metabolomics. *Philosophical Transactions of the Royal Society A: Mathematical, Physical and Engineering Sciences*, *374*(2079), 20150370.

Kaneria, M. J., Rakholiya, K. D., Marsonia, L. R., Dave, R. A., & Golakiya, B. A. (2018). Nontargeted metabolomics approach to determine metabolites profile and antioxidant study of Tropical Almond (*Terminalia catappa* L.) fruit peels using GC-QTOF-MS and LC-QTOF-MS. *Journal of Pharmaceutical and Biomedical Analysis*, *160*, 415–427. https://doi.org/10.1016/j.jpba.2018.08.026.

Khakimov, B., Mongi, R. J., Sørensen, K. M., Ndabikunze, B. K., Chove, B. E., & Engelsen, S. B. (2016). A comprehensive and comparative GC-MS metabolomics study of non-volatiles in Tanzanian grown mango, pineapple, jackfruit, baobab and tamarind fruits. *Food Chemistry*, *213*, 691–699. https://doi.org/10.1016/j.foodchem.2016.07.005.

Kim, H.-S., Park, S. J., Hyun, S.-H., Yang, S.-O., Lee, J., Auh, J.-H., et al. (2011). Biochemical monitoring of black raspberry (*Rubus coreanus* Miquel) fruits according to maturation stage by 1H NMR using multiple solvent systems. *Food Research International*, *44*(7), 1977–1987. https://doi.org/10.1016/j.foodres.2011.01.023.

Kim, M. J., Lee, M. Y., Shon, J. C., Kwon, Y. S., Liu, K.-H., Lee, C. H., et al. (2019). Untargeted and targeted metabolomics analyses of blackberries—Understanding post-harvest red drupelet disorder. *Food Chemistry*, *300*, 125169. https://doi.org/10.1016/j.foodchem.2019.125169.

Kusano, M., Redestig, H., Hirai, T., Oikawa, A., Matsuda, F., Fukushima, A., et al. (2011). Covering chemical diversity of genetically-modified tomatoes using metabolomics for objective substantial equivalence assessment. *Plos One*, *6*(2). https://doi.org/10.1371/journal.pone.0016989.

Kwon, C.-T., Heo, J., Lemmon, Z. H., Capua, Y., Hutton, S. F., Van Eck, J., et al. (2020). Rapid customization of Solanaceae fruit crops for urban agriculture. *Nature Biotechnology*, *38*(2), 182–188. https://doi.org/10.1038/s41587-019-0361-2.

Lecourieux, D., Kappel, C., Claverol, S., Pieri, P., Feil, R., Lunn, J. E., et al. (2020). Proteomic and metabolomic profiling underlines the stage- and time-dependent effects of high temperature on grape berry metabolism. *Journal of Integrative Plant Biology*, *62*(8), 1132–1158. https://doi.org/10.1111/jipb.12894.

Lemaire-Chamley, M., Mounet, F., Deborde, C., Maucourt, M., Jacob, D., & Moing, A. (2019). NMR-based tissular and developmental metabolomics of tomato fruit. *Metabolites*, *9*, 93. https://doi.org/10.3390/metabo9050093.

Li, Y., Fang, J., Qi, X., Lin, M., Zhong, Y., Sun, L., et al. (2018). Combined analysis of the fruit metabolome and transcriptome reveals candidate genes involved in flavonoid biosynthesis in *Actinidia arguta*. *International Journal of Molecular Sciences*, *19*(5), 1471. https://doi.org/10.3390/ijms19051471.

Li, T., Yun, Z., Wu, Q., Qu, H., Duan, X., & Jiang, Y. (2019). Combination of transcriptomic, proteomic, and metabolomic analysis reveals the ripening mechanism of banana pulp. *Biomolecules*, *9*(10), 523. https://doi.org/10.3390/biom9100523.

Lisec, J., Schauer, N., Kopka, J., Willmitzer, L., & Fernie, A. R. (2006). Gas chromatography mass spectrometry–based metabolite profiling in plants. *Nature Protocols*, *1*(1), 387–396. https://doi.org/10.1038/nprot.2006.59.

Lloyd, A. J., William Allwood, J., Winder, C. L., Dunn, W. B., Heald, J. K., Cristescu, S. M., et al. (2011). Metabolomic approaches reveal that cell wall modifications play a major role in ethylene-mediated resistance against *Botrytis cinerea*. *The Plant Journal*, *67*(5), 852–868. https://doi.org/10.1111/j.1365-313X.2011.04639.x.

López, M. G., Zanor, M. I., Pratta, G. R., Stegmayer, G., Boggio, S. B., Conte, M., et al. (2015). Metabolic analyses of interspecific tomato recombinant inbred lines for fruit quality improvement. *Metabolomics*, *11*(5), 1416–1431. https://doi.org/10.1007/s11306-015-0798-3.

Markley, J. L., Brüschweiler, R., Edison, A. S., Eghbalnia, H. R., Powers, R., Raftery, D., et al. (2017). The future of NMR-based metabolomics. *Current Opinion in Biotechnology*, *43*, 34–40. https://doi.org/10.1016/j.copbio.2016.08.001.

Maulidiani, M., Mediani, A., Abas, F., Park, Y. S., Park, Y.-K., Kim, Y. M., et al. (2018). 1H NMR and antioxidant profiles of polar and non-polar extracts of persimmon (*Diospyros kaki* L.)—Metabolomics study based on cultivars and origins. *Talanta*, *184*, 277–286. https://doi.org/10.1016/j.talanta.2018.02.084.

Moing, A., Aharoni, A., Biais, B., Rogachev, I., Meir, S., Brodsky, L., et al. (2011). Extensive metabolic cross-talk in melon fruit revealed by spatial and developmental combinatorial metabolomics. *New Phytologist*, *190*(3), 683–696. https://doi.org/10.1111/j.1469-8137.2010.03626.x.

Mori, K., Beauvoit, B. P., Biais, B., Chabane, M., Allwood, J. W., Deborde, C., et al. (2019). Central metabolism is tuned to the availability of oxygen in developing melon fruit. *Frontiers in Plant Science*, *10*. https://doi.org/10.3389/fpls.2019.00594.

Mounet, F., Moing, A., Garcia, V., Petit, J., Maucourt, M., Deborde, C., et al. (2009). Gene and metabolite regulatory network analysis of early developing fruit tissues highlights new candidate genes for the control of tomato fruit composition and development. *Plant Physiology*, *149*(3), 1505–1528. https://doi.org/10.1104/pp.108.133967.

Muñoz, P., & Munné-Bosch, S. (2018). Photo-oxidative stress during leaf, flower and fruit development. *Plant Physiology*, *176*(2), 1004–1014. https://doi.org/10.1104/pp.17.01127.

Musse, M., Quellec, S., Cambert, M., Devaux, M.-F., Lahaye, M., & Mariette, F. (2009). Monitoring the postharvest ripening of tomato fruit using quantitative MRI and NMR relaxometry. *Postharvest Biology and Technology*, *53*(1), 22–35. https://doi.org/10.1016/j.postharvbio.2009.02.004.

Nagana Gowda, G. A., & Raftery, D. (2017). Recent advances in NMR-based metabolomics. *Analytical Chemistry*, *89*(1), 490–510. https://doi.org/10.1021/acs.analchem.6b04420.

Nakamura, J., Morikawa-Ichinose, T., Fujimura, Y., Hayakawa, E., Takahashi, K., Ishii, T., et al. (2017). Spatially resolved metabolic distribution for unraveling the physiological change and responses in tomato fruit using matrix-assisted laser desorption/ionization–mass spectrometry imaging (MALDI-MSI). *Analytical and Bioanalytical Chemistry*, *409*(6), 1697–1706. https://doi.org/10.1007/s00216-016-0118-4.

Nile, S. H., & Park, S. W. (2014). Edible berries: Bioactive components and their effect on human health. *Nutrition*, *30*(2), 134–144. https://doi.org/10.1016/j.nut.2013.04.007.

Obata, T., Witt, S., Lisec, J., Palacios-Rojas, N., Florez-Sarasa, I., Yousfi, S., et al. (2015). Metabolite profiles of maize leaves in drought, heat, and combined stress field trials reveal the relationship between metabolism and grain yield. *Plant Physiology*, *169*(4), 2665–2683. https://doi.org/10.1104/pp.15.01164.

Okazaki, Y., & Saito, K. (2018). Plant lipidomics Using UPLC-QTOF-MS. In C. António (Ed.), *Plant metabolomics: Methods and protocols* (pp. 157–169). New York, NY: Springer New York.

Omena-Garcia, R. P., de Ávila Silva, L., Vital, C. E., Araújo, W. L., Ramos, H. J. O., & Nunes-Nesi, A. (2019). Identification of metabolite traits from the current metabolomic approaches. *Theoretical and Experimental Plant Physiology*, *31*(1), 1–19. https://doi.org/10.1007/s40626-019-00147-9.

Orth, J. D., Thiele, I., & Palsson, B.Ø. (2010). What is flux balance analysis? *Nature Biotechnology*, *28*(3), 245–248. https://doi.org/10.1038/nbt.1614.

Osorio, S., Carneiro, R. T., Lytovchenko, A., McQuinn, R., Sørensen, I., Vallarino, J. G., et al. (2020). Genetic and metabolic effects of ripening mutations and vine detachment on tomato fruit quality. *Plant Biotechnology Journal, 18*(1), 106–118. https://doi.org/10.1111/pbi.13176.

Osorio, S., Ruan, Y.-L., & Fernie, A. R. (2014). An update on source-to-sink carbon partitioning in tomato. *Frontiers in Plant Science, 5*(516). https://doi.org/10.3389/fpls.2014.00516.

Osorio, S., Vallarino, J. G., Szecowka, M., Ufaz, S., Tzin, V., Angelovici, R., et al. (2013). Alteration of the interconversion of pyruvate and malate in the plastid or cytosol of ripening tomato fruit invokes diverse consequences on sugar but similar effects on cellular organic acid, metabolism, and transitory starch accumulation. *Plant Physiology, 161*(2), 628–643. https://doi.org/10.1104/pp.112.211094.

Pascual, L., Xu, J., Biais, B., Maucourt, M., Ballias, P., Bernillon, S., et al. (2013). Deciphering genetic diversity and inheritance of tomato fruit weight and composition through a systems biology approach. *Journal of Experimental Botany, 64*(18), 5737–5752. https://doi.org/10.1093/jxb/ert349.

Peng, J., Chen, Y.-T., Chen, C.-L., & Li, L. (2014). Development of a universal metabolome-standard method for long-term LC-MS metabolome profiling and its application for bladder cancer urine-metabolite-biomarker discovery. *Analytical Chemistry, 86*(13), 6540–6547. https://doi.org/10.1021/ac5011684.

Peters, K., Bradbury, J., Bergmann, S., Capuccini, M., Cascante, M., de Atauri, P., et al. (2018). PhenoMeNal: Processing and analysis of metabolomics data in the cloud. *GigaScience, 8*(2). https://doi.org/10.1093/gigascience/giy149.

Podda, R., Delli Castelli, D., Digilio, G., Gullino, M. L., & Aime, S. (2015). Asparagine in plums detected by CEST-MRI. *Food Chemistry, 169*, 1–4. https://doi.org/10.1016/j.foodchem.2014.07.041.

Pott, D. M., Osorio, S., & Vallarino, J. G. (2019). From central to specialized metabolism: An overview of some secondary compounds derived from the primary metabolism for their role in conferring nutritional and organoleptic characteristics to fruit. *Frontiers in Plant Science, 10*. https://doi.org/10.3389/fpls.2019.00835.

Quadrana, L., Almeida, J., Asís, R., Duffy, T., Dominguez, P. G., Bermúdez, L., et al. (2014). Natural occurring epialleles determine vitamin E accumulation in tomato fruits. *Nature Communications, 5*(1), 4027. https://doi.org/10.1038/ncomms5027.

Rai, A., Yamazaki, M., & Saito, K. (2019). A new era in plant functional genomics. *Current Opinion in Systems Biology, 15*, 58–67. https://doi.org/10.1016/j.coisb.2019.03.005.

Ramadan, S., Lin, A., & Stanwell, P. (2013). Glutamate and glutamine: A review of in vivo MRS in the human brain. *NMR in Biomedicine, 26*(12), 1630–1646. https://doi.org/10.1002/nbm.3045.

Ratcliffe, R. G., & Shachar-Hill, Y. (2006). Measuring multiple fluxes through plant metabolic networks. *The Plant Journal, 45*(4), 490–511. https://doi.org/10.1111/j.1365-313X.2005.02649.x.

Razzaq, A., Sadia, B., Raza, A., Khalid Hameed, M., & Saleem, F. (2019). Metabolomics: A way forward for crop improvement. *Metabolites, 9*(12), 303. https://doi.org/10.3390/metabo9120303.

Redestig, H., Szymanski, J., Hirai, M. Y., Selbig, J., Willmitzer, L., Nikoloski, Z., et al. (2018). Data integration, metabolic networks and systems biology. *Annual Plant Reviews online*, 261–316. https://doi.org/10.1002/9781119312994.apr0469.

Reshef, N., Agam, N., & Fait, A. (2018). Grape berry acclimation to excessive solar irradiance leads to repartitioning between major flavonoid groups. *Journal of Agricultural and Food Chemistry, 66*(14), 3624–3636. https://doi.org/10.1021/acs.jafc.7b04881.

Riedelsheimer, C., Czedik-Eysenberg, A., Grieder, C., Lisec, J., Technow, F., Sulpice, R., et al. (2012). Genomic and metabolic prediction of complex heterotic traits in hybrid maize. *Nature Genetics, 44*(2), 217–220. https://doi.org/10.1038/ng.1033.

Roch, L., Dai, Z., Gomès, E., Bernillon, S., Wang, J., Gibon, Y., et al. (2019). Fruit salad in the lab: Comparing botanical species to help deciphering fruit primary metabolism. *Frontiers in Plant Science*, *10*, 836. https://doi.org/10.3389/fpls.2019.00836.

Roch, L., Prigent, S., Klose, H., Capko, C. B., Beauvoit, B., Deborde, C., et al. (2020). Biomass composition explains fruit relative growth rate and discriminates climacteric from non-climacteric species. *Journal of Experimental Botany*, *71*(19), 5823–5836. https://doi.org/10.1093/jxb/eraa302.

Rodriguez-Casado, A. (2016). The health potential of fruits and vegetables phytochemicals: Notable examples. *Critical Reviews in Food Science and Nutrition*, *56*(7), 1097–1107. https://doi.org/10.1080/10408398.2012.755149.

Roessner-Tunali, U., Hegemann, B., Lytovchenko, A., Carrari, F., Bruedigam, C., Granot, D., et al. (2003). Metabolic profiling of transgenic tomato plants overexpressing hexokinase reveals that the influence of hexose phosphorylation diminishes during fruit development. *Plant Physiology*, *133*(1), 84–99. https://doi.org/10.1104/pp.103.023572.

Rohwer, J. M. (2012). Kinetic modelling of plant metabolic pathways. *Journal of Experimental Botany*, *63*(6), 2275–2292. https://doi.org/10.1093/jxb/ers080.

Rothan, C., Diouf, I., & Causse, M. (2019). Trait discovery and editing in tomato. *The Plant Journal*, *97*(1), 73–90. https://doi.org/10.1111/tpj.14152.

Ru, L., Osorio, S., Wang, L., Fernie, A. R., Patrick, J. W., & Ruan, Y.-L. (2017). Transcriptomic and metabolomics responses to elevated cell wall invertase activity during tomato fruit set. *Journal of Experimental Botany*, *68*(15), 4263–4279. https://doi.org/10.1093/jxb/erx219.

Ruan, Y.-L. (2014). Sucrose metabolism: Gateway to diverse carbon use and sugar signaling. *Annual Review of Plant Biology*, *65*(1), 33–67. https://doi.org/10.1146/annurev-arplant-050213-040251.

Ruttkies, C., Neumann, S., & Posch, S. (2019). Improving MetFrag with statistical learning of fragment annotations. *BMC Bioinformatics*, *20*(1), 376. https://doi.org/10.1186/s12859-019-2954-7.

Saini, R. K., & Keum, Y.-S. (2018). Carotenoid extraction methods: A review of recent developments. *Food Chemistry*, *240*, 90–103. https://doi.org/10.1016/j.foodchem.2017.07.099.

Sajitz-Hermstein, M., Töpfer, N., Kleessen, S., Fernie, A. R., & Nikoloski, Z. (2016). iReMet-flux: Constraint-based approach for integrating relative metabolite levels into a stoichiometric metabolic models. *Bioinformatics*, *32*(17), i755–i762. https://doi.org/10.1093/bioinformatics/btw465.

Salzano, A. M., Renzone, G., Sobolev, A. P., Carbone, V., Petriccione, M., Capitani, D., et al. (2019). Unveiling kiwifruit metabolite and protein changes in the course of post-harvest cold storage. *Frontiers in Plant Science*, *10*(71). https://doi.org/10.3389/fpls.2019.00071.

Sangster, T., Major, H., Plumb, R., Wilson, A. J., & Wilson, I. D. (2006). A pragmatic and readily implemented quality control strategy for HPLC-MS and GC-MS-based metabonomic analysis. *Analyst*, *131*(10), 1075–1078. https://doi.org/10.1039/B604498K.

Savoi, S., Wong, D. C. J., Arapitsas, P., Miculan, M., Bucchetti, B., Peterlunger, E., et al. (2016). Transcriptome and metabolite profiling reveals that prolonged drought modulates the phenylpropanoid and terpenoid pathway in white grapes (*Vitis vinifera* L.). *BMC Plant Biology*, *16*(1), 67. https://doi.org/10.1186/s12870-016-0760-1.

Shammai, A., Petreikov, M., Yeselson, Y., Faigenboim, A., Moy-Komemi, M., Cohen, S., et al. (2018). Natural genetic variation for expression of a SWEET transporter among wild species of *Solanum lycopersicum* (tomato) determines the hexose composition of ripening tomato fruit. *The Plant Journal*, *96*(2), 343–357. https://doi.org/10.1111/tpj.14035.

Shi, H., & Schwender, J. (2016). Mathematical models of plant metabolism. *Current Opinion in Biotechnology*, *37*, 143–152. https://doi.org/10.1016/j.copbio.2015.10.008.

Šimura, J., Antoniadi, I., Široká, J., Tarkowská, D., Strnad, M., Ljung, K., et al. (2018). Plant hormonomics: Multiple phytohormone profiling by targeted metabolomics. *Plant Physiology*, *177*(2), 476–489. https://doi.org/10.1104/pp.18.00293.

Sobolev, A. P., Mannina, L., Proietti, N., Carradori, S., Daglia, M., Giusti, A. M., et al. (2015). Untargeted NMR-based methodology in the study of fruit metabolites. *Molecules*, *20*(3), 4088–4108.

Srivastava, R. K., Talluri, S., Beebi, S. K., & Kumar, B. R. (2018). Magnetic resonance imaging for quality evaluation of fruits: A review. *Food Analytical Methods*, *11*(10), 2943–2960. https://doi.org/10.1007/s12161-018-1262-6.

Stanstrup, J., Broeckling, C. D., Helmus, R., Hoffmann, N., Mathé, E., Naake, T., et al. (2019). The metaRbolomics Toolbox in Bioconductor and beyond. *Metabolites*, *9*(10), 200. https://doi.org/10.3390/metabo9100200.

Stevens, R. G., Baldet, P., Bouchet, J.-P., Causse, M., Deborde, C., Deschodt, C., et al. (2018). A systems biology study in tomato fruit reveals correlations between the ascorbate pool and genes involved in ribosome biogenesis, translation, and the heat-shock response. *Frontiers in Plant Science*, *9*(137). https://doi.org/10.3389/fpls.2018.00137.

Stewart, D., Shepherd, L. V. T., Hall, R. D., & Fraser, P. D. (2018). Crops and tasty, nutritious food—How can metabolomics help? In R. D. Hall (Ed.), *Annual plant reviews* (pp. 181–217). Wiley. online.

Stitt, M., & Gibon, Y. (2014). Why measure enzyme activities in the era of systems biology? *Trends in Plant Science*, *19*(4), 256–265. https://doi.org/10.1016/j.tplants.2013.11.003.

Sumner, L. W., Yang, D. S., Bench, B. J., Watson, B. S., Li, C., & Jones, A. D. (2018). Spatially resolved plant metabolomics. *Annual Plant Reviews*, 343–366. https://doi.org/10.1002/9781119312994.apr0471. online.

Sun, X., Zhu, A., Liu, S., Sheng, L., Ma, Q., Zhang, L., et al. (2013). Integration of metabolomics and subcellular organelle expression microarray to increase understanding the organic acid changes in post-harvest Citrus fruit. *Journal of Integrative Plant Biology*, *55*(11), 1038–1053. https://doi.org/10.1111/jipb.12083.

Sysi-Aho, M., Katajamaa, M., Yetukuri, L., & Orešič, M. (2007). Normalization method for metabolomics data using optimal selection of multiple internal standards. *BMC Bioinformatics*, *8*(1), 93. https://doi.org/10.1186/1471-2105-8-93.

Tadeo, F. R., Terol, J., Rodrigo, M. J., Licciardello, C., & Sadka, A. (2020). Chapter 12—Fruit growth and development. In M. Talon, M. Caruso, & F. G. Gmitter (Eds.), *The genus citrus* (pp. 245–269). Woodhead Publishing.

Tang, H., Zhang, X., Gong, B., Yan, Y., & Shi, Q. (2020). Proteomics and metabolomics analysis of tomato fruit at different maturity stages and under salt treatment. *Food Chemistry*, *311*, 126009. https://doi.org/10.1016/j.foodchem.2019.126009.

Tietel, Z., Feldmesser, E., Lewinsohn, E., Fallik, E., & Porat, R. (2011). Changes in the transcriptome of 'Mor' mandarin flesh during storage: Emphasis on molecular regulation of fruit flavor deterioration. *Journal of Agricultural and Food Chemistry*, *59*(8), 3819–3827. https://doi.org/10.1021/jf104614s.

Tikunov, Y. M., Molthoff, J., de Vos, R. C. H., Beekwilder, J., van Houwelingen, A., van der Hooft, J. J. J., et al. (2013). NON-SMOKY GLYCOSYLTRANSFERASE1 prevents the release of smoky aroma from tomato fruit. *Plant Cell*, *25*(8), 3067–3078. https://doi.org/10.1105/tpc.113.114231.

Tohge, T., & Fernie, A. R. (2015). Metabolomics-inspired insight into developmental, environmental and genetic aspects of tomato fruit chemical composition and quality. *Plant and Cell Physiology*, *56*(9), 1681–1696. https://doi.org/10.1093/pcp/pcv093.

Tsugawa, H. (2018). Advances in computational metabolomics and databases deepen the understanding of metabolisms. *Current Opinion in Biotechnology*, *54*, 10–17. https://doi.org/10.1016/j.copbio.2018.01.008.

Tummler, K., Lubitz, T., Schelker, M., & Klipp, E. (2014). New types of experimental data shape the use of enzyme kinetics for dynamic network modeling. *The FEBS Journal*, *281*(2), 549–571. https://doi.org/10.1111/febs.12525.

Uarrota, G. V., Fuentealba, C., Hernández, I., Defilippi-Bruzzone, B., Meneses, C., Campos-Vargas, R., et al. (2019). Integration of proteomics and metabolomics data of early and middle season Hass avocados under heat treatment. *Food Chemistry*, *289*, 512–521. https://doi.org/10.1016/j.foodchem.2019.03.090.

Uekane, T. M., Nicolotti, L., Griglione, A., Bizzo, H. R., Rubiolo, P., Bicchi, C., et al. (2017). Studies on the volatile fraction composition of three native Amazonian-Brazilian fruits: Murici (*Byrsonima crassifolia* L., Malpighiaceae), bacuri (*Platonia insignis* M., Clusiaceae), and sapodilla (*Manilkara sapota* L., Sapotaceae). *Food Chemistry*, *219*, 13–22. https://doi.org/10.1016/j.foodchem.2016.09.098.

Vallarino, J. G., de Abreu e Lima, F., Soria, C., Tong, H., Pott, D. M., Willmitzer, L., et al. (2018). Genetic diversity of strawberry germplasm using metabolomic biomarkers. *Scientific Reports*, *8*(1), 14386. https://doi.org/10.1038/s41598-018-32212-9.

Vallarino, J. G., Osorio, S., Bombarely, A., Casañal, A., Cruz-Rus, E., Sánchez-Sevilla, J. F., et al. (2015). Central role of FaGAMYB in the transition of the strawberry receptacle from development to ripening. *New Phytologist*, *208*(2), 482–496. https://doi.org/10.1111/nph.13463.

Vallarino, J. G., Pott, D. M., Cruz-Rus, E., Miranda, L., Medina-Minguez, J. J., Valpuesta, V., et al. (2019). Identification of quantitative trait loci and candidate genes for primary metabolite content in strawberry fruit. *Horticulture Research*, *6*. https://doi.org/10.1038/s41438-018-0077-3.

van der Knaap, E., & Østergaard, L. (2018). Shaping a fruit: Developmental pathways that impact growth patterns. *Seminars in Cell & Developmental Biology*, *79*, 27–36. https://doi.org/10.1016/j.semcdb.2017.10.028.

Veličković, D., Herdier, H., Philippe, G., Marion, D., Rogniaux, H., & Bakan, B. (2014). Matrix-assisted laser desorption/ionization mass spectrometry imaging: A powerful tool for probing the molecular topology of plant cutin polymer. *The Plant Journal*, *80*(5), 926–935. https://doi.org/10.1111/tpj.12689.

Venter, P., Muller, M., Vestner, J., Stander, M. A., Tredoux, A. G. J., Pasch, H., et al. (2018). Comprehensive three-dimensional LC × LC × ion mobility spectrometry separation combined with high-resolution MS for the analysis of complex samples. *Analytical Chemistry*, *90*(19), 11643–11650. https://doi.org/10.1021/acs.analchem.8b03234.

Viant, M. R., Kurland, I. J., Jones, M. R., & Dunn, W. B. (2017). How close are we to complete annotation of metabolomes? *Current Opinion in Chemical Biology*, *36*, 64–69. https://doi.org/10.1016/j.cbpa.2017.01.001.

Wang, Z., Cui, Y., Vainstein, A., Chen, S., & Ma, H. (2017). Regulation of fig (*Ficus carica* L.) fruit color: Metabolomic and transcriptomic analyses of the flavonoid biosynthetic pathway. *Frontiers in Plant Science*, *8*, 1990. https://doi.org/10.3389/fpls.2017.01990.

Wang, Z., Du, H., Zhai, R., Song, L., Ma, F., & Xu, L. (2017). Transcriptome analysis reveals candidate genes related to color fading of 'red bartlett' (*Pyrus communis* L.). *Frontiers in Plant Science*, *8*(455). https://doi.org/10.3389/fpls.2017.00455.

Wang, J.-G., Gao, X.-M., Ma, Z.-L., Chen, J., Liu, Y.-N., & Shi, W.-Q. (2019). Metabolomic and transcriptomic profiling of three types of litchi pericarps reveals that changes in the hormone balance constitute the molecular basis of the fruit cracking susceptibility of *Litchi chinensis* cv. Baitangying. *Molecular Biology Reports*, *46*(5), 5295–5308. https://doi.org/10.1007/s11033-019-04986-2.

Wang, Y., Zhang, X., Yang, S., & Yuan, Y. (2018). Lignin involvement in programmed changes in peach-fruit texture indicated by metabolite and transcriptome analyses. *Journal of Agricultural and Food Chemistry*, *66*(48), 12627–12640. https://doi.org/10.1021/acs.jafc.8b04284.

Wiseman, J. M., Ifa, D. R., Song, Q., & Cooks, R. G. (2006). Tissue imaging at atmospheric pressure using desorption electrospray ionization (DESI) mass spectrometry. *Angewandte Chemie International Edition*, *45*(43), 7188–7192. https://doi.org/10.1002/anie.200602449.

Wu, X., Jiang, L., Yu, M., An, X., Ma, R., & Yu, Z. (2016). Proteomic analysis of changes in mitochondrial protein expression during peach fruit ripening and senescence. *Journal of Proteomics*, *147*, 197–211. https://doi.org/10.1016/j.jprot.2016.06.005.

Yang, W., Ma, X., Laaksonen, O., He, W., Kallio, H., & Yang, B. (2019). Effects of latitude and weather conditions on proanthocyanidins in blackcurrant (*Ribes nigrum*) of Finnish commercial cultivars. *Journal of Agricultural and Food Chemistry*, *67*(51), 14038–14047. https://doi.org/10.1021/acs.jafc.9b06031.

Ying, H., Shi, J., Zhang, S., Pingcuo, G., Wang, S., Zhao, F., et al. (2019). Transcriptomic and metabolomic profiling provide novel insights into fruit development and flesh coloration in *Prunus mira* Koehne, a special wild peach species. *BMC Plant Biology*, *19*(1), 463. https://doi.org/10.1186/s12870-019-2074-6.

Yuan, Y., Zhao, Y., Yang, J., Jiang, Y., Lu, F., Jia, Y., et al. (2017). Metabolomic analyses of banana during postharvest senescence by 1H-high resolution-NMR. *Food Chemistry*, *218*, 406–412. https://doi.org/10.1016/j.foodchem.2016.09.080.

Zampieri, M., Sekar, K., Zamboni, N., & Sauer, U. (2017). Frontiers of high-throughput metabolomics. *Current Opinion in Chemical Biology*, *36*, 15–23. https://doi.org/10.1016/j.cbpa.2016.12.006.

Zanor, M. I., Osorio, S., Nunes-Nesi, A., Carrari, F., Lohse, M., Usadel, B., et al. (2009). RNA Interference of LIN5 in tomato confirms its role in controlling brix content, uncovers the influence of sugars on the levels of fruit hormones, and demonstrates the importance of sucrose cleavage for normal fruit development and fertility. *Plant Physiology*, *150*(3), 1204–1218. https://doi.org/10.1104/pp.109.136598.

Zhang, W.-F., Gong, Z.-H., Wu, M.-B., Chan, H., Yuan, Y.-J., Tang, N., et al. (2019). Integrative comparative analyses of metabolite and transcript profiles uncovers complex regulatory network in tomato (*Solanum lycopersicum* L.) fruit undergoing chilling injury. *Scientific Reports*, *9*(1), 4470. https://doi.org/10.1038/s41598-019-41065-9.

Zhang, Q., Wang, L., Liu, Z., Zhao, Z., Zhao, J., Wang, Z., et al. (2020). Transcriptome and metabolome profiling unveil the mechanisms of *Ziziphus jujuba* Mill. peel coloration. *Food Chemistry*, *312*, 125903. https://doi.org/10.1016/j.foodchem.2019.125903.

Zhao, G., Lian, Q., Zhang, Z., Fu, Q., He, Y., Ma, S., et al. (2019). A comprehensive genome variation map of melon identifies multiple domestication events and loci influencing agronomic traits. *Nature Genetics*, *51*(11), 1607–1615. https://doi.org/10.1038/s41588-019-0522-8.

Zhao, X., Yuan, X., Chen, S., Fu, D.-Q., & Jiang, C.-Z. (2019). Metabolomic and transcriptomic analyses reveal that a MADS-box transcription factor TDR4 regulates tomato fruit quality. *Frontiers in Plant Science*, *10*(792). https://doi.org/10.3389/fpls.2019.00792.

Zhao, X. D., Yuan, X. Y., Chen, S., Meng, L. H., & Fu, D. Q. (2018). Role of the tomato TAGL1 gene in regulating fruit metabolites elucidated using RNA sequence and metabolomics analyses. *Plos One*, *13*(6). https://doi.org/10.1371/journal.pone.0199083.

Zhu, G., Wang, S., Huang, Z., Zhang, S., Liao, Q., Zhang, C., et al. (2018). Rewiring of the fruit metabolome in tomato breeding. *Cell*, *172*(1). https://doi.org/10.1016/j.cell.2017.12.019. 249-261.e212.

CHAPTER TWO

Specialized metabolites in seeds

Massimiliano Corso, François Perreau, Loïc Rajjou, Rawen Ben Malek, Loïc Lepiniec, and Grégory Mouille*

Institut Jean-Pierre Bourgin, Université Paris-Saclay, INRAE, AgroParisTech, Versailles, France
*Corresponding author: e-mail address: gregory.mouille@inrae.fr

Contents

Abstract

Seed quality is considered as a major agricultural issue with respect to food and non-food uses, biodiversity preservation and environmental protection. Germination efficiency and seed vigor are key factors to ensure correct plant production and yield. Germination can be limited by seed dormancy and/or suboptimal environmental conditions. In addition, seed viability is a fundamental trait to consider in the management of genetic resources and market chain. Seed quality is largely determined by the accumulation of various storage compounds including specialized metabolites (SMs) with protective functions. Therefore, a comprehensive investigation of seed SMs is of paramount importance. During the last years, developing multiomics approaches has provided a wealth of data highlighting the key biological functions and major molecular/metabolic hubs of seed specialized metabolites. This chapter will give an updated overview of the biodiversity of these metabolites and the regulatory networks that control their production and storage. We will also discuss the spatial distribution and structure/function relationships of these SMs. We will focus on the following four major groups detected in seeds, phenylpropanoids (more than about 10,000 compounds), alkaloids (about 20,000 compounds), sulfur-containing compounds (e.g., glucosinolates), and terpenoids (more than 40,000 compounds).

Advances in Botanical Research, Volume 98
ISSN 0065-2296
https://doi.org/10.1016/bs.abr.2020.11.001

1. Introduction

Seeds are the main vectors for spermaphyte dissemination and the foundation of agriculture. Plant seeds comprise three genetically distinct compartments namely the zygotic embryo, the endosperm, and maternal tissues including the seed coat. In many species that produce indehiscent fruits, the pericarp surrounding the seed is adjoined to the seed coat. Seed quality is of major importance since it determines crop yield and other physiological parameters. During seed maturation, the incoming material flow is plentiful while the outcoming material flow is residual, thus allowing the accumulation of storage compounds. Therefore, seeds are considered as a sink organ storing carbon, nitrogen and sulfur reserves in proteins, lipids, carbohydrates and specialized metabolites (Ohto, Stone, & Harada, 2007). The late maturation program is associated with the acquisition of seed physiological properties (e.g., dormancy, longevity, stress resistance and germination potential) (Leprince, Pellizzaro, Berriri, & Buitink, 2017). During maturation, seeds are subjected to a large range of complex biochemical changes and metabolic transitions from autotrophic to heterotrophic state with a significant reduction of the metabolic activity, the loss of photosynthetic ability, loss of water leading to metabolic quiescence and dormancy. The so-called orthodox seeds form an intracellular glass that is required for long-term anhydrobiotic survival. The dry mature seeds are full of storage nutrients that act as a fuel for the germination and the emergence of the future seedling. The metabolic quiescence requires the limitation of oxidative metabolism by a diversity of antioxidant systems and the inhibition of protease, lipase and amylase activities (Zahedi, Karimi, & Venditti, 2019). The metabolic switches occurring from maturation to germination are associated with specific gene expression programs related to specialized metabolism. In both monocots and dicots, storage compounds and specialized metabolites are located in the endosperm, the embryo or maternal tissues (e.g., the nucellus or seed coat). In most of the non-endospermic dicots, cotyledons act as the main storage compartment, while in endospermic monocots the endosperm ensures this function. The seed coat provides physical and chemical barriers to protect the embryo (Rajjou & Debeaujon, 2008). The accumulation of the main seed storage compounds has been largely described (Baud, Dubreucq, Miquel, Rochat, & Lepiniec, 2008; Shewry, Napier, & Tatham, 1995; Tetlow, 2011). Moreover, seeds accumulate a wide range of specialized metabolites namely, phenolic compounds (or phenylpropanoids), alkaloids and N-containing metabolites and terpenes (Corso, Perreau, Mouille, & Lepiniec, 2020), which are accumulated at different levels depending on the considered species. The accumulation of these SMs in seeds is strongly influenced by the environment of the mother plant. For instance, under low temperatures, Arabidopsis seed accumulates a large amount of procyanidins (polyphenolic polymers) in the seed coat triggering a deeper primary dormancy (Macgregor et al., 2015). In addition, glucosinolate accumulation is enhanced in *Brassica napus* seed, when the mother plant is subjected to drought stress (Mailer & Cornish, 1987). Similarly, drought and high-temperature stresses increase alkaloid content in seeds of *Lupinus angustifolius* (Christiansen, Joørnsgård, Buskov, & Olsen, 1997; Frick et al., 2018; Jansen, Jürgens, & Ordon, 2009). Besides their role in abiotic stress responses, flavonoids, cinnamic acids and coumarins are important in seed defence against biotic attacks (Corso et al., 2020; Shirley, 1998), as testified by the antipathogenic activity of anthocyanidins and isoflavones against several pathogenic microorganisms in model and crop plants (Shirley, 1998). These examples emphasize the induction of seed SM accumulation by biotic and abiotic stresses, allowing the seed to cope with harmful environments.

In addition to the physiological role, SMs abundance impacts the nutritional and organoleptic qualities of seeds. Interestingly, plant specialized metabolites are largely recognized as health-beneficial compounds (Dillard & Bruce German, 2000). However, some of these metabolites also display undesirable food properties such as anti-nutritional effects, anti-palatability or toxicity (Emire, Jha, & Mekam, 2013). Therefore, during plant domestication, seed composition has been mainly oriented towards the elimination of the anti-nutritional and toxic SMs (e.g., Sánchez-Pérez et al., 2019).

The accumulation of plant SMs is driven by regulatory genes that intervene at the transcriptional and epigenetic levels (Jacobowitz & Weng, 2020; Kooke et al., 2019; Lacchini & Goossens, 2020; Vom Endt, Kijne, & Memelink, 2002; Zhao & Rhee, 2019). A major role in these processes is played by transcription factors (TFs) that are able to coordinate specialized metabolisms through the regulation of key enzymes of the biosynthetic pathways or transporters (Broun, 2005; Colinas & Goossens, 2018). Seed SMs pathways are widely, but not exclusively, regulated by members of the MYB and bHLH families (Corso et al., 2020; Dubos et al., 2010; Hirai et al., 2007; Kroc et al., 2019; Kroc, Czepiel, Wilczura, Mokrzycka, & Święcicki, 2019; Lepiniec et al., 2006; Liu, Osbourn, & Ma, 2015; Xu, Dubos, & Lepiniec, 2015).

Seeds undergo two highly-specific developmental phases (i.e., maturation and germination) impacting dramatically SM distribution, as shown for oilseed rape (Bhandari et al., 2015; Bhandari, Jo, & Lee, 2015), barley (Gorzolka, Kölling, Nattkemper, & Niehaus, 2016; Gupta et al., 2019), sorghum (Montini et al., 2020), soybean (Gupta et al., 2019) or *Maytenus senegalensis* (Eckelmann, Kusari, & Spiteller, 2017). SMs localization in plant and seed tissues can be performed via in vitro isolation and extraction of specific tissues, or imaging in situ (Boughton, Thinagaran, Sarabia, Bacic, & Roessner, 2016). While separation of the different seed sections and analysis of SMs accumulation using chemical techniques can be a very laborious and inaccurate process, cytological and histological were more efficient techniques to perform in situ localization of SMs. In particular, Mass Spectrometry Imaging (MSI) has been shown to provide a powerful tool to investigate the spatiotemporal distribution of the SM. MSI allowed the analyses of SM accumulation between and within tissues, which is particularly interesting for plant seeds that are characterized by highly differentiated tissues (seed coat, endosperm, embryo and, within the embryo, cotyledon, radicle, and hypocotyl). Matrix-Assisted Laser Desorption Ionization (MALDI) is the most widely used technique for seed imaging. Despite its high resolution, MALDI is not suitable for identifying molecules smaller than 1000 Da due to matrix interference with the signal (Bednarz, Rolo, & Niehaus, 2019; Bhandari, Jo, & Lee, 2015; Bhandari, Wang, et al., 2015; Gorzolka et al., 2016; Gorzolka, Bednarz, & Niehaus, 2014; Gupta et al., 2019; He et al., 2019; Sagara, Bhandari, Spengler, & Vollmann, 2020). Desorption Electrospray Ionization (DESI) is another high-resolution MSI technique adapted to polar compounds (Bjarnholt, Li, D'Alvise, & Janfelt, 2014), which was already used to study seed SMs distribution (Mohana Kumara, Srimany, Ravikanth, Uma Shaanker, & Pradeep, 2015; Srimany et al., 2016). TOF-SIMS was also used to visualise seed SMs distribution (Seyer, Einhorn, Brunelle, & Laprévote, 2010). Finally, Laser desorption Ionization (LDI) and Laser-Assisted Electrospray Ionization (LAESI) are other promising, but never employed, techniques that could be used to perform SMs MSI (Bjarnholt et al., 2014).

The above mentioned MSI techniques could be potentially used to monitor SMs accumulation dynamics in seeds at different developmental stages, and give further information regarding metabolites fluxes, locations and about their putative function during seed maturation. Although few time-course MSI analyses have been published on developing and maturing seeds (Bednarz et al., 2019; Mohana Kumara et al., 2015;

Srimany et al., 2016), this technique has been employed to analyse metabolites profiles during seed development (Bhandari, Jo, & Lee, 2015; Bhandari, Wang, et al., 2015).

A main challenge for metabolomic analyses remains SMs annotation and identification. Although untargeted metabolomic analysis by chromatography and tandem mass spectrometry generates thousands of MS and MS/MS spectra, this technique does not rely on automatic and reliable annotation of the compounds detected. Molecular networking is an innovative and efficient approach that allowed the clustering of compounds with a high degree of chemical similarity according to their MS^2 spectra and therefore identified clusters of unknown metabolites using the MSMS information of annotated compounds (Olivon et al., 2017, 2018; Wang et al., 2016). It improves the understanding of chemical diversity and allows annotation by proximity with the already identified or annotated compounds. Molecular networking has already been widely used in pharmacognosy studies to explore chemical diversity in several plant species (Wolfender, Litaudon, Touboul, & Queiroz, 2019). Despite its great potential, it has never been adopted to characterize the metabolic diversity in seeds and potentially relate the accumulation of a date metabolite to its function, for example in response to environmental perturbations.

2. Phenylpropanoids

Phenylpropanoids from plants include more than 10,000 molecules (Tungmunnithum, Thongboonyou, Pholboon, & Yangsabai, 2018). Anthocyanins, flavonols, flavan-3-ols and proanthocyanidins (PAs; also called condensed tannins polymers of flavan-3-ols) are abundant phenylpropanoid classes found in seed tissues. PAs and anthocyanins are preferentially located in seed coats. In food/feed supplies, condensed PAs and hydrolysable tannins are capable of binding and precipitating the proteins, thereby altering their digestibility and the bioavailability of amino acids. The use of tannin-rich seeds and grains in animal feed is often not recommended and can affect nutritional qualities of food products for human consumption. By binding to salivary proteins, the tannins give an astringent taste. Mild astringency can enhance the taste of a number of foods and beverages, although an excess of tannins leads to extreme astringency that can make seed-based food unpleasant or even repulsive. Most seeds have a very high tannin content in the outer integument, this has certainly been a selective advantage during evolution, limiting the consumption of seeds by animals until they are able to be spread and germinate. Thus, the history of plant domestication and breeding has driven humans to select more light-coloured seeds, with low tannin content, for most seed crops such as rice (Zhu et al., 2011), wheat (Kuraparthy, Sood, & Gill, 2008), Brassica (Nesi, Delourme, Brégeon, Falentin, & Renard, 2008; Rahman, Joersbo, & Poulsen, 2001), amaranth (Stetter, Vidal-Villarejo, Schmid, & Wright, 2020), or pulses (Zhou et al., 2015), in order to improve protein bioavailability, palatability and industrial processing of seed. PAs of the seed coat are not essential for embryo and endosperm development. Nevertheless, the Arabidopsis mutants impaired in flavonoid metabolism (a.k.a. *transparent testa* mutants) exhibit a reduction of both seed longevity and dormancy (Debeaujon et al., 2003; Debeaujon, Léon-Kloosterziel, & Koornneef, 2000) resulting in better germination performance under optimal conditions (Rajjou et al., 2004). In rice, the *Rc* locus encoding for a bHLH transcription factor is involved in the regulation of both ABA (Abscisic acid) content in seeds and flavonoid accumulation in the pericarp resulting in the control of seed dormancy (Gu et al., 2011; Liu, Li, & Xing, 2018). The hypothesis to explain the close relationship between flavonoids, seed longevity and seed dormancy is related to the control of metabolic activities and cellular redox homeostasis. Flavonoids act as effective scavengers of reactive oxygen species (ROS). For instance, the flavonol aglycones, namely quercetin, and its derivatives were showed to inhibit trypsin-like serine proteases (Xue et al., 2017), showed metal chelation properties (Leopoldini, Russo, Chiodo, & Toscano, 2006) and displayed strong antioxidant activity (Zhang et al., 2011). It has been suggested that seed germination is possible only when the ROS content reaches a threshold to enter the appropriate "oxidative window" (Arc et al., 2011; Bailly, El-Maarouf-Bouteau, & Corbineau, 2008). Below this threshold, the amount of ROS in dry or imbibed seed remains too low for dormancy release. Thus, overaccumulation of flavonoids promotes dormancy by lowering the redox potential. During dry storage, the photo-protective role of flavonoids in the seed coat prevents light-induced oxidative damage and the accumulation of age-associated oxidative damage. Therefore, seeds with low flavonoids content exhibit a reduced longevity. Because of their described antibacterial, antiviral, and antifungal properties (Orhan, Özçelik, Özgen, & Ergun, 2010), flavonoids probably act as major regulators of beneficial and pathogenic seed-microbe interactions during germination and seedling establishment (Fig. 1) but only a few works have been conducted on this issue.

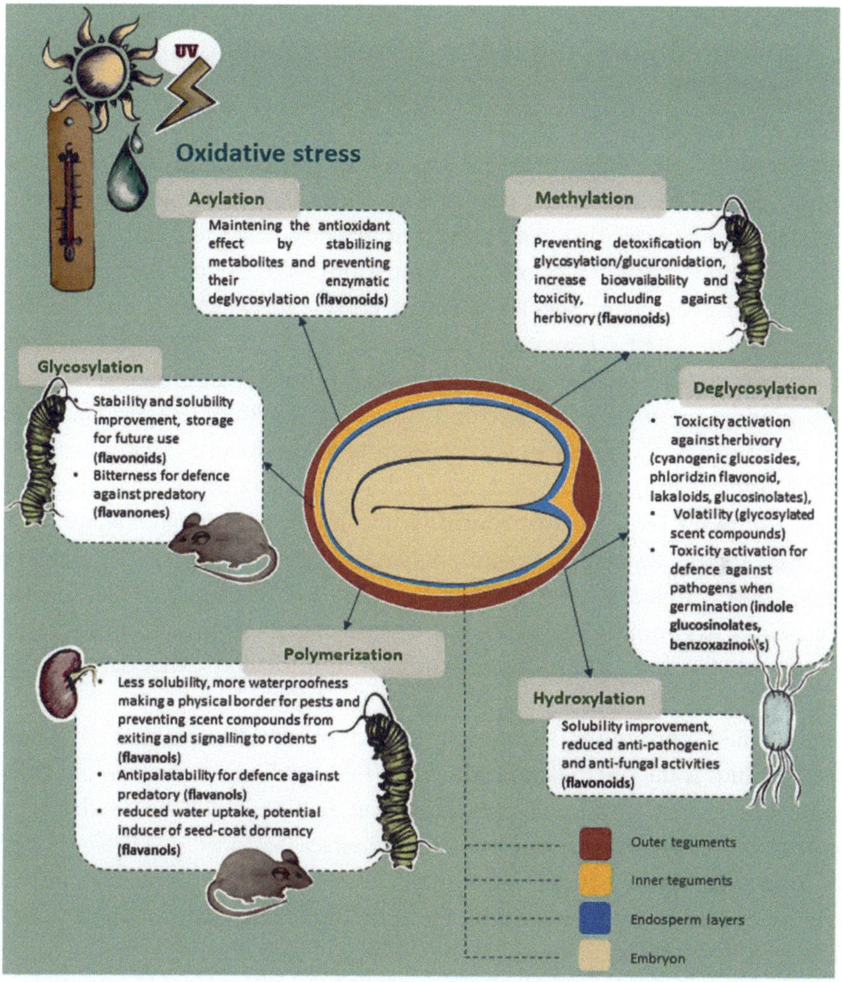

Fig. 1 Functions of chemical modifications of seed specialized metabolites, based on knowledge about flavonoids (Alseekh, Perez de Souza, Benina, & Fernie, 2020) and other compounds (Wang et al., 2019).

Given the function of PAs and flavonols as phytonutrients with a wide range of positive health benefits on human health (Martin & Li, 2017) and the involvement in seed tolerance to biotic and abiotic stresses (Corso et al., 2020; Paauw, Koes, & Quattrocchio, 2019), selection of crops with reduced levels of these metabolites in seeds might have generated undesired effects.

2.1 Genetic, transcriptional and post-transcriptional spatiotemporal regulation of seed phenylpropanoid synthesis

Phenylpropanoids are largely accumulated in plant seeds, including benzoic and cinnamic acids, coumarins and flavonoids (Corso et al., 2020). While the transcriptional regulators of coumarin, benzoic and cinnamic acids biosynthesis in seeds are poorly characterized or unknown, some transcription factors associated with flavonoid accumulation have been characterized in model and crop plant species (Baudry et al., 2004; Corso et al., 2020; Lepiniec et al., 2006). Flavonoids, in particular PA and flavonols, constitute the largest class of phenylpropanoids and are widely distributed in seeds of Arabidopsis and other plant species (Lepiniec et al., 2006). PAs and flavonols accumulation in the seed can vary according to the species. For example, In Arabidopsis, legume species and rapeseed, PAs are accumulated in the innermost tegument layer (endothelium), while in *Medicago truncatula* are accumulated in the outermost teguments layer (Corso et al., 2020; Lepiniec et al., 2006; Smýkal, Vernoud, Blair, Soukup, & Thompson, 2014). In addition, other flavonoid metabolites, such as the anthocyanins, accumulated in the seeds of maize, rice, wheat, barley, sorghum and other Brassicaceae species, but not in Arabidopsis (Corso et al., 2020).

The first committed enzyme involved in the biosynthesis of phenylpropanoids is the phenylammonia-lyase (PAL), which cleaves the phenylalanine to produce ammonia trans-cinnamic acid that will be used for the synthesis of coumarins, flavonoids and stilbenes (Saito et al., 2013; Tohge, de Souza, & Fernie, 2017). The genes involved in flavonoid biosynthesis are divided into two categories: the structural Early Biosynthetic Genes (EBGs) and Late Biosynthetic Genes (LBGs) (Gonzalez, Zhao, Leavitt, & Lloyd, 2008; Lepiniec et al., 2006). *CHALCONE SYNTHASE* (*CHS*), *CHALCONE ISOMERASE* (*CHI*), *FLAVANONE 3-HYDROXYLASE* (*F3H*), *FLAVONOID 3'-HYDROXYLASE* (*F3'H*) and *FLAVONOL SYNTHASE* (*FLS*) are the EBGs, involved in the biosynthesis of flavanone (e.g., naringenin) and flavonols (e.g., kaempferol and quercetin) (Lepiniec et al., 2006; Saito et al., 2013; Tohge et al., 2017). Anthocyanins (e.g., cyanidin) and proanthocyanidins (PAs, e.g., epicatechin polymer) biosynthesis is driven by the LBGs, namely *DIHYDROFLAVONOL REDUCTASE* (*DFR*), *LEUCOANTHOCYANIDIN DIOXYGENASE/ANTHOCYANIDIN SYNTHASE* (*LDOX/ANS*), and *ANTHOCYANIDIN REDUCTASE* (*ANR*), the latter encoded by the *BANYULS* (*BAN*) gene in *Arabidopsis thaliana* (Lepiniec et al., 2006).

Flavonoids biosynthetic enzymes are weakly bound to each other and form protein complexes that are associated with specific plant organelles. These ER-associated "multienzyme complexes" or "metabolons" play an important role in flavonoid accumulation, bioactivity and structural diversification (Nakayama, Takahashi, & Waki, 2019). The existence of flavonoids metabolons was firstly proposed in 1974 by Stafford (1974) and subsequently confirmed by other studies (see Obata, 2019 and references inside). The interaction between flavonoid biosynthetic enzymes was described as a mechanism affecting the canalization of metabolic fluxes and therefore allowing a quick and efficient production of these compounds in response to environmental perturbations (Obata, 2019). One major challenge in the characterization of plant metabolons is the choice of the methodology/technique to identify protein-protein interactions, otherwise named "enzyme–enzyme assemblies" (Zhang & Fernie, 2020). In the case of flavonoids, the interactions between biosynthetic proteins were detected using Förster resonance energy transfer (FRET), co-immunoprecipitation (coIP), tandem affinity purification–MS (AP–MS) and yeast-two-hybrid (Y2H) assays techniques (Zhang & Fernie, 2020). Interactions between flavonoids biosynthetic enzymes have been detected in several plant species (Jørgensen et al., 2005; Nakayama et al., 2019). In *Arabidopsis thaliana*, CHS protein interacts with CHI, F3H, DFR, FLS and has been identified as the central hub for flavonoid metabolons formation and regulation of flavonoids level (Nakayama et al., 2019; Obata, 2019). In addition, a recent example of metabolons and metabolic channelling has been shown by Waki et al. (2020), demonstrating that the interaction between CHS and chalcone isomerase-like proteins (CHILs) increases the efficiency of flavonoid biosynthesis through the enhancement of 2′,4,4′,6′-tetrahydroxychalcone production (specific to the flavonoid pathway) and decrease of the synthesis of other polyketides, as p-coumaroyltriacetic acid lactone. Besides CHS, the interaction between other flavonoid biosynthetic enzymes has been demonstrated between DFR and LAR (Diharce, Golebiowski, Fiorucci, & Antonczak, 2016), CHI-DFR, FLS-DFR, CHI-F3H and FLS1-F3H (Nakayama et al., 2019; Obata, 2019). A major role for flavonoid metabolons has been highlighted in plant species other than Arabidopsis, as in soybean or snapdragon and torenia species of the *Lamiales* genus (Fujino et al., 2018; Nakayama et al., 2019).

Besides metabolons, the regulation and expression of flavonoid EBGs and LBGs in seeds is mainly controlled by members of MYBs and bHLHs TF families, which are regulated by several environmental signals

(Airoldi, Hearn, Brockington, Webb, & Glover, 2019; Hichri et al., 2011; Li, 2014; Xu et al., 2015). *Arabidopsis MYB11, MYB12* and *MYB111* TFs are the master regulators of flavonoid EBGs, while the R2R3-MYB/bHLH/WD40 (MBW) complex regulates the expression of the LBGs (Baudry et al., 2004; Dubos et al., 2010; Stracke et al., 2007, 2010; Xu et al., 2015, 2014). The LBGs are involved in the production of the seed-specific PAs and their expression is controlled by MYB (TRANSPARENT TESTA 2, TT2 and MYB5) (Nesi, Jond, Debeaujon, Caboche, & Lepiniec, 2001), bHLH (TT8/GL3/EGL3) (Nesi et al., 2000) and WDR (TRANSPARENT TESTA GLABRA1, TTG1) (Zhang & Schrader, 2017) proteins, which constitute the MBW complex (Baudry et al., 2004; Gonzalez, Mendenhall, Huo, & Lloyd, 2009; Xu et al., 2015). MYB5–TT8–TTG1 and TT2–EGL3/GL3–TTG1 regulate the expression of LBGs in the seed endothelium and chalaza, respectively, while TT8-TTG1-TT2 complex is active in both seed tissues (Xu et al., 2015, 2014).

Within the MBW TFs, *TTG1* works on a wide range of seed developmental/differentiation processes (Gonzalez et al., 2009; Wei et al., 2019; Zhang & Schrader, 2017). Despite its main importance, little is known about TTG1 regulation during flavonoid accumulation and in response to environmental stimuli.

Members of the *MYB, bHLH* and *WDR* gene families control the accumulation of flavonoid in seeds or grains in crop species (Zhu et al., 2020), including rapeseed (Hong et al., 2017; Jiang et al., 2019), *Medicago truncatula* (Liu, Hyung Jun, & Dixon, 2014) and tomato (Gao et al., 2018). While in Arabidopsis these transcription factors control the accumulation of PAs, in other species they can regulate the synthesis of other classes of flavonoids. In maize, the MYBs *COLORED ALEURONE1* (*C1*) and *PURPLE LEAF* (*PL*), the bHLHs *R* and *B1* (*BOOSTER1*), and the WDR gene *PAC1* (*PALE ALEURONE COLOR1*) control the accumulation of anthocyanins (Liu et al., 2015). Contrary to the MBW complexes found in Arabidopsis, these TFs could work independently on the regulation of anthocyanin pigmentation in maize kernels (C1). Metabolic engineering approaches showed that the maize C1 and R TFs expressed in soybean were able to modulate isoflavone accumulation in seeds, by increasing the levels of daidzein and decreasing the genistein accumulation (Yu et al., 2003). P1 (pericarp color1) control phlobaphenes (3-deoxy flavonoid / flavan-4-ol) biosynthesis in pericarp and floral tissues (Grotewold, Drummond, Bowen, & Peterson, 1994; Quattrocchio, Baudry, Lepiniec, & Grotewold, 2006). Another bHLH transcription factor, namely *Noemi*, interacts with *Ruby* transcriptional activator to control PAs production in citrus seeds and anthocyanin accumulation in the fruit, with consequences on fruit acidity (Butelli et al., 2019; Zhu, Wen, & Fernie, 2019).

In addition to the *MYB*, *bHLH* and *WDR* TFs, the GARP gene family has been characterized as a regulator of seed flavonoid synthesis (Gao et al., 2010). Despite some similarities with the MYB, it has been shown that the GARP family is phylogenetically different from MYB TFs (Safi et al., 2017). Within these TFs, *KAN4* has been characterized as a flavonoid regulator. *KAN4* loss-of-function mutant showed lower flavonols and PAs levels and decreased expression of related genes in seeds (Gao et al., 2010). In addition, Gao et al. (2010) demonstrate the ability of KAN4 to control the expression of flavonoids EBGs and LBGs. Despite that, it is not clear whether KAN4 plays a direct role in the control of flavonoid biosynthetic genes. KAN4 indeed is encoded by the ABERRANT TESTA SHAPE (ATS) locus and play also a role in integument development (McAbee et al., 2006). When ectopically expressed, it can bind the promoters of TT2, TT8, and TTG1 to inhibit PA accumulation. Nevertheless, the physiological significance of this inhibition and the interactions of ATS with TT1, TT16, or TTG2 still need further investigation (McAbee et al., 2006).

The accumulation of seed phenylpropanoids is controlled by various epigenetic mechanisms including chromatin remodelling, histone modifications and RNA silencing. For instance, RNA silencing has been demonstrated to control paramutations of regulatory genes in maize seeds (Chandler, 2007) and silencing of CHS in soybean seeds (Cho, Jones, & Vodkin, 2017; Tuteja, Zabala, Varala, Hudson, & Vodkin, 2009). More recently other mechanisms have been elucidated. For example, *SEEDSTICK* (*STK*) regulates the expression of *BAN* gene through the suppression of histone deacetylases (HDACs) activity (Mizzotti et al., 2014). The accumulation of H2A.Z variant is associated with lower flavonoid gene expression and weak H3K4me3 deposition (Cai et al., 2019). On the contrary, the ectopic expression of *SGD2* increases both H3K4me3 deposition and flavonoid gene expression in seedlings and seeds. Consistent with these results, the inhibition of H3K4me3 deposition in *sgd2* mutant reduces flavonoid gene expression. Finally, epistatic analysis demonstrated that SGD2 (and thus H3K4me3) acts downstream of H2A.Z deposition (Cai et al., 2019). In poplar, H3K9me2 and DNA demethylation repress anthocyanin biosynthesis, through the induction of *MYB182* expression that encodes a repressor of flavonoid genes (Fan et al., 2018). Although epigenetic mechanisms play an important role in the regulation of seed phenylpropanoid and most likely of other SMs, epigenetic controls have been described mainly in the vegetative organs (Cai et al., 2019; Kooke et al., 2019) and further studies are still needed to elucidate their impact in seeds of model and crop plants.

2.2 Spatiotemporal distribution of phenylpropanoids in seeds

Mass Spectrometry Imaging techniques have been used to localize flavonoids in seeds. Seyer et al. (2010) used Cluster-TOF-SIMS imaging technique on several genotypes of Arabidopsis dry seeds (WT and *tt7* mutant) differing in their flavonoid biosynthesis and composition. They confirmed the previously observed localization of many flavonoids in the seed coat (Routaboul et al., 2006), including quercetin-3-O-rhamnoside, being 10 to 20 times more abundant in seed coat than in cotyledons. MSI imaging techniques applied to *Pisum sativum* seeds, allowed the detection of eight phenolic species distributed in the seed coat, radicle or cotyledons according to their chemical composition (Seyer et al., 2010). Isoflavones (daidzein, glycitein, genistein and their derivatives) and polyamines were the main specialized metabolites detected by MALDI imaging in soybean seeds (Sagara et al., 2020), being more highly accumulated in the germ than in the cotyledons. Inside the germ, polyamines were found to be abundant in the apical and root meristem and the outer layers of root parenchyma, whereas isoflavones were observed especially in root parenchyma tissues. In another work, high accumulation of a putative gancaonin metabolite has been detected in the embryo of barley seeds from different accessions using MALDI imaging. Gancaonin highlighted different accumulation kinetics according to the barley accession and showed more pronounced accumulation in response to salt stress (Gupta et al., 2019). In addition, Gupta et al. (2019) showed that flavonoid accumulation rate in response to salt stress is positively correlated with the germination efficiency, suggesting that the flavonoids detected in the embryo may play a role in salt stress resistance during germination.

3. Alkaloids

Alkaloids are nitrogen-containing compounds mostly derived from amino acids. This large class of low molecular weight metabolites comprises approximately 20,000 annotated molecules that are mostly used as pharmaceuticals, stimulants, poisons and narcotics (Schläger & Dräger, 2016; Wangchuk, 2019). Among the daily-ingested alkaloids by humans there is caffeine (1,3,7-trimethylxanthine) and theobromine (3,7-dimethyl-1H-purine-2,6-dione), both purine alkaloids highly accumulated in coffee

(*Coffea arabica*) and cacao (*Theobroma cacao*) fruits and seeds (Ashihara & Suzuki, 2004), but also present in other plant species. Besides, alkaloids possess allelopathic activities during germination and defensive role against herbivores and pathogens (Table 1). In particular, an allelopathic function of seed coat purine alkaloids has been suggested. These alkaloids are released into the soil upon

Table 1 Few examples of abundant alkaloids in seeds (glucosinolates excluded).

Metabolite	Chemical family	Main seed sources	Main human perception or use (https:/www.drugbank.ca)	Physiological function for seeds
Caffeine	Purine alkaloid	*Coffea arabica*	Stimulant; energy supplement; bitter-tasting components	Allelopathy, elicitor antimicrobial activity, act against plant predators
Theobromine	Purine alkaloid	*Theobroma cacao*	Vasodilator; a diuretic, and heart stimulant; bitter-tasting components	Allelopathy
Hyoscyamine/ Daturine	Tropane alkaloid	*Datura stramonium*	Gastrointestinal disorders; control the symptoms of Parkinson's disease and rhinitis	Allelopathy
Lupanine	Quinolizidine alkaloid	*Lupinus spp.*	Hepatotoxicity; neurotoxicity	Allelopathy, act against plant predators, insecticide activity
Physostigmine	Indole alkaloid	*Physostigma venenosum*	Cholinesterase inhibitor; used to treat glaucoma; used to treat the central nervous system effects of anticholinergic drug overdoses	

imbibition inhibiting seed germination nearby the seedlings (Baumann & Gabriel, 1984; Suzuki & Waller, 1987). In addition, caffeine is able to stimulate plant defense mechanisms against herbivores and pathogens by activating the expression of defense-related genes and is involved in the control of diseases caused by plant-pathogenic bacteria (Ashihara, Sano, & Crozier, 2008; Kim, Choi, & Sano, 2010).

Quinolizidine alkaloids (QAs) are found in various *Leguminosae* seeds, particularly in the genus Lupinus (Frick, Kamphuis, Siddique, Singh, & Foley, 2017). These alkaloids can protect seeds by acting as antinutritional compounds and inhibiting germination of neighboring plants (Wink, 1983). Although these plants can grow in poor soils and produce protein-rich seeds, their use for feed/food is limited by the abundance and the toxicity of these SMs which have to be eliminated by soaking seeds in water for extended periods prior to consumption. The most abundant SMs, such as the QAs lupinine, angustifoline, cystisine, lupanine, sparteine or anagyrine, can cause neurological damages and teratogenicity (Cheeke & Kelly, 1989). Genetics, genomics and breeding approaches allowed to substantially reduce the alkaloid content in modern lupin varieties (Abraham et al., 2019; Frick et al., 2017; Rychel & Książkiewicz, 2019).

3.1 Spatiotemporal regulation of alkaloids and N-containing compounds content in seeds

Glucosinolates (GLSs) are N-containing SMs widely accumulated in Brassicaceae such as *B. napus* or *B. juncea*, although levels vary among species (Frerigmann, 2016; Li et al., 2013; Seo & Kim, 2017). GLSs have well-known anti-nutritional properties and are associated to increased astringency (Augustine, Mukhopadhyay, & Bisht, 2013). Glucosinolates are synthesized in roots and leaves, then transported to seeds (Nour-Eldin et al., 2012), where they accumulate at high levels (Bhandari, Jo, & Lee, 2015; Bhandari, Wang, et al., 2015). GLSs accumulation in plants is mainly regulated by MYB and bHLH TFs (Li et al., 2018; Li et al., 2013; Seo & Kim, 2017). For instance, eight R2R3 MYB TFs with two MYB repeats (R2R3) are involved in the regulation of GLSs in Arabidopsis (Seo & Kim, 2017). Surprisingly, some of these TFs also play a role in the regulation of palmitoleic acid (omega-7 monounsaturated fatty acid) (Troncoso-Ponce et al., 2016) suggesting their involvement in a more complex regulatory network. Interestingly, GLSs composition in *Arabidopsis thaliana* seeds were characterized by lower accumulation of aliphatic glucosinolates, and higher levels of gluconasturtiin (2-phenylethyl glucosinolate) and benzoyloxy glucosinolates than in leaves and roots, suggesting that at least benzoylation could take place in seeds as the last steps of biosynthesis (Brown, Tokuhisa, Reichelt, & Gershenzon, 2003).

Similarly to GLSs, QAs are synthetized in plant aerial tissues, then transported and highly accumulated in pods and then in seeds during the drying phase (up to 5% of their dry weight) (Otterbach, Yang, Kato, Janfelt, & Geu-Flores, 2019). Similarly, coniine, the toxic alkaloid of poison hemlock (*Conium maculatum*), was shown to be synthesized in maturing leaves, and then transported in seeds (Hotti & Rischer, 2017). Finally, in *Nigella sativa* seeds, monoterpenoids and indazole alkaloids were only found in the seed coat, whereas dopamine and the saponin α-hederin differentially accumulated in the seed coat and internal tissues (Botnick et al., 2012), suggesting differential roles in seed defense.

QAs biosynthetic genes have been only partially identified and characterized (Frick et al., 2018, 2017; Otterbach et al., 2019). Genomic and transcriptomics analyses allowed the identification of *APETALA2/ethylene response transcription factor, RAP2-7,* which co-segregates with the *iucundus* locus, associated to lower QAs content in seeds (Kroc, Czepiel, et al., 2019; Kroc, Koczyk, et al., 2019; Oram, 1983). In addition, RAP2-7 is associated with a single-nucleotide polymorphism linked to QAs alkaloid content in seeds (Kroc, Czepiel, et al., 2019; Kroc, Koczyk, et al., 2019). RAP2-7 belongs to the *AP2/ERF* TF, a large family in the plant kingdoms involved in abiotic and biotic stress responses, and that plays a central role in the regulation of diverse specialized metabolites synthesis, like the steroidal glycoalkaloids, nicotine and terpenoid indole alkaloids (Dietz, Vogel, & Viehhauser, 2010; Paul et al., 2020; Riechmann & Meyerowitz, 1998).

Alkaloid distribution in seeds has been characterized in various species by MSI (i.e., ESI, DESI and MALDI) imaging (Bednarz et al., 2019; Gorzolka et al., 2014, 2016; Mohana Kumara et al., 2015). For example, in *Areca catechu* seed (betel nut) alkaloids level are specifically accumulated in two regions according to their chemical identity and showed an overall reduction during ripening (Srimany et al., 2016). The major tomato saponin alkaloid α-tomatine is localized in the endosperm, whereas trisaccharide β-tomatine was specifically found in some cotyledon structures, complementary to tetrasaccharide α-tomatine (Bednarz et al., 2019). The chromone alkaloid rohitukine is preferentially accumulated in cotyledons of *Dysoxylum binectariferum*, while it is less abundant in the embryo and in the seed coat (Mohana Kumara et al., 2015). As demonstrated by MALDI imaging, barley dry and early germinating seed contain few alkaloid compounds (hordatines) (Gorzolka et al., 2016) with no specific accumulation in the different tissues of the embryo (Gorzolka et al., 2014). Finally, two major tropane alkaloids—acetyltropine and protopine—were specifically found in the internal seed coat of germinating *Taxus chinensis* seed (He et al., 2019).

4. Terpenoids

Terpenoids have historically been used as flavors, fragrances, pharmaceuticals and pesticides. More than 40,000 different terpenoid molecules have been described (Misawa, 2011). In plants two independent pathways are involved in the production of the basic isoprene units, namely isopentenyl diphosphate (IDP) and dimethylallyl diphosphate (DMADP), required for terpenoids biosynthesis: the methylerythritol phosphate (MEP) pathway, which takes place in the plastids, and the mevalonic acid (MVA) pathway, which is localized to the cytosol, endoplasmic reticulum (ER) and peroxisomes (Zhou, 2018). In seed physiology, the control of MEP pathway is of paramount importance because it synthetizes the basic units used to produce monoterpenoids, diterpenoids and tetraterpenoids. Diterpenoids are the precursors of gibberellins (GAs) and vitamin E (tocopherols and tocotrienols). The Arabidopsis mutants affected in vitamin E biosynthesis (tocopherol/tocotrienol cyclase *tc/vte1* and homogentisate phytyltransferase *hpt/vte2* mutants) allowed to show that the primary function of tocopherols in the seed is to limit nonenzymatic lipid oxidation during dry storage, germination and early seedling establishment. As a consequence, reduced storability has been observed for seeds with low vitamin E content (Sattler, Gilliland, Magallanes-Lundback, Pollard, & DellaPenna, 2004). In monocot seeds, the tocotrienol synthesis is initiated by the homogentisate geranylgeranyl transferase (HGGT) enzyme, which is related to homogentisate phytyltransferase (HPT) that catalyzes the prenylation step in tocopherol synthesis. Rice knockout lines for *TC/VTE1*, *HPT/VTE2* and *HGGT* genes displayed low germination potential and a higher proportion of abnormal seedlings than the wild-type plants. It has been suggested that tocopherols may protect the embryo from ROS under both artificial aging and abiotic stress conditions, whereas tocotrienols in the pericarp may exclusively help in reducing the metabolic activity of the seed during artificial aging (Chen, Li, Fang, Shi, & Chen, 2016). Interestingly, the overexpression of the barley *HGGT* gene in corn seeds resulted in an increase in tocotrienol and tocopherol content of as much as sixfold (Cahoon et al., 2003; Schuy et al., 2019). Hence, the synthesis of diterpenoids in seeds appears to be essential for seed longevity and germination performance.

The MVA pathway predominantly provides the precursors for the cytosolic biosynthesis of primary terpenoids such as, phytosterols, saponins or brassinosteroids and trans-FPP synthases (FPSs) that catalyze the formation of C15-trans,trans-farnesyl diphosphate ((E,E)-FPP), as a central precursor. In Arabidopsis, both *FPS* paralogue genes, namely *FPS1* and *FPS2*, are highly expressed in seeds. *FPS1* gene is specifically expressed in the maternal seed coat, while, *FPS2* is predominantly expressed in mature seeds and during early seedling establishment (Keim et al., 2012). In addition, *fps1fps2* double mutants displayed defects during early embryo development (Closa et al., 2010). The availability of (E,E)-FPP as a precursor for sterols/brassinosteroids in seeds may promote germination for many species (Leubner-Metzger, 2003) and contribute to modulate seed coat permeability by cell wall modifications that might be involved in the loss of seed longevity during seed priming (Sano et al., 2017).

Many seeds used in food as spices or for their essential oils (i.e., Caraway, *Carum carvi*; Dill, *Anethum graveolens*) are rich in terpenoids, particularly, the monoterpenes carvone, limonene, and α-phellandrene (Al-Snafi, 2014; De Carvalho & Da Fonseca, 2006). These metabolites are abundant seed SMs and are affect the organoleptic characteristics of flavors. Interestingly, carvone was largely described for its antimicrobial effects and sprout suppression in stored potatoes (Hartmans, Diepenhorst, Bakker, & Gorris, 1995), suggesting terpenes involvement in seed stress response and control of germination. In coriander seed (*Coriandrum sativum* L.) the accumulation of another terpene, namely linalool, is positively influenced by salicylic acid (SA), which is a key hormone in plant innate immunity, and is involved in seedling vigour (Rahimi, Mashayekhi, Ghasemnezhad, Ebrahimi, & Hemmati, 2009).

Well-known molecules such as Paclitaxel/Taxol (a diterpenoid produced by *Taxus brevifolia*), Artemisinin (a sesquiterpenoids lactone derived from *Artemisia annua*), and ginkgolides (diterpenoids family produced by *Ginkgo biloba*) are terpenoids with pharmacological effects. These molecules and their derivatives were detected in seeds although their function in seed physiology remains unclear. During germination and seedlings growth of Taxus baccata, two genes of the paclitaxel biosynthetic pathway are highly expressed and match with the increase of taxanes, allowing the seed to cope with various herbivores and pathogens (Bruňáková, Košuth, Katkovčinová, Lázárová, & Čellárová, 2010). Artemisinin and its derivatives showed allelopathic activity with promising bioherbicide potential. These sesquiterpenes inhibit seed germination of weed species through crosstalk with ABA signalling. Transgenic Arabidopsis plants over-expressing ABA receptor from *Artemisia annua* (namely AaPYL9) enhanced ABA sensitivity in the seed germination, resulting in more artemisinin production (Zhang et al., 2015, 2013).

4.1 Spatiotemporal regulation of terpenoids content in seeds

Very little information is available on the tissue-specific accumulation of terpenes in seeds and their transcriptional regulation. It has been recently shown that the bHLH *TRITERPENE SAPONIN ACTIVATION REGULATOR3* (*TSAR3*) TF controls the synthesis of hemolytic saponin in *Medicago truncatula* seeds (Ribeiro et al., 2020). *TSAR3* is specifically and highly expressed in seeds during development, differently from *TSAR1* and *2* that are expressed in vegetative tissues (Mertens et al., 2016). In addition, metabolomic analysis of *TSAR3* loss-of-function lines highlighted a lower accumulation of several hemolytic triterpene saponins compared to wild-type plants. Finally, co-expression analyses allowed the identification of *CYP88A13*, a cytochrome P450 co-regulated with *TSAR3*, which is involved in a final oxidation step of hemolytic saponin biosynthesis in *Medicago truncatula* seeds (Ribeiro et al., 2020).

5. Structure/function relationship of specialized metabolites in seeds

Chemical structure cannot be the sole explanation to the localization of metabolites in plant organs, including seeds. Structure/function relationships of several SMs classes have been indeed mainly investigated for their use as pharmaceuticals (e.g., anticancer) and their functions in plants, as exemplified for flavonoids (Alseekh et al., 2020) or other SMs (Wang, Alseekh, Fernie, & Luo, 2019). Nevertheless, the information available is rather scarce for seed SMs (Fig. 1).

5.1 Chemical modifications of specialized metabolites: structure/function relationships

Specialized metabolites chemical modifications (i.e., hydroxylation, glycosylation, methylation and acylation) give a substantial contribution to metabolites diversity, affect their physicochemical properties and hence their bioactivity and transport (Table 2).

Table 2 Modifications of physicochemical properties induced by chemical modifications of metabolites.

Modifications from quercetin	Name	logP (hydrophobicity)[a]	Water solubility (mg/mL)[a]	Redox potential (mV)[b]
None	Quercetin	1.81	0.26	398
Dehydroxylation on C-ring	Luteolin	2.73	0.132	299
Dehydroxylation on A-ring	Fisetin	2.03	0.151	214
Dehydroxylation on B-ring	Kaempferol	1.99	0.18	209
Deoxyhexosylation (rhamnose) on C-ring	Quercitrin (Quercetin 3-O-rhamnose)	1.31	1.79	/
Hexosylation (galactose) on C-ring	Hyperin (Quercetin 3-O-galactose)	0.47	1.95	/
Hexosylation (glucose) on C-ring	Isoquercitrin (Quercetin 3-O-glucose)	0.47	1.95	/
Diglycosylation (rutinose) on C-ring	Rutin (Quercetin-3-O-hexose-rhamnose)	0.15	3.54	275
Hexosylation (glucose) on B-ring	Quercetin-4′-O-glucose	0.63	1.59	/
Uronidation (glucuronic acid) on B-ring	Quercetin-4′-O-glucuronide	1.05	1.63	/
Hexosylation (glucose) and galloylation, on C-ring	Quercetin-3-(2-galloylglucoside)	1.88	0.95	/
Pentosylation (arabinose) on C-ring	Guaiaverin (Quercetin-3-O-arabinoside)	0.75	1.15	/

Continued

Table 2 Modifications of physicochemical properties induced by chemical modifications of metabolites.—cont'd

Modifications from quercetin	Name	logP (hydrophobicity)[a]	Water solubility (mg/mL)[a]	Redox potential (mV)[b]
Hexosylation (glucose) and acetylation on C-ring	Quercetin-3-(6″-acetylglucoside)	1.41	1.42	/
O-methylation on B-ring	Isorhamnetin	1.96	0.15	/
Sulfation on C-ring	Quercetin-3-O-sulfate	0.87	0.6	/
Sulfation on B-ring	Quercetin-3′-O-sulfate	0.65	0.52	/

[a]As calculated (http://www.vcclab.org/lab/alogps/).
[b]As measured (Bors, Mici-Ml, & Schikora, 1995) on a series of flavonoids.

5.1.1 Hydroxylation

Hydroxylation highly affects SMs properties, flavonoids in particular, by impacting water solubility and hydrophilic characteristics (Wang et al., 2019) and increasing their vacuolar storage (Le Roy, Huss, Creach, Hawkins, & Neutelings, 2016). For instance, it has been shown that quercetin hydroxylation reduced hydrophobicity by 9—33% and increased water solubility by a factor of 1.5—2 (Table 2). Hydroxylation can also increase the antioxidant capacity of flavonols (Treml & Šmejkal, 2016). More specifically, when comparing quercetin with its dehydroxylated counterparts (Table 2), the redox potential was twice lower with dehydroxylation on the A or B rings (quercetin vs. kaempferol, 185—189 mV) than on the C ring (quercetin vs. luteolin, 99 mV). These results highlighted the major role of hydroxyl group on the C ring antioxidant potential and the higher antioxidant capacity of flavonols compared to flavones. *Nigella sativa* immature seeds contain specific monoterpenes that were gradually replaced during maturation by monoterpenoids (i.e., oxygenated monoterpenes) including the efficient antioxidants thymohydroquinone and thymoquinone (Botnick et al., 2012). These results highlighted a putative role of oxygenation and hydroxylation in the activation of flavonoids and other SMs against abiotic stresses in seeds, such as oxidative and light stresses.

5.1.2 Glycosylation

Glycosylation occurs in most SMs classes and several examples are available for flavonoids. Similarly to hydroxylation, O-glycosylation increases quercetin water solubility and hydrophilic properties of numerous metabolites, with some differential effects according to the type of sugars (Table 2) (Wang et al., 2019), thus increasing their vacuolar storage (Le Roy et al., 2016). However, O-glycosylation can be hydrolysed either enzymatically or under acidic conditions, whereas C-glycosylation, a modification frequently found in flavanones of cereal seeds (e.g., wheat, rice) is known to produce more stable glycosides (Brazier-Hicks et al., 2009). In Almond from *Prunus dulcis*, the cyanogenic glucoside prunasin, which is synthesized in the integuments, can be transported to the cotyledons, where it is glycosylated and stored as amygdalin (Thodberg et al., 2018), suggesting that this conjugation may enhance metabolite inter-tissue transport, storage and stability.

Glycosidically-bound volatiles are non-volatile and odourless, and most of them are β-D-glucosides and β-primeverosides (Song, Härtl, McGraphery, Hoffmann, & Schwab, 2018). For instance, in dry soybean seed, 1-Octen-3-ol is liberated as a volatile compound after hydration and deglycosylation of its primeveroside, presumably by a glucosidase. This diglycosylated precursor is stored in seed, and contained in green immature seeds in high amounts (Matsui, Takemoto, Koeduka, & Ohnishi, 2018), which could support its role in defence during maturation.

Many glycosylated SMs playing a role in plant defense against bioagressors are localized in compartments or cells other than their vacuolar substrates. For instance, GLSs are stored in the vacuole whereas their glycoside hydrolase myrosinase accumulates in the neighboring cells. Cyanogenic glycosides in *Phaseolus lunatus* or in Prunus seeds are accumulated in vacuoles whereas their related glucosidase linamarase is localized in the apoplasm. Benzoxazinoids glycosides are located in plastids, cytoplasm and cell walls out of reach of their vacuolar substrates (Morant et al., 2008). Hence, inactive compounds are stored in vacuoles when seeds are not subjected to stress and are quickly transported to other compartments and activated when the seeds are subjected to biotic or abiotic stresses.

When comparing the redox properties of quercetin with its glycosylated metabolite quercetin-O-rutinoside (rutin), the rutinosyl group reduces the redox potential of quercetin by more than 100 mV. O-glycosylation decreases the antioxidant capacity of flavonoids by masking hydroxyl groups (Plaza et al., 2014). The hydrolysis of sugar moieties by glucosidases could therefore improve the antioxidant capacity of metabolites. As a general observation, glycosylation reduces the activity of the nucleophilic groups that are conjugated and prevents them from strongly reacting with other components of the cell (Jones & Vogt, 2001).

Benzoxazinoids differentially accumulate in Rye embryo, where they are subjected to a wide range of modifications during seed germination. While diglycosylated DIBOA was the main benzoxazinoid present in mature rye seeds (Tanwir, Dionisio, Adhikari, Fomsgaard, & Gregersen, 2017), during germination benzoxazinoids lose their glycosylated moieties, producing benzoxazinoids with one or no glycosylation, which might be related to a role of the aglycone in seed defense against pathogens or herbivory.

Glycosylation of flavanones is known to increase the bitterness and play a role in plant defense (Wang et al., 2019). As an example, a "bitter" glycosylated flavanone was found in seeds of *Garcinia kola* Heckel (Okwu & Morah, 2007) and may contribute to deterring seed predation.

While glycosylation is a way to stabilize compounds and avoid autotoxicity, deglycosylation increases the reactivity of the compounds by unprotecting the sugar moiety. More generally, deglycosylation is the activating step for phytoanticipins and other preformed defense compounds (Morant et al., 2008), such as benzoxazinoids, glycosylated alkaloids, cyanogenic glucosides and glucosinolates, that undergo the action of specific glucosidases or hydrolases to be active and/or toxic (Dick et al., 2012; Ketudat Cairns, Mahong, Baiya, & Jeon, 2015). Rather than phytoalexins, preformed defense products and their deglycosylated form are more likely playing a role in seeds protection against pathogen and herbivore attacks.

Brassica napus seeds contain sinapoyl glucose that is turned into sinapoylcholine or sinapine, antinutritional bitter compounds being the main specialized metabolite in these seeds, localized in embryo but not in the seed coat (Fang, Reichelt, Hidalgo, Agnolet, & Schneider, 2012). Sinapine is then hydrolysed during germination, sinapate and choline being used later in other biosynthetic pathways.

In addition to their ROS scavenging activity, flavonoids exhibit metal-binding activity (Corso & de la Torre, 2020; Kasprzak, Erxleben, & Ochocki, 2015), whose efficiency is also influenced by glycosylation. In particular, stoichiometry of flavonols : metal complexes is strongly affected by the addition of a sugar moiety and metal-binding ability is reduced in glycosylated flavonols compared to their aglycone counterpart (Guo et al., 2007). Given the essential contribute of metals (mainly iron and zinc) homeostasis in plant seed quality, a deeper understanding the impact of glycosylation on flavonoids (or other SMs) metal binding activity will be of major importance.

5.1.3 Acylation

Acylation play also an important role in seed tolerance to stresses and interaction with the environment. Tohge et al. (2016) found that phenylacylation has a positive effect on the UV absorption of some flavonols found in different Arabidopsis accessions (Tohge et al., 2016). Interestingly these phenylacylated flavonols were detected only in organs exposed to light stress (flowers, leaf, stamen and siliques), and not in seeds, roots and pollen. Furthermore, acylation protects glycosides from being hydrolysed, giving them some stability against light and thermodegradation. Phenylacetylated flavonoids often exhibit increased solubility and higher stability by the interaction between aromatic cycles (Tohge et al., 2016), and that could contribute to their accumulation.

5.1.4 Methylation

Methylation of OH groups increases hydrophobicity and modifies molecular conformation in flavonoids (Alseekh et al., 2020), for instance, isorhamnetin that is an O-methylated quercetin has a slightly higher hydrophobicity (+0.15 for log P) than quercetin and reduced water solubility (by a 1.7 factor) (Table 2). While O-methylated flavonoids could have herbivores deterrence functions in the vegetative tissues, their role in seeds remains elusive.

5.1.5 Sulfation

Sulfation of quercetin resulted in a similar effect as glycosylation, since it increases water solubility. O-sulfated flavonoids are however less stable than O-glycosylated flavonoids (Teles, Souza, & De Souza, 2018). A sulfated isorhamnetin hexoside was found to be the most accumulated flavonoid in the seed coat of developing and matures seeds of black-seeded *Brassica napus* genotypes (Auger et al., 2010). The sulfation might increase flavonoids vacuolar storage and establish a sulfur source for the future seedling as well.

6. Conclusions

Here, we have summarized and exemplified the current knowledge on the diversity, distribution, regulations and structure/function relationships of seed SMs. There is a growing interest for seed SMs because of their potential impact on agroecology and the identification of alternative solutions to pesticides and herbicides. SMs can improve seed quality and seedling establishment, and are widely used for feed/food or as pharmaceuticals. Recent advances in molecular plant -biotic and -abiotic stress interactions are increasingly highlighting the role of inducible specialized metabolism in driving important aspects of plant ecological interactions and, more generally, plant adaptation to the environment. Several studies have focused on the characterization in plant vegetative tissues of the biosynthetic pathways and regulation by many SMs in response to biotic and abiotic stresses, while little is known about these processes in seeds, which are particularly important in many species where they constitute the final product for industrial or nutritional purposes (e.g., oilseed *Brassicaceae*). The integration of multiomics and MSI approaches will be essential to characterize seed SMs environmental regulation and could be a relevant lever to use in new plant breeding programs.

References

Abraham, E. M., Ganopoulos, I., Madesis, P., Mavromatis, A., Mylona, P., Nianiou-Obeidat, I., et al. (2019). The use of lupin as a source of protein in animal feeding: Genomic tools and breeding approaches. *International Journal of Molecular Sciences*, *20*, 851. https://doi.org/10.3390/ijms20040851.

Airoldi, C. A., Hearn, T. J., Brockington, S. F., Webb, A. A. R., & Glover, B. J. (2019). TTG1 proteins regulate circadian activity as well as epidermal cell fate and pigmentation. *Nature Plants*, *5*(11), 1145–1153. https://doi.org/10.1038/s41477-019-0544-3.

Alseekh, S., Perez de Souza, L., Benina, M., & Fernie, A. R. (2020). The style and substance of plant flavonoid decoration; towards defining both structure and function. *Phytochemistry*, *174*, 112347. https://doi.org/10.1016/j.phytochem.2020.112347.

Al-Snafi, A. E. (2014). The pharmacological importance of Anethum graveolens. A review. *International Journal of Pharmacy and Pharmaceutical Sciences*, *6*(4), 11–13.

Arc, E., Galland, M., Cueff, G., Godin, B., Lounifi, I., Job, D., et al. (2011). Reboot the system thanks to protein post-translational modifications and proteome diversity: How quiescent seeds restart their metabolism to prepare seedling establishment. *Proteomics*, *11*(9), 1606–1618. https://doi.org/10.1002/pmic.201000641.

Ashihara, H., Sano, H., & Crozier, A. (2008). Caffeine and related purine alkaloids: Biosynthesis, catabolism, function and genetic engineering. *Phytochemistry*, *69*(4), 841–856. https://doi.org/10.1016/j.phytochem.2007.10.029.

Ashihara, H., & Suzuki, T. (2004). Distribution and biosynthesis of caffeine in plants. *Frontiers in Bioscience*, *9*, 1864–1876. https://doi.org/10.2741/1367.

Auger, B., Marnet, N., Gautier, V., Maia-Grondard, A., Leprince, F., Renard, M., et al. (2010). A detailed survey of seed coat flavonoids in developing seeds of *Brassica napus* L. *Journal of Agricultural and Food Chemistry*, *58*(10), 6246–6256. https://doi.org/10.1021/jf903619v.

Augustine, R., Mukhopadhyay, A., & Bisht, N. C. (2013). Targeted silencing of BjMYB28 transcription factor gene directs development of low glucosinolate lines in oilseed *Brassica juncea*. *Plant Biotechnology Journal*, *11*(7), 855–866. https://doi.org/10.1111/pbi.12078.

Bailly, C., El-Maarouf-Bouteau, H., & Corbineau, F. (2008). From intracellular signaling networks to cell death: the dual role of reactive oxygen species in seed physiology. *Comptes Rendus Biologies*, *331*(10), 806–814. https://doi.org/10.1016/j.crvi.2008.07.022.

Baud, S., Dubreucq, B., Miquel, M., Rochat, C., & Lepiniec, L. (2008). Storage reserve accumulation in Arabidopsis: Metabolic and developmental control of seed filling. *The Arabidopsis Book*, *6*, e0113. https://doi.org/10.1199/tab.0113.

Baudry, A., Heim, M. A., Dubreucq, B., Caboche, M., Weisshaar, B., & Lepiniec, L. (2004). TT2, TT8, and TTG1 synergistically specify the expression of *BANYULS* and proanthocyanidin biosynthesis in *Arabidopsis thaliana*. *The Plant Journal*, *39*(3), 366–380. https://doi.org/10.1111/j.1365-313X.2004.02138.x.

Baumann, T. W., & Gabriel, H. (1984). Metabolism and excretion of caffeine during germination of *Coffea arabica* L. *Plant and Cell Physiology*, *25*(8), 1431–1436. https://doi.org/10.1093/oxfordjournals.pcp.a076854.

Bednarz, H., Rolo, N., & Niehaus, K. (2019). Mass spectrometry imaging of the spatial and temporal localization of alkaloids in nightshades. *Journal of Agricultural and Food Chemistry*, *67*(49), 13470–13477. https://doi.org/10.1021/acs.jafc.9b01155.

Bhandari, S. R., Jo, J. S., & Lee, J. G. (2015). Comparison of glucosinolate profiles in different tissues of nine brassica crops. *Molecules*, *20*(9), 15827–15841. https://doi.org/10.3390/molecules200915827.

Bhandari, D. R., Wang, Q., Friedt, W., Spengler, B., Gottwald, S., & Römpp, A. (2015). High resolution mass spectrometry imaging of plant tissues: Towards a plant metabolite atlas. *Analyst*, *140*(22), 7696–7709. https://doi.org/10.1039/c5an01065a.

Bjarnholt, N., Li, B., D'Alvise, J., & Janfelt, C. (2014). Mass spectrometry imaging of plant metabolites-principles and possibilities. *Natural Product Reports*, *31*, 818–837. https://doi.org/10.1039/c3np70100j.

Bors, W., Mici-Ml, C., & Schikora, B. (1995). Interaction of flavonoids with ascorbate and determination of their univalent redox potentials: A pulse radiolysis study. *Free Radical Biology & Medicine*, *19*(1), 45–52. https://doi.org/10.1016/0891-5849(95)00011-L.

Botnick, I., Xue, W., Bar, E., Ibdah, M., Schwartz, A., Joel, D. M., et al. (2012). Distribution of primary and specialized metabolites in *Nigella sativa* seeds, a spice with vast traditional and historical uses. *Molecules*, *17*(9), 10159–10177. https://doi.org/10.3390/molecules170910159.

Boughton, B. A., Thinagaran, D., Sarabia, D., Bacic, A., & Roessner, U. (2016). Mass spectrometry imaging for plant biology: a review. *Phytochemistry Reviews*, *15*, 445–488. https://doi.org/10.1007/s11101-015-9440-2.

Brazier-Hicks, M., Evans, K. M., Gershater, M. C., Puschmann, H., Steel, P. G., & Edwards, R. (2009). The C-glycosylation of flavonoids in cereals. *Journal of Biological Chemistry*, *284*(27), 17926–17934. https://doi.org/10.1074/jbc.M109.009258.

Broun, P. (2005). Transcriptional control of flavonoid biosynthesis: A complex network of conserved regulators involved in multiple aspects of differentiation in Arabidopsis. *Current Opinion in Plant Biology*, *8*, 272–279. https://doi.org/10.1016/j.pbi.2005.03.006.

Brown, P. D., Tokuhisa, J. G., Reichelt, M., & Gershenzon, J. (2003). Variation of glucosinolate accumulation among different organs and developmental stages of *Arabidopsis thaliana*. *Phytochemistry*, *62*, 471–481. https://doi.org/10.1016/S0031-9422(02)00549-6.

Bruňáková, K., Košuth, J., Katkovčinová, Z., Lázárová, M., & Čellárová, E. (2010). Expression of two genes of paclitaxel biosynthetic pathway during germination of *Taxus baccata* zygotic embryos. *Biologia Plantarum*, *54*, 515–519. https://doi.org/10.1007/s10535-010-0090-3.

Butelli, E., Licciardello, C., Ramadugu, C., Recupero, G. R., Froelicher, Y., & Correspondence, C. M. (2019). Noemi controls production of flavonoid pigments and fruit acidity and illustrates the domestication routes of modern citrus varieties. *Current Biology*, *29*(1), 158–164. https://doi.org/10.1016/j.cub.2018.11.040.

Cahoon, E. B., Hall, S. E., Ripp, K. G., Ganzke, T. S., Hitz, W. D., & Coughlan, S. J. (2003). Metabolic redesign of vitamin E biosynthesis in plants for tocotrienol production and increased antioxidant content. *Nature Biotechnology*, *21*(9), 1082–1087. https://doi.org/10.1038/nbt853.

Cai, H., Zhang, M., Chai, M., He, Q., Huang, X., Zhao, L., et al. (2019). Epigenetic regulation of anthocyanin biosynthesis by an antagonistic interaction between H2A.Z and H3K4me3. *New Phytologist*, *221*(1), 295–308. https://doi.org/10.1111/nph.15306.

Chandler, V. L. (2007). Paramutation: From Maize to Mice. *Cell*, *128*(4), 641–645. https://doi.org/10.1016/j.cell.2007.02.007.

Cheeke, P. R., & Kelly, J. D. (1989). Metabolism, toxicity and nutritional implications of quinolizidine (lupin) alkaloids. Recent Advances of Research in Antinutritional Factors in Legume Seeds. In *Proceedings of the First International Workshop on "Antinutritional Factors (ANF) in Legume Seeds"*.

Chen, D., Li, Y., Fang, T., Shi, X., & Chen, X. (2016). Specific roles of tocopherols and tocotrienols in seed longevity and germination tolerance to abiotic stress in transgenic rice. *Plant Science*, *244*, 31–39. https://doi.org/10.1016/j.plantsci.2015.12.005.

Cho, Y. B., Jones, S. I., & Vodkin, L. O. (2017). Mutations in Argonaute5 illuminate epistatic interactions of the K1 and I loci leading to saddle seed color patterns in *Glycine max*. *Plant Cell*, *29*(4), 708–725. https://doi.org/10.1105/tpc.17.00162.

Christiansen, J. L., Joørnsgård, B., Buskov, S., & Olsen, C. E. (1997). Effect of drought stress on content and composition of seed alkaloids in narrow-leafed lupin, *Lupinus angustifolius* L. *European Journal of Agronomy*, 7(4), 307–314. https://doi.org/10.1016/S1161-0301(97)00017-8.

Closa, M., Vranová, E., Bortolotti, C., Bigler, L., Arró, M., Ferrer, A., et al. (2010). The Arabidopsis thaliana FPP synthase isozymes have overlapping and specific functions in isoprenoid biosynthesis, and complete loss of FPP synthase activity causes early developmental arrest. *Plant Journal*, *63*(3), 512–525. https://doi.org/10.1111/j.1365-313X.2010.04253.x.

Colinas, M., & Goossens, A. (2018). Combinatorial Transcriptional Control of Plant Specialized Metabolism. *Trends in Plant Science*, *23*(4), 324–336. https://doi.org/10.1016/J.TPLANTS.2017.12.006.

Corso, M., & de la Torre, V. S. G. (2020). Biomolecular approaches to understanding metal tolerance and hyperaccumulation in plants. *Metallomics*, *12*(6), 840–859. https://doi.org/10.1039/d0mt00043d.

Corso, M., Perreau, F., Mouille, G., & Lepiniec, L. (2020). Specialized Phenolic compounds in seeds: structures, functions, and regulations. *Plant Science*, *296*, 110471. https://doi.org/10.1016/j.plantsci.2020.110471.

De Carvalho, C. C. C. R., & Da Fonseca, M. M. R. (2006). Carvone: Why and how should one bother to produce this terpene. *Food Chemistry*, *95*(3), 413–422. https://doi.org/10.1016/j.foodchem.2005.01.003.

Debeaujon, I., Léon-Kloosterziel, K. M., & Koornneef, M. (2000). Influence of the testa on seed dormancy, germination, and longevity in Arabidopsis. *Plant Physiology*, *22*(2), 403–414. https://doi.org/10.1104/pp.122.2.403.

Debeaujon, I., Nesi, N., Perez, P., Devic, M., Grandjean, O., Caboche, M., et al. (2003). Proanthocyanidin-accumulating cells in arabidopsis testa: regulation of differentiation and role in seed development. *Plant Cell*, *15*(11), 2514–2531. https://doi.org/10.1105/tpc.014043.

Dick, R., Rattei, T., Haslbeck, M., Schwab, W., Gierl, A., & Frey, M. (2012). Comparative analysis of benzoxazinoid biosynthesis in monocots and dicots: Independent recruitment of Stabilization and activation functions. *Plant Cell*, *24*(3), 915–928. https://doi.org/10.1105/tpc.112.096461.

Dietz, K.-J., Vogel, M. O., & Viehhauser, A. (2010). AP2/EREBP transcription factors are part of gene regulatory networks and integrate metabolic, hormonal and environmental signals in stress acclimation and retrograde signalling. *Protoplasma*, *245*, 3–14. https://doi.org/10.1007/s00709-010-0142-8.

Diharce, J., Golebiowski, J., Fiorucci, S., & Antonczak, S. (2016). Fine-tuning of microsolvation and hydrogen bond interaction regulates substrate channelling in the course of flavonoid biosynthesis. *Physical Chemistry Chemical Physics*, *18*(15), 10337–10345. https://doi.org/10.1039/c5cp05059f.

Dillard, C. J., & Bruce German, J. (2000). Phytochemicals: Nutraceuticals and human health. *Journal of the Science of Food and Agriculture*, *80*, 1744–1756. https://doi.org/10.1002/1097-0010(20000915)80:12 < 1744::AID-JSFA725 > 3.0.CO;2-W.

Dubos, C., Stracke, R., Grotewold, E., Weisshaar, B., Martin, C., & Lepiniec, L. (2010). MYB transcription factors in Arabidopsis. *Trends in Plant Science*, *15*(10), 573–581. https://doi.org/10.1016/j.tplants.2010.06.005.

Eckelmann, D., Kusari, S., & Spiteller, M. (2017). Spatial profiling of maytansine during the germination process of Maytenus senegalensis seeds. *Fitoterapia*, *119*, 51–56. https://doi.org/10.1016/j.fitote.2017.03.014.

Emire, S. A., Jha, Y. K., & Mekam, F. (2013). *Role of Anti-nutritional Factors in Food Industry Role of Anti-nutritional Factors*. Beverage & Food World.

Fan, D., Wang, X., Tang, X., Ye, X., Ren, S., Wang, D., et al. (2018). Histone H3K9 demethylase JMJ25 epigenetically modulates anthocyanin biosynthesis in poplar. *The Plant Journal*, *96*(6), 1121–1136. https://doi.org/10.1111/tpj.14092.

Fang, J., Reichelt, M., Hidalgo, W., Agnolet, S., & Schneider, B. (2012). Tissue-specific distribution of secondary metabolites in rapeseed (*Brassica napus* L.). *PLoS One*, 7(10), e48006. https://doi.org/10.1371/journal.pone.0048006.

Frerigmann, H. (2016). Glucosinolate regulation in a complex relationship – MYC and MYB – No one can act without each other. *Advances in Botanical Research*, *80*, 57–97. https://doi.org/10.1016/bs.abr.2016.06.005.

Frick, K. M., Foley, R. C., Kamphuis, L. G., Siddique, K. H. M., Garg, G., & Singh, K. B. (2018). Characterization of the genetic factors affecting quinolizidine alkaloid biosynthesis and its response to abiotic stress in narrow-leafed lupin (*Lupinus angustifolius* L.). *Plant, Cell and Environment*, *41*(9), 2155–2168. https://doi.org/10.1111/pce.13172.

Frick, K. M., Kamphuis, L. G., Siddique, K. H. M., Singh, K. B., & Foley, R. C. (2017). Quinolizidine alkaloid biosynthesis in lupins and prospects for grain quality improvement. *Frontiers in Plant Science*, *8*, 87. https://doi.org/10.3389/fpls.2017.00087.

Fujino, N., Tenma, N., Waki, T., Ito, K., Komatsuzaki, Y., Sugiyama, K., et al. (2018). Physical interactions among flavonoid enzymes in snapdragon and torenia reveal the diversity in the flavonoid metabolon organization of different plant species. *Plant Journal*, *94*(2), 372–392. https://doi.org/10.1111/tpj.13864.

Gao, P., Li, X., Cui, D., Wu, L., Parkin, I., & Gruber, M. Y. (2010). A new dominant Arabidopsis transparent testa mutant, *sk21-D*, and modulation of seed flavonoid biosynthesis by KAN4. *Plant Biotechnology Journal*, *8*(9), 979–993. https://doi.org/10.1111/j.1467-7652.2010.00525.x.

Gao, Y., Liu, J., Chen, Y., Tang, H., Wang, Y., He, Y., et al. (2018). Tomato SlAN11 regulates flavonoid biosynthesis and seed dormancy by interaction with bHLH proteins but not with MYB proteins. *Horticulture Research*, *5*(1), 1–18. https://doi.org/10.1038/s41438-018-0032-3.

Gonzalez, A., Mendenhall, J., Huo, Y., & Lloyd, A. (2009). TTG1 complex MYBs, MYB5 and TT2, control outer seed coat differentiation. *Developmental Biology*, *325*(2), 412–421. https://doi.org/10.1016/j.ydbio.2008.10.005.

Gonzalez, A., Zhao, M., Leavitt, J. M., & Lloyd, A. M. (2008). Regulation of the anthocyanin biosynthetic pathway by the TTG1/bHLH/Myb transcriptional complex in Arabidopsis seedlings. *The Plant Journal*, *53*(5), 814–827. https://doi.org/10.1111/j.1365-313X.2007.03373.x.

Gorzolka, K., Bednarz, H., & Niehaus, K. (2014). Detection and localization of novel hordatine-like compounds and glycosylated derivates of hordatines by imaging mass spectrometry of barley seeds. *Planta*, *239*(6), 1321–1335. https://doi.org/10.1007/s00425-014-2061-y.

Gorzolka, K., Kölling, J., Nattkemper, T. W., & Niehaus, K. (2016). Spatio-Temporal metabolite profiling of the barley germination process by MALDI MS imaging. *PLoS One*, *11*(3), e0150208. https://doi.org/10.1371/journal.pone.0150208.

Grotewold, E., Drummond, B. J., Bowen, B., & Peterson, T. (1994). The myb-homologous P gene controls phlobaphene pigmentation in maize floral organs by directly activating a flavonoid biosynthetic gene subset. *Cell*, *76*(3), 543–553. https://doi.org/10.1016/0092-8674(94)90117-1.

Gu, X. Y., Foley, M. E., Horvath, D. P., Anderson, J. V., Feng, J., Zhang, L., et al. (2011). Association between seed dormancy and pericarp color is controlled by a pleiotropic gene that regulates abscisic acid and flavonoid synthesis in weedy red rice. *Genetics*, *189*(4), 1515–1524. https://doi.org/10.1534/genetics.111.131169.

Guo, M., Perez, C., Wei, Y., Rapoza, E., Su, G., Bou-Abdallah, F., et al. (2007). Iron-binding properties of plant phenolics and cranberry's bio-effects. *Dalton Transactions*, *21*(43), 4951–4961. https://doi.org/10.1039/b705136k.

Gupta, S., Rupasinghe, T., Callahan, D. L., Natera, S. H. A., Smith, P. M. C., Hill, C. B., et al. (2019). Spatio-temporal metabolite and elemental profiling of salt stressed barley seeds during initial stages of germination by MALDI-MSI and μ-XRF spectrometry. *Frontiers in Plant Science*, *10*, 1139. https://doi.org/10.3389/fpls.2019.01139.

Hartmans, K. J., Diepenhorst, P., Bakker, W., & Gorris, L. G. M. (1995). The use of carvone in agriculture: sprout suppression of potatoes and antifungal activity against potato tuber and other plant diseases. *Industrial Crops and Products*, *4*(1), 3–13. https://doi.org/10.1016/0926-6690(95)00005-W.

He, H., Qin, L., Zhang, Y., Han, M., Li, J., Liu, Y., et al. (2019). 3,4-Dimethoxycinnamic acid as a novel matrix for enhanced in situ detection and imaging of low-molecular-weight compounds in biological tissues by MALDI-MSI. *Analytical Chemistry*, *91*(4), 2634–2643. https://doi.org/10.1021/acs.analchem.8b03522.

Hichri, I., Barrieu, F., Bogs, J., Kappel, C., Delrot, S., & Lauvergeat, V. (2011). Recent advances in the transcriptional regulation of the flavonoid biosynthetic pathway. *Journal of Experimental Botany*, *62*(8), 2465–2483. https://doi.org/10.1093/jxb/erq442.

Hirai, M. Y., Sugiyama, K., Sawada, Y., Tohge, T., Obayashi, T., Suzuki, A., et al. (2007). Omics-based identification of *Arabidopsis* Myb transcription factors regulating aliphatic glucosinolate biosynthesis. *Proceedings of the National Academy of Sciences of the United States of America*, *104*(15), 6478–6483. https://doi.org/10.1073/pnas.0611629104.

Hong, M., Hu, K., Tian, T., Li, X., Chen, L., Zhang, Y., et al. (2017). Transcriptomic analysis of seed coats in yellow-seeded *Brassica napus* reveals novel genes that influence proanthocyanidin biosynthesis. *Frontiers in Plant Science*, *8*, 1674. https://doi.org/10.3389/fpls.2017.01674.

Hotti, H., & Rischer, H. (2017). The killer of socrates: Coniine and related alkaloids in the plant Kingdom. *Molecules*, *22*(11), 1962. https://doi.org/10.3390/molecules22111962.

Jacobowitz, J. R., & Weng, J. (2020). Exploring uncharted territories of plant specialized metabolism in the postgenomic era. *Annual Review of Plant Biology*, *71*, 631–658. https://doi.org/10.1146/annurev-arplant-081519-035634.

Jansen, G., Jürgens, H. U., & Ordon, F. (2009). Effects of temperature on the alkaloid content of seeds of *Lupinus angustifolius* cultivars. *Journal of Agronomy and Crop Science*, *195*, 172–177. https://doi.org/10.1111/j.1439-037X.2008.00356.x.

Jiang, J., Zhu, S., Yuan, Y., Wang, Y., Zeng, L., Batley, J., et al. (2019). Transcriptomic comparison between developing seeds of yellow- and black- seeded *Brassica napus* reveals that genes influence seed quality. *BMC Plant Biology, 19*, 203. https://doi.org/10.1186/s12870-019-1821-z.

Jones, P., & Vogt, T. (2001). Glycosyltransferases in secondary plant metabolism: Tranquilizers and stimulant controllers. *Planta, 213*(2), 164–174. https://doi.org/10.1007/s004250000492.

Jørgensen, K., Rasmussen, A. V., Morant, M., Nielsen, A. H., Bjarnholt, N., Zagrobelny, M., et al. (2005). Metabolon formation and metabolic channeling in the biosynthesis of plant natural products. *Current Opinion in Plant Biology, 8*(3), 280–291. https://doi.org/10.1016/j.pbi.2005.03.014.

Kasprzak, M. M., Erxleben, A., & Ochocki, J. (2015). Properties and applications of flavonoid metal complexes. *RSC Advances, 5*(57), 45853–45877. https://doi.org/10.1039/C5RA05069C.

Keim, V., Manzano, D., Fernández, F. J., Closa, M., Andrade, P., Caudepón, D., et al. (2012). Characterization of arabidopsis FPS isozymes and FPS gene expression analysis provide insight into the biosynthesis of isoprenoid precursors in seeds. *PLoS One,* 7(11), e49109. https://doi.org/10.1371/journal.pone.0049109.

Ketudat Cairns, J. R., Mahong, B., Baiya, S., & Jeon, J. S. (2015). β-Glucosidases: Multitasking, moonlighting or simply misunderstood? *Plant Science, 241*, 246–259. https://doi.org/10.1016/j.plantsci.2015.10.014.

Kim, Y. S., Choi, Y. E., & Sano, H. (2010). Plant vaccination: Stimulation of defense system by caffeine production in planta. *Plant Signaling & Behavior, 5*(5), 489–493. https://doi.org/10.4161/psb.11087.

Kooke, R., Morgado, L., Becker, F., Van Eekelen, H., Hazarika, R., Zheng, Q., et al. (2019). Epigenetic mapping of the Arabidopsis metabolome reveals mediators of the epigenotype-phenotype map. *Genome Research, 29*(1), 96–106. https://doi.org/10.1101/gr.232371.117.

Kroc, M., Czepiel, K., Wilczura, P., Mokrzycka, M., & Święcicki, W. (2019). Development and validation of a gene-targeted dCAPS marker for marker-assisted selection of low-alkaloid content in seeds of narrow-leafed lupin (*Lupinus angustifolius* L.). *Genes, 10*(6), 428. https://doi.org/10.3390/genes10060428.

Kroc, M., Koczyk, G., Kamel, K. A., Czepiel, K., Fedorowicz-Strońska, O., Krajewski, P., et al. (2019). Transcriptome-derived investigation of biosynthesis of quinolizidine alkaloids in narrow-leafed lupin (*Lupinus angustifolius* L.) highlights candidate genes linked to iucundus locus. *Scientific Reports, 9*(1), 2231. https://doi.org/10.1038/s41598-018-37701-5.

Kuraparthy, V., Sood, S., & Gill, B. S. (2008). Targeted genomic mapping of a red seed color gene (R-A1) in wheat. *Crop Science, 48*, 37–48. https://doi.org/10.2135/cropsci2007.08.0488tpg.

Lacchini, E., & Goossens, A. (2020). Combinatorial control of plant specialized metabolism: mechanisms, functions, and consequences. *Annual Review of Cell and Developmental Biology, 36*(1), 1–23.

Le Roy, J., Huss, B., Creach, A., Hawkins, S., & Neutelings, G. (2016). Glycosylation is a major regulator of phenylpropanoid availability and biological activity in plants. *Frontiers in Plant Science,* 7, 735. https://doi.org/10.3389/fpls.2016.00735.

Leopoldini, M., Russo, N., Chiodo, S., & Toscano, M. (2006). Iron chelation by the powerful antioxidant flavonoid quercetin. *Journal of Agricultural and Food Chemistry, 54*(17), 6343–6351. https://doi.org/10.1021/jf060986h.

Lepiniec, L., Debeaujon, I., Routaboul, J.-M., Baudry, A., Pourcel, L., Nesi, N., et al. (2006). Genetics and biochemistry of seed flavonoids. *Annual Review of Plant Biology, 57*(1), 405–430. https://doi.org/10.1146/annurev.arplant.57.032905.105252.

Leprince, O., Pellizzaro, A., Berriri, S., & Buitink, J. (2017). Late seed maturation: Drying without dying. *Journal of Experimental Botany*, *68*(4), 827–841. https://doi.org/10.1093/jxb/erw363.

Leubner-Metzger, G. (2003). Brassinosteroids promote seed germination. *Brassinosteroids*, 119–128. https://doi.org/10.1007/978-94-017-0948-4_5.

Li, S. (2014). Transcriptional control of flavonoid biosynthesis: Fine-tuning of the MYB-bHLH-WD40 (MBW) complex. *Plant Signaling & Behavior*, *9*(1), e27522. https://doi.org/10.4161/psb.27522.

Li, Y., Sawada, Y., Hirai, A., Sato, M., Kuwahara, A., Yan, X., et al. (2013). Novel insights into the function of arabidopsis R2R3-MYB transcription factors regulating aliphatic glucosinolate biosynthesis. *Plant & Cell Physiology*, *54*(8), 1335–1344. https://doi.org/10.1093/pcp/pct085.

Li, B., Tang, M., Nelson, A., Caligagan, H., Zhou, X., Clark-Wiest, C., et al. (2018). Network-guided discovery of extensive epistasis between transcription factors involved in aliphatic glucosinolate biosynthesis. *Plant Cell*, *30*(1), 178–195. https://doi.org/10.1105/tpc.17.00805.

Liu, C., HyungJun, J., & Dixon, R. A. (2014). MYB5 and MYB14 play pivotal roles in seed coat polymer biosynthesis in *Medicago truncatula*. *Plant Physiology*, *165*, 1424–1439. https://doi.org/10.1104/pp.114.241877.

Liu, H., Li, Q., & Xing, Y. (2018). Genes contributing to domestication of rice seed traits and its global expansion. *Genes*, *9*(10), 489. https://doi.org/10.3390/genes9100489.

Liu, J., Osbourn, A., & Ma, P. (2015). MYB transcription factors as regulators of phenylpropanoid metabolism in plants. *Molecular Plant*, *8*, 689–708. https://doi.org/10.1016/j.molp.2015.03.012.

Macgregor, D. R., Kendall, S. L., Florance, H., Fedi, F., Moore, K., Paszkiewicz, K., et al. (2015). Seed production temperature regulation of primary dormancy occurs through control of seed coat phenylpropanoid metabolism. *New Phytologist*, *205*(2), 642–652. https://doi.org/10.1111/nph.13090.

Mailer, R. J., & Cornish, P. S. (1987). Effects of water stress on glucosinolate and oil concentrations in the seeds of rape (*Brassica Napus* L.) and turnip rape (*Brassica Rapa* L. Var. Silvestris [Lam.] Briggs). *Australian Journal of Experimental Agriculture*, *27*(5), 707–711. https://doi.org/10.1071/EA9870707.

Martin, C., & Li, J. (2017). Medicine is not health care, food is health care: plant metabolic engineering, diet and human health. *New Phytologist*, *216*(3), 699–719. https://doi.org/10.1111/nph.14730.

Matsui, K., Takemoto, H., Koeduka, T., & Ohnishi, T. (2018). 1-Octen-3-ol is formed from its glycoside during processing of soybean [Glycine max (L.) Merr.] seeds. *Journal of Agricultural and Food Chemistry*, *66*(28), 7409–7416. https://doi.org/10.1021/acs.jafc.8b01950.

McAbee, J. M., Hill, T. A., Skinner, D. J., Izhaki, A., Hauser, B. A., Meister, R. J., et al. (2006). *ABERRANT TESTA SHAPE* encodes a KANADI family member, linking polarity determination to separation and growth of Arabidopsis ovule integuments. *The Plant Journal*, *46*(3), 522–531. https://doi.org/10.1111/j.1365-313X.2006.02717.x.

Mertens, J., Pollier, J., Bossche, R. V., Lopez-Vidriero, I., Franco-Zorrilla, J. M., & Goossens, A. (2016). The bHLH transcription factors TSAR1 and TSAR2 regulate triterpene saponin biosynthesis in Medicago truncatula. *Plant Physiology*, *170*(1), 194–210. https://doi.org/10.1104/pp.15.01645.

Misawa, N. (2011). Pathway engineering for functional isoprenoids. *Current Opinion in Biotechnology*, *22*(5), 627–633. https://doi.org/10.1016/j.copbio.2011.01.002.

Mizzotti, C., Ezquer, I., Paolo, D., Rueda-Romero, P., Guerra, R. F., Battaglia, R., et al. (2014). SEEDSTICK is a master regulator of development and metabolism in the Arabidopsis seed coat. *PLoS Genetics*, *10*(12). https://doi.org/10.1371/journal.pgen.1004856.

Mohana Kumara, P., Srimany, A., Ravikanth, G., Uma Shaanker, R., & Pradeep, T. (2015). Ambient ionization mass spectrometry imaging of rohitukine, a chromone anti-cancer alkaloid, during seed development in *Dysoxylum binectariferum* Hook.f (Meliaceae). *Phytochemistry*, *116*, 104–110. https://doi.org/10.1016/j.phytochem.2015.02.031.

Montini, L., Crocoll, C., Gleadow, R., Motawia, M. S., Janfelt, C., & Bjarnholt, N. (2020). Matrix-assisted laser desorption/ionization-mass spectrometry imaging of metabolites during sorghum germination. *Plant Physiology*, *183*(3), 925–942. https://doi.org/10.1104/pp.19.01357.

Morant, A. V., Jørgensen, K., Jørgensen, C., Paquette, S. M., Sánchez-Pérez, R., Møller, B. L., et al. (2008). β-Glucosidases as detonators of plant chemical defense. *Phytochemistry*, *69*, 1795–1813. https://doi.org/10.1016/j.phytochem.2008.03.006.

Nakayama, T., Takahashi, S., & Waki, T. (2019). Formation of flavonoid metabolons: Functional significance of protein-protein interactions and impact on flavonoid chemodiversity. *Frontiers in Plant Science*, *10*, 821. https://doi.org/10.3389/fpls.2019.00821.

Nesi, N., Debeaujon, I., Jond, C., Pelletier, G., Caboche, M., & Lepiniec, L. (2000). The TT8 gene encodes a basic helix-loop-helix domain protein required for expression of DFR and BAN genes in arabidopsis siliques. *The Plant Cell*, *12*(10), 1863. https://doi.org/10.2307/3871198.

Nesi, N., Delourme, R., Brégeon, M., Falentin, C., & Renard, M. (2008). Genetic and molecular approaches to improve nutritional value of *Brassica napus* L. seed. *Comptes Rendus Biologies*, *331*(10), 763–771. https://doi.org/10.1016/j.crvi.2008.07.018.

Nesi, N., Jond, C., Debeaujon, I., Caboche, M., & Lepiniec, L. (2001). The arabidopsis TT2 gene encodes an R2R3 MYB domain protein that acts as a key determinant for proanthocyanidin accumulation in developing seed. *The Plant Cell*, *13*, 2099–2114. https://doi.org/10.1105/tpc.010098.

Nour-Eldin, H. H., Andersen, T. G., Burow, M., Madsen, S. R., Jørgensen, M. E., Olsen, C. E., et al. (2012). NRT/PTR transporters are essential for translocation of glucosinolate defence compounds to seeds. *Nature*, *488*(7412), 531–534. https://doi.org/10.1038/nature11285.

Obata, T. (2019). Metabolons in plant primary and secondary metabolism. *Phytochemistry Reviews*, *18*(6), 1483–1507. https://doi.org/10.1007/s11101-019-09619-x.

Ohto, M. A., Stone, S. L., & Harada, J. J. (2007). Genetic control of seed development and seed mass. In *Seed Development, Dormancy and Germination*. Blackwell Publishing. https://doi.org/10.1002/9780470988848.ch1.

Okwu, D. E., & Morah, F. N. I. (2007). Isolation and characterization of flavanone glycoside 4',5,7-trihydroxy flavanone rhamnoglucose from Garcinia kola seed. *Journal of Applied Sciences*, 7(2), 306–309. https://doi.org/10.3923/jas.2007.306.309.

Olivon, F., Allard, P.-M., Koval, A., Righi, D., Genta-Jouve, G., Neyts, J., et al. (2017). Bioactive natural products prioritization using massive multi-informational molecular networks. *ACS Chemical Biology*, *12*, 2644–2651. https://doi.org/10.1021/acschembio.7b00413.

Olivon, F., Elie, N., Grelier, G., Roussi, F., Litaudon, M., & Touboul, D. (2018). MetGem software for the generation of molecular networks based on the t-SNE algorithm. *Analytical Chemistry*, *90*(23), 13900–13908. https://doi.org/10.1021/acs.analchem.8b03099.

Oram, R. N. (1983). Selection for higher seed yield in the presence of the deleterious low alkaloid allele iucundus in *Lupinus angustifolius* L. *Field Crops Research*, 7(C), 169–180. https://doi.org/10.1016/0378-4290(83)90020-5.

Orhan, D. D., Özçelik, B., Özgen, S., & Ergun, F. (2010). Antibacterial, antifungal, and antiviral activities of some flavonoids. *Microbiological Research*, *165*(6), 496–504. https://doi.org/10.1016/j.micres.2009.09.002.

Otterbach, S. L., Yang, T., Kato, L., Janfelt, C., & Geu-Flores, F. (2019). Quinolizidine alkaloids are transported to seeds of bitter narrow-leafed lupin. *Journal of Experimental Botany*, *70*(20), 5799–5808. https://doi.org/10.1093/jxb/erz334.

Paauw, M., Koes, R., & Quattrocchio, F. M. (2019). Alteration of flavonoid pigmentation patterns during domestication of food crops. *Journal of Experimental Botany*, *70*(15), 3719–3735. https://doi.org/10.1093/jxb/erz141.

Paul, P., Singh, S. K., Patra, B., Liu, X., Pattanaik, S., & Yuan, L. (2020). Mutually regulated AP2/ERF gene clusters modulate biosynthesis of specialized metabolites in plants. *Plant Physiology*, *182*(2), 840–856. https://doi.org/10.1104/pp.19.00772.

Plaza, M., Pozzo, T., Liu, J., Gulshan Ara, K. Z., Turner, C., & Nordberg Karlsson, E. (2014). Substituent effects on in vitro antioxidizing properties, stability, and solubility in flavonoids. *Journal of Agricultural and Food Chemistry*, *62*, 3321–3333. https://doi.org/10.1021/jf405570u.

Quattrocchio, F., Baudry, A., Lepiniec, L., & Grotewold, E. (2006). The regulation of flavonoid biosynthesis. *The Science of Flavonoids*, 97–122. https://doi.org/10.1007/978-0-387-28822-2_4.

Rahimi, A. R., Mashayekhi, K., Ghasemnezhad, A., Ebrahimi, P., & Hemmati, K. (2009). Salicylic acid plus nutritional minerals influence the essential oil production and linalool accumulation of coriander seeds (*Coriandrum sativum* L.). *Zeitschrift Fur Arznei- Und Gewurzpflanzen*, *14*(4), 167–171.

Rahman, M. H., Joersbo, M., & Poulsen, M. H. (2001). Development of yellow-seeded *Brassica napus* of double low quality. *Plant Breeding*, *120*(6), 473–478. https://doi.org/10.1046/j.1439-0523.2001.00639.x.

Rajjou, L., & Debeaujon, I. (2008). Seed longevity: Survival and maintenance of high germination ability of dry seeds. *Comptes Rendus Biologies*, *331*(10), 796–805. https://doi.org/10.1016/j.crvi.2008.07.021.

Rajjou, L., Gallardo, K., Debeaujon, I., Vandekerckhove, J., Job, C., & Job, D. (2004). The effect of α-amanitin on the arabidopsis seed proteome highlights the distinct roles of stored and neosynthesized mRNAs during germination. *Plant Physiology*, *134*, 1598–1613. https://doi.org/10.1104/pp.103.036293.

Ribeiro, B., Lacchini, E., Bicalho, K., Mertens, J., Arendt, P., Vanden Bossche, R., et al. (2020). A seed-specific regulator of triterpene saponin biosynthesis in *Medicago truncatula*. *The Plant Cell*. tpc.00609.2019. https://doi.org/10.1105/tpc.19.00609.

Riechmann, J. L., & Meyerowitz, E. M. (1998). The AP2/EREBP family of plant transcription factors. *Biological Chemistry*, *379*, 633–646. https://doi.org/10.1515/bchm.1998.379.6.633.

Routaboul, J. M., Kerhoas, L., Debeaujon, I., Pourcel, L., Caboche, M., Einhorn, J., et al. (2006). Flavonoid diversity and biosynthesis in seed of *Arabidopsis thaliana*. *Planta*, *224*(1), 96–107. https://doi.org/10.1007/s00425-005-0197-5.

Rychel, S., & Książkiewicz, M. (2019). Development of gene-based molecular markers tagging low alkaloid pauper locus in white lupin (*Lupinus albus* L.). *Journal of Applied Genetics*, *60*, 269–281. https://doi.org/10.1007/s13353-019-00508-9.

Safi, A., Medici, A., Szponarski, W., Ruffel, S., Lacombe, B., & Krouk, G. (2017). The world according to GARP transcription factors. *Current Opinion in Plant Biology*, *39*, 159–167. https://doi.org/10.1016/j.pbi.2017.07.006.

Sagara, T., Bhandari, D. R., Spengler, B., & Vollmann, J. (2020). Spermidine and other functional phytochemicals in soybean seeds: Spatial distribution as visualized by mass spectrometry imaging. *Food Science & Nutrition*, *8*(1), 675–682. https://doi.org/10.1002/fsn3.1356.

Saito, K., Yonekura-Sakakibara, K., Nakabayashi, R., Higashi, Y., Yamazaki, M., Tohge, T., et al. (2013). The flavonoid biosynthetic pathway in Arabidopsis: Structural and genetic diversity. *Plant Physiology and Biochemistry*, *72*, 21–34. https://doi.org/10.1016/j.plaphy.2013.02.001.

Sánchez-Pérez, R., Pavan, S., Mazzeo, R., Moldovan, C., Cigliano, R. A., Cueto, J. D., et al. (2019). Mutation of a bHLH transcription factor allowed almond domestication. *Science*, *364*, 1095–1098.

Sano, N., Kim, J. S., Onda, Y., Nomura, T., Mochida, K., Okamoto, M., et al. (2017). RNA-Seq using bulked recombinant inbred line populations uncovers the importance of brassinosteroid for seed longevity after priming treatments. *Scientific Reports*, 7, 8095. https://doi.org/10.1038/s41598-017-08116-5.

Sattler, S. E., Gilliland, L. U., Magallanes-Lundback, M., Pollard, M., & DellaPenna, D. (2004). Vitamin E is essential for seed longevity and for preventing lipid peroxidation during germination. *Plant Cell*, *16*(6), 1419–1432. https://doi.org/10.1105/tpc.021360.

Schläger, S., & Dräger, B. (2016). Exploiting plant alkaloids. *Current Opinion in Biotechnology*, *37*, 155–164. https://doi.org/10.1016/j.copbio.2015.12.003.

Schuy, C., Groth, J., Ammon, A., Eydam, J., Baier, S., Schweizer, G., et al. (2019). Deciphering the genetic basis for vitamin E accumulation in leaves and grains of different barley accessions. *Scientific Reports*, *9*, 9470. https://doi.org/10.1038/s41598-019-45572-7.

Seo, M.-S., & Kim, J. S. (2017). Understanding of MYB Transcription Factors Involved in Glucosinolate Biosynthesis in Brassicaceae. *Molecules*, *22*, 1549. https://doi.org/10.3390/molecules22091549.

Seyer, A., Einhorn, J., Brunelle, A., & Laprévote, O. (2010). Localization of flavonoids in seeds by cluster time-of-flight secondary ion mass spectrometry imaging. *Analytical Chemistry*, *82*(6), 2326–2333. https://doi.org/10.1021/ac902528t.

Shewry, P. R., Napier, J. A., & Tatham, A. S. (1995). Seed storage proteins: Structures and biosynthesis. *Plant Cell*, 7, 945–956. https://doi.org/10.1105/tpc.7.7.945.

Shirley, B. W. (1998). Flavonoids in seeds and grains: physiological function, agronomic importance and the genetics of biosynthesis. *Seed Science Research*, *8*, 415–422. https://doi.org/10.1017/S0960258500004372.

Smýkal, P., Vernoud, V., Blair, M. W., Soukup, A., & Thompson, R. D. (2014). The role of the testa during development and in establishment of dormancy of the legume seed. *Frontiers in Plant Science*, *5*, 1–19. https://doi.org/10.3389/fpls.2014.00351.

Song, C., Härtl, K., McGraphery, K., Hoffmann, T., & Schwab, W. (2018). Attractive but toxic: Emerging roles of glycosidically bound volatiles and glycosyltransferases involved in their formation. *Molecular Plant*, *11*, 1225–1236. https://doi.org/10.1016/j.molp.2018.09.001.

Srimany, A., George, C., Naik, H. R., Pinto, D. G., Chandrakumar, N., & Pradeep, T. (2016). Developmental patterning and segregation of alkaloids in areca nut (seed of *Areca catechu*) revealed by magnetic resonance and mass spectrometry imaging. *Phytochemistry*, *125*, 35–42. https://doi.org/10.1016/j.phytochem.2016.02.002.

Stafford, H. A. (1974). The metabolism of aromatic compounds. *Annual Review of Plant Physiology*, *25*, 459–486. https://doi.org/10.1146/annurev.pp.25.060174.002331.

Stetter, M. G., Vidal-Villarejo, M., Schmid, K. J., & Wright, S. (2020). Parallel seed color adaptation during multiple domestication attempts of an ancient new world grain. *Molecular Biology and Evolution*, *37*(5), 1407–1419. https://doi.org/10.1093/molbev/msz304.

Stracke, R., Ishihara, H., Huep, G., Barsch, A., Mehrtens, F., Niehaus, K., et al. (2007). Differential regulation of closely related R2R3-MYB transcription factors controls flavonol accumulation in different parts of the *Arabidopsis thaliana* seedling. *Plant Journal*, *50*(4), 660–677. https://doi.org/10.1111/j.1365-313X.2007.03078.x.

Stracke, R., Jahns, O., Keck, M., Tohge, T., Niehaus, K., Fernie, A. R., et al. (2010). Analysis of production of flavonol glycosides-dependent flavonol glycoside accumulation in *Arabidopsis thaliana* plants reveals MYB11-, MYB12- and MYB111-independent flavonol glycoside accumulation. *New Phytologist*, *188*(4), 985–1000. https://doi.org/10.1111/j.1469-8137.2010.03421.x.

Suzuki, T., & Waller, G. R. (1987). Allelopathy due to purine alkaloids in tea seeds during germination. *Plant and Soil*, *98*, 131–136. https://doi.org/10.1007/BF02381733.

Tanwir, F., Dionisio, G., Adhikari, K. B., Fomsgaard, I. S., & Gregersen, P. L. (2017). Biosynthesis and chemical transformation of benzoxazinoids in rye during seed germination and the identification of a rye Bx6-like gene. *Phytochemistry*, *140*, 95–107. https://doi.org/10.1016/j.phytochem.2017.04.020.

Teles, Y. C. F., Souza, M. S. R., & De Souza, M.d. F. V. (2018). Sulphated flavonoids: Biosynthesis, structures, and biological activities. *Molecules*, *23*(2), 1–11. https://doi.org/10.3390/molecules23020480.

Tetlow, I. J. (2011). Starch biosynthesis in developing seeds. *Seed Science Research*, *21*(1), 5–32. https://doi.org/10.1017/S0960258510000292.

Thodberg, S., Cueto, J. D., Mazzeo, R., Pavan, S., Lotti, C., Dicenta, F., et al. (2018). Elucidation of the amygdalin pathway reveals the metabolic basis of bitter and sweet almonds prunus. *178*, 1096–1111. https://doi.org/10.1104/pp.18.00922.

Tohge, T., de Souza, L. P., & Fernie, A. R. (2017). Current understanding of the pathways of flavonoid biosynthesis in model and crop plants. *Journal of Experimental Botany*, *68*(15), 4013–4028. https://doi.org/10.1093/jxb/erx177.

Tohge, T., Wendenburg, R., Ishihara, H., Nakabayashi, R., Watanabe, M., Sulpice, R., et al. (2016). Characterization of a recently evolved flavonol-phenylacyltransferase gene provides signatures of natural light selection in Brassicaceae. *Nature Communications*, *7*, 12399. https://doi.org/10.1038/ncomms12399.

Treml, J., & Šmejkal, K. (2016). Flavonoids as potent scavengers of hydroxyl radicals. *Comprehensive Reviews in Food Science and Food Safety*, *15*(4), 720–738. https://doi.org/10.1111/1541-4337.12204.

Troncoso-Ponce, M. A., Barthole, G., Tremblais, G., To, A., Miquel, M., Lepiniec, L., et al. (2016). Transcriptional activation of two delta-9 palmitoyl-ACP desaturase genes by MYB115 and MYB118 is critical for biosynthesis of omega-7 monounsaturated fatty acids in the endosperm of Arabidopsis seeds. *The Plant Cell*, *28*, 2666–2682. https://doi.org/10.1105/tpc.16.00612.

Tungmunnithum, D., Thongboonyou, A., Pholboon, A., & Yangsabai, A. (2018). Flavonoids and other phenolic compounds from medicinal plants for pharmaceutical and medical aspects: An overview. *Medicines*, *5*(3), 93. https://doi.org/10.3390/medicines5030093.

Tuteja, J. H., Zabala, G., Varala, K., Hudson, M., & Vodkin, L. O. (2009). Endogenous, tissue-specific short interfering RNAs silence the chalcone synthase gene family in *Glycine max* seed coats. *Plant Cell*, *21*(10), 3063–3077. https://doi.org/10.1105/tpc.109.069856.

Vom Endt, D., Kijne, J. W., & Memelink, J. (2002). Transcription factors controlling plant secondary metabolism: What regulates the regulators? *Phytochemistry*, *61*(2), 107–114. https://doi.org/10.1016/S0031-9422(02)00185-1.

Waki, T., Mameda, R., Nakano, T., Yamada, S., Terashita, M., Ito, K., et al. (2020). A conserved strategy of chalcone isomerase-like protein to rectify promiscuous chalcone synthase specificity. *Nature Communications*, *11*(1), 1–14. https://doi.org/10.1038/s41467-020-14558-9.

Wang, S., Alseekh, S., Fernie, A. R., & Luo, J. (2019). The structure and function of major plant metabolite modifications. *Molecular Plant*, *12*, 899–919. https://doi.org/10.1016/j.molp.2019.06.001.

Wang, M., Carver, J. J., Phelan, V. V., Sanchez, L. M., Garg, N., Peng, Y., et al. (2016). Sharing and community curation of mass spectrometry data with Global Natural Products Social Molecular Networking. *Nature Biotechnology*, *34*, 828–837. https://doi.org/10.1038/nbt.3597.

Wangchuk, P. (2019). Plant alkaloids: Classification, isolation, and drug development. In *Medicinal Plants: Chemistry, Pharmacology, and Therapeutic Applications* (pp. 131–137). CRC Press.

Wei, Z., Cheng, Y., Zhou, C., Li, D., Gao, X., Zhang, S., et al. (2019). Genome-wide identification of direct targets of the TTG1-bHLH-MYB complex in regulating trichome formation and flavonoid accumulation in *Arabidopsis thaliana*. *International Journal of Molecular Sciences*, *20*(20), 1–16. https://doi.org/10.3390/ijms20205014.

Wink, M. (1983). Inhibition of seed germination by quinolizidine alkaloids. *Planta*, *158*, 365–368. https://doi.org/10.1007/bf00397339.

Wolfender, J. L., Litaudon, M., Touboul, D., & Queiroz, E. F. (2019). Innovative omics-based approaches for prioritisation and targeted isolation of natural products-new strategies for drug discovery. *Natural Product Reports*, *36*, 855–868. https://doi.org/10.1039/c9np00004f.

Xu, W., Dubos, C., & Lepiniec, L. (2015). Transcriptional control of flavonoid biosynthesis by MYB-bHLH-WDR complexes. *Trends in Plant Science*, *20*(3), 176–185. https://doi.org/10.1016/j.tplants.2014.12.001.

Xu, W., Grain, D., Bobet, S., Le Gourrierec, J., Thévenin, J., Kelemen, Z., et al. (2014). Complexity and robustness of the flavonoid transcriptional regulatory network revealed by comprehensive analyses of MYB-bHLH-WDR complexes and their targets in Arabidopsis seed. *New Phytologist*, *202*(1), 132–144. https://doi.org/10.1111/nph.12620.

Xue, G., Gong, L., Yuan, C., Xu, M., Wang, X., Jiang, L., et al. (2017). A structural mechanism of flavonoids in inhibiting serine proteases. *Food & Function*, *8*, 2437–2443. https://doi.org/10.1039/c6fo01825d.

Yu, O., Shi, J., Hession, A. O., Maxwell, C. A., McGonigle, B., & Odell, J. T. (2003). Metabolic engineering to increase isoflavone biosynthesis in soybean seed. *Phytochemistry*, *63*(7), 753–763. https://doi.org/10.1016/S0031-9422(03)00345-5.

Zahedi, S. M., Karimi, M., & Venditti, A. (2019). Plants adapted to arid areas: specialized metabolites. *Natural Product Research*. https://doi.org/10.1080/14786419.2019.1689500.

Zhang, F., Fu, X., Lv, Z., Lu, X., Shen, Q., Zhang, L., et al. (2015). A basic leucine zipper transcription factor, AabZIP1, connects abscisic acid signaling with artemisinin biosynthesis in *Artemisia annua*. *Molecular Plant*, *8*(1), 163–175. https://doi.org/10.1016/j.molp.2014.12.004.

Zhang, F., Lu, X., Lv, Z., Zhang, L., Zhu, M., Jiang, W., et al. (2013). Overexpression of the artemisia orthologue of ABA Receptor, AaPYL9, Enhances ABA sensitivity and improves artemisinin content in *Artemisia annua* L. *PLoS One*, *8*(2), e56697. https://doi.org/10.1371/journal.pone.0056697.

Zhang, B., & Schrader, A. (2017). Transparent testa glabra 1-dependent regulation of flavonoid biosynthesis. *Plants*, *6*(4), 65. https://doi.org/10.3390/plants6040065.

Zhang, M., Swarts, S. G., Yin, L., Liu, C., Tian, Y., Cao, Y., et al. (2011). Antioxidant properties of quercetin. *Advances in Experimental Medicine and Biology*. https://doi.org/10.1007/978-1-4419-7756-4_38.

Zhang, Y., & Fernie, A. R. (2020). Metabolons, enzyme–enzyme assemblies that mediate substrate channeling, and their roles in plant metabolism. *Plant Communications*, *100081*. In press. https://doi.org/10.1016/j.xplc.2020.100081.

Zhao, K., & Rhee, S. Y. (2019). Epigenomic landscape of *Arabidopsis thaliana* metabolism reveals bivalent chromatin on specialized metabolic genes. *BioRxiv*. https://doi.org/10.1101/589036.

Zhou, Y. J. (2018). Expanding the terpenoid kingdom. *Nature Chemical Biology*, *14*(12), 1069–1070. https://doi.org/10.1038/s41589-018-0167-4.

Zhou, Z., Jiang, Y., Wang, Z., Gou, Z., Lyu, J., Li, W., et al. (2015). Resequencing 302 wild and cultivated accessions identifies genes related to domestication and improvement in soybean. *Nature Biotechnology*, *33*, 408–414. https://doi.org/10.1038/nbt.3096.

Zhu, B. F., Si, L., Wang, Z., Zhou, Y., Zhu, J., Shangguan, Y., et al. (2011). Genetic control of a transition from black to straw-white seed hull in rice domestication. *Plant Physiology*, *155*(3), 1301–1311. https://doi.org/10.1104/pp.110.168500.

Zhu, Q., Wang, B., Tan, J., Liu, T., Li, L., & Liu, Y.-G. (2020). Plant synthetic metabolic engineering for enhancing crop nutritional quality. *Plant Communications*, *1*(1), 100017. https://doi.org/10.1016/j.xplc.2019.100017.

Zhu, F., Wen, W., & Fernie, A. R. (2019). Finding Noemi: The transcription factor mutations underlying trait differentiation amongst citrus. *Trends in Plant Science*, *24*(5), 384–386. https://doi.org/10.1016/j.tplants.2019.03.001.

SECTION 2

Metabolomics and plant-environment interactions

CHAPTER THREE

Untangling plant immune responses through metabolomics

Alex Williams[a], Jordi Gamir[b], Antoine Gravot[c], and Pierre Pétriacq[d,*]

[a]School of Earth and Environmental Sciences, The University of Manchester, Manchester, United Kingdom
[b]Metabolic Integration and Cell Signaling Laboratory, Plant Physiology Section, Dept Ciencias Agrarias y del Medio Natural, Universitat Jaume I, Castellón, Spain
[c]IGEPP, AGROCAMPUS OUEST, INRAE, Univ Rennes, Le Rheu, France
[d]UMR1332, Université de Bordeaux, INRAE, Bordeaux Metabolome, MetaboHUB, PHENOME, Villenave d'Ornon, France
*Corresponding author: e-mail address: pierre.petriacq@inrae.fr

Contents

Abstract

Plant metabolomics is a set of fast-moving, analytical and chemometric tools and methods for plant functional genomics, phenotyping and systems biology. This multidisciplinary "omics" science can deliver qualitative and quantitative data that provide a detailed description of biochemical systems that are influenced by environmental changes, such as plant–pathogen interactions. This chapter examines key insights and recent outputs in the field of phytopathological metabolomics with a specific focus on the perturbations of primary compounds involved in central metabolism and infection-related metabolites including phytohormones. Priming of plant immune responses is also considered through the angle of metabolomics. Furthermore, the chapter highlights the recurring challenges in the distinction of host and pathogen metabolomes and the hurdles of metabolome annotation. The chapter concludes with

Advances in Botanical Research, Volume 98
ISSN 0065-2296
https://doi.org/10.1016/bs.abr.2020.09.017

perspectives indicating key avenues in ongoing efforts to decode metabolic landscapes of plants under biotic stress conditions.

1. Introduction

In the last two decades, the advent of metabolomics as a systems biology strategy has greatly contributed to rapidly enhancing our knowledge of complex biological systems such as plants. As an integrative, functional methodology, metabolomics gathers insight from the different omics scales including the transcriptome, proteome or enzyme activome (Hall, 2011). In this sense, the metabolome (i.e., the complete set of low molecular weight compounds) embodies not only the eventual phenotype influenced by gene expression, and the alteration of protein function from genetic and environmental origins, but also the feedback and response that these cellular processes constantly undergo (Hong, Yang, Zhang, & Shi, 2016).

Nowadays, the term "metabolomics" refers to both untargeted and targeted approaches that describe metabolomes through global pictures, or precise concentrations of metabolites, characterizing organismal chemistries. To untangle this great chemical diversity, metabolomics relays on two major analytical technologies that historically originate from Physics, and are now widespread within (bio)chemistry and biology laboratories: nondestructive nuclear magnetic resonance spectroscopy (NMR) and mass spectrometry (MS)-based techniques, including GC/LC/CE-MS (gas/liquid chromatography/capillary electrophoresis-MS) and FT-ICR–MS (Fourier transform-ion cyclotron resonance-MS; Deborde et al., 2017; Perez de Souza, Alseekh, Naake, & Fernie, 2019). While it is yet not possible to attain an exact and complete depiction of the whole plant metabolome, multiple combined approaches are being increasingly used in metabolomics investigations to identify, quantify and localize plant metabolites. The combination of complex techniques warrants sensitivity, selectivity and repeatability to increase metabolome coverage, providing highly multivariate datasets that require computational tools. As such, alongside analytical advances, bioinformatics and biostatistics have progressed to offer tools and methods to depict the metabolomics complexity, reveal discriminant biomarkers and even predict metabolic fluxes based on mathematical modeling (Fernandez et al., 2016). These technological developments are beyond the scope of this chapter and have been covered in recent reviews (Castro-Moretti, Gentzel, Mackey, & Alonso, 2020) and volume 67 of Advances in Botanical Research (ABR; Rolin, 2013).

With a swiftly growing human population setting to exceed 9 billion by 2050, it will be vital to develop improved crop varieties toward better tolerance to abiotic and biotic stress in order to address the ever-increasing challenge of food security aggravated by increasing climate change. This task relies on a better comprehension of the metabolic drivers for increased plant performance with resilience to stress and diseases that have adverse effects on crop yields. Indeed, plants suffer a fixed life which prevents them from escaping or hiding from a plethora of latent biotic foes (i.e. bacteria, fungi, viruses, herbivores, nematodes) and abiotic environmental constraints. To defend themselves against these hostile stress conditions, plants have developed unique chemical features, approximately 200,000 molecules, thanks to a very dynamic metabolism that participates in a fine adjustment of plant developmental processes and their responses to environmental stimuli (Kessler & Kalske, 2018). Furthermore, since ancient times, mankind has used plants as natural biochemists owing to their chemically complex metabolomes, with diversified structures and abundance, which have already delivered effective medication and will continue to provide an untapped resource for the discovery of original pharmaceuticals (Guerriero et al., 2018).

Like other omics, metabolomics has made manifest contributions to the recent advancements in our comprehension of plant immune responses (Castro-Moretti et al., 2020; Tugizimana, Mhlongo, Piater, & Dubery, 2018). A typical metabolomics design and workflow are presented in Fig. 1 depicting both targeted and untargeted strategies. Upon microbial challenges, metabolomics has revealed specific and common metabolic signatures as hallmarks of infection processes, antimicrobial metabolite counterattack or metabolic signals to rearrange plant physiology and adjust to the stress. In the present chapter, we will inspect the major and recent contributions of metabolomics to plant pathology studies with a detailed focus on the reorchestration of central and stress-related metabolic pathways in response to biotic stimuli and immune priming. Finally, we will emphasize how phytopathological metabolomics can contribute to take up the new challenges and perspectives on plant–pathogen threats that are emerging in the context of climate change.

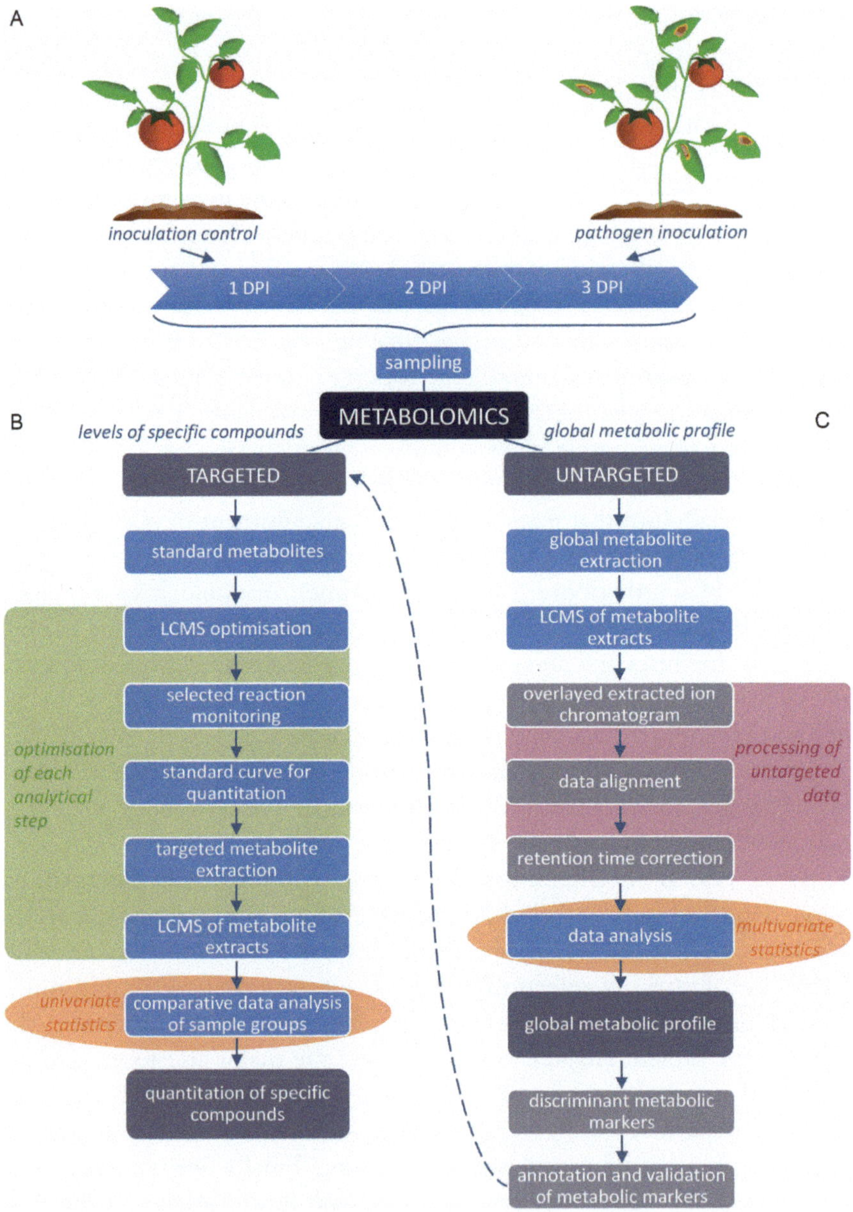

Fig. 1 Typical LCMS-based metabolomics workflow in phytopathological studies. (A) Plant tissues (here, tomato leaves) are inoculated without (i.e., mock) or with the pathogenic microbe (here, a fungus causing necrotic lesions) or subjected to another relevant biotic stress (e.g., elicitors). Following a time course expressed as day post-inoculation (DPI), plant tissues are sampled in liquid nitrogen for metabolic quenching and stored until targeted (B) and/or untargeted (C) metabolomics can be performed. *(Continued)*

2. Deciphering the Janus faces of primary metabolic responses in plant–pathogen interactions

The importance of carbohydrate metabolism for the outcome of plant–pathogen interactions has been studied since the 1960s (Long, Fung, McGEE, Cooke, & Lewis, 1975; Smith, Muscatine, & Lewis, 1969). In those early works, the use of chromatographic profiling and radio-tracing approaches enabled the sketching of some hallmarks of pathogen-triggered alterations in primary metabolism. This includes (1) the downregulation of photosynthesis and the rise of respiratory activities, (2) the promotion of nutrient sink in infected tissues, in relation with (3) Cell-Wall Invertase (CWI)-mediated release of monosaccharides in the apoplast compartment. Pivotal questioning has emerged from these early studies, concerning positive or negative influences exerted by photo-assimilate allocation/partitioning responses onto pathogen nutrient uptake and disease susceptibility. Those issues have thereafter been overshadowed by outstanding sagas focused on defensive specialized metabolites and *R*-gene cloning.

However, the development of omic approaches over the last two decades has once more stimulated interest on the importance of primary metabolism regulation during biotic interactions (Berger, Sinha, & Roitsch, 2007; Kanwar & Jha, 2019). Targeted and quantitative profiling

Fig. 1—Cont'd Targeted metabolomics (B) is based on the quantitative measurement of specific compounds, using metabolite standards to optimize the LC and MS analysis and extraction. The resulting targeted data are processed by univariate statistical methods to provide comparative quantitation (or absolute quantification when using isotopically labeled standards) of the compound of interest between the control and infected samples at different time points. Untargeted metabolomics (C) provides a comprehensive metabolic overview of several thousands of metabolic features from a global extraction that are analyzed by untargeted LCMS profiling. The processing of untargeted data relies on sophisticated bioinformatics workflow (including extraction, alignment, grouping and retention time correction) on all the detected LCMS features. Multivariate statistical methods are subsequently applied to the dataset to visualize how metabolic profiles respond to the conditions (i.e., biotic stress) and unravel discriminant metabolic markers that link to the stress. These relevant markers are putatively annotated using metabolite databases, then confirmed by comparing MSMS and retention time data with metabolite standards, thereby returning to a targeted approach for verification. Targeted and untargeted metabolomics are complementary in addressing how relevant metabolites and metabolism respond or comprehensively reacts to stress.

of specific families of compounds, and NMR- or GC/LC-MS based untargeted metabolomics methods, have been extensively used to complement our views on the regulation of major primary metabolic pathways. Beyond that, the combination of transcriptomics and metabolomics has resulted in a series of holistic datasets covering a wealth of pathosystems. Together with automated methods to map both metabolites and transcripts against genome-scale networks, this has provided global views of regulatory processes during pathogen infection (Aliferis & Jabaji, 2012; Copley, Aliferis, Kliebenstein, & Jabaji, 2017; Deeken et al., 2006; Doehlemann et al., 2008; Duplessis et al., 2011; Gonzalez-Mula et al., 2019; Rudd et al., 2015). Fascinatingly, these works also brought credence to the idea that pathogens could (deliberately) trigger genome-scale reprogramming of plant primary metabolism to fulfill their nutrient acquisition strategies.

The functional interpretation of primary metabolism responses to infection can, however, be a delicate task. Both pathogen and host plants depend on the energy and carbon skeletons of primary metabolites, and both organisms are also potentially impacted by the signaling information conveyed by those compounds. Infection-triggered regulations of primary metabolism can thus have multiple roles: (1) promoting pathogen nutrition ("metabolic high-jacking" by pathogen effectors), (2) fuelling plant defense responses ("fuel for the fire" hypothesis developed by Bolton (2009), and/or (3) promoting plant defense via signaling processes (Kanwar & Jha, 2019). The aforementioned combined metabolomic and genetic/transcriptomic approaches have then become paramount in deciphering these multiple putative roles of primary metabolic pathways during plant–pathogen interactions.

Following the discovery that certain fungi can detoxify and potentially utilize plant defense chemicals (glucosinolates; Chen et al., 2010), further research has investigated the roles of sugar transporters families in plant–pathogen competition for soluble carbohydrates (Bezrutczyk et al., 2018; Kanwar & Jha, 2019). Contrasted conclusions from this research have highlighted the importance of trophic lifestyles of different pathogen species; the mechanism involved in the import of hexose into the plant cytoplasm promotes the nutrition of a pathogen with intracellular or haustorium-based lifestyle (Huai et al., 2019) while impairing access to nutriment for pathogens with apoplastic lifestyles (Lemonnier et al., 2014; Yamada, Saijo, Nakagami, & Takano, 2016). Alternatively, the sequestration of sugar in vacuoles has also been suggested to enhance resistance to *Verticillium* (Meteier et al., 2019). Thus, measuring metabolite dynamics at a subcellular level is crucial for the validation of those models. A few studies have developed profiling methods to investigate rapid metabolic changes in apoplastic fractions from infected plants (O'Leary et al., 2016). However, this approach did not always match with hypothesized roles of transporter induction, possibly as a result of technical reasons (discussed in Bezrutczyk et al., 2018). Such inadequacies encourage the development of complementary approaches, such as the use of metabolic sensors (Toyota et al., 2018), or metabolomics investigations of subcellular fractions obtained using nonaqueous purification methods (Gagneul et al., 2007; Hossain, Persicke, ElSayed, Kalinowski, & Dietz, 2017). Matrix-assisted laser desorption ionization imaging (MALDI-I) techniques (reviewed in chapter "Metabolite imaging by mass spectrometry: A new discovery tool" by Walker) also offer promising perspectives for spatial dissection of metabolic responses to infection. These techniques allowed validating the role of SWEET transporters in the accumulation of diholosides in roots of Arabidopsis infected by *Plasmodiophora brassicae* (Walerowski et al., 2018). Spatial metabolomics strategy in plant-pathogen interactions might be implemented in the future by methodological developments to assess colocalizations between pathogen structures and metabolic patterns at the subcellular scale (Geier et al., 2020).

As mentioned earlier, the typical induction of CWI-triggered sucrose hydrolysis in infected tissues, (lowering sucrose/hexose ratio) can also initiate systemic signaling that contributes to the induction of major defense processes such as salicylic acid (SA) signaling, *Pathogen Related* (*PR*) genes transcription, phytoalexin biosynthesis, and photosynthesis downregulation (Gebauer et al., 2017; Tauzin & Giardina, 2014). Besides "SWEET signaling," a series of nitrogen primary compounds, initially identified in plant–pathogen metabolomics investigations, was also revealed to play important immune roles. A good example is the importance of pipecolic acid during plant immunity, initially identified by amino acid profiling (Návarová, Bernsdorff, Döring, & Zeier, 2012), and the role of proline or polyamine in the regulation of reactive oxygen species (ROS)-mediated defense signaling (Fabro, Kovács, Pavet, Szabados, & Alvarez, 2004; Fabro, Rizzi, & Alvarez, 2016; Liu, Atanasov, Tiburcio, & Alcázar, 2019; Lou, Bor, Yan, Preuss, & Jander, 2016). Recently, it was reported that damage-induced glutamate concentration in apoplast could increase by up to 50 times, mediating long-distance defense responses by the activation of glutamate receptors (Toyota et al., 2018). These recent observations suggest that stress-induced subcellular partitioning of primary metabolites could thus play a range of still uncovered signaling roles in plant defense.

It is clear that plant–pathogen interactions typically lead to large perturbations of plant primary metabolism. Different modules of these responses can result from diverse plant or pathogen genetic factors, and possibly trigger multiple layers of above-described processes, with contradictory consequences on the fate of pathogen infection and plant defense. Quantitative genetics is a potentially auspicious option to investigate this complexity. This has been implemented with some success, for instance, a colocalization study between metabolic and resistance Quantitative Trait Loci pointed to the genetic-based deconvolution of links (or absence of) between different modules of primary metabolism and pathogen development (Wagner et al., 2019). The implementation of such strategies may benefit in the future from the advancement of high-throughput profiling protocols capable of dealing with large germplasm collections for genome-wide association analyses and generate links between primary metabolism and specific defense mechanisms and metabolites, that are still not fully established.

3. Special preparation for the battle: Stress-related specialized metabolites

Generally, specialized metabolites (SM) are biochemical compounds originating almost solely from plants and fungi (Kessler & Kalske, 2018). When referred to as secondary metabolites, their definition implies that these compounds have *stricto sensu* no direct function in growth and developmental processes but rather in interactions with the environment. This appears somewhat erroneous as some specific SM (e.g., phytohormones, antioxidant metabolites) are master regulators of plant growth and participate in critical functions like photosynthesis, respiration or nutrient assimilation. Compared to primary compounds, SM are usually restricted to one species or a group of taxonomically related plants (Bednarek & Osbourn, 2009). Besides, due to their diversified chemical structure and ornamentation, SM are assumed to be costly from a production perspective, which could imbalance growth and reproductive processes (Guerriero et al., 2018). Despite this paradigm, some plant species are capable of accumulating large amounts of SM, thereby contributing to stress mitigation, and particularly immune responses. Nevertheless, this trade-off between development and defense requires much more attention to understand better how growth-immune balance could be manipulated to maximize crop quality and yield upon microbial and pest challenges (Huot, Yao, Montgomery, & He, 2014).

Plant SM can be distinguished in three chemically distinct major classes of molecules, namely phenolics, terpenes and nitrogen (N)- or sulfur (S)-containing biochemicals. SM have well-known defensive roles including deterrent, antifeedant or toxic activities, and could act as precursors to physical immune barriers (Zaynab et al., 2018). Due to their antioxidant properties, SM are also associated with the reduction–oxidation (redox) metabolism that is particularly altered by stressful conditions like biotic challenges (Decros et al., 2019). Functionally, defense-related SM are divided into either phytoanticipins which are constitutive molecules, or phytoalexins induced upon biotic challenges thus requiring *de novo* enzyme synthesis. Metabolomics has proven particularly crucial to detect, annotate, identify and quantify SM in phytopathological systems (Castro-Moretti et al., 2020). In this regard, targeted approaches have focused on metabolites of interest involved in stress responses, while untargeted strategies have evaluated the global metabolomic responses (Fig. 1). This section will present some recent examples of the major classes of SM where metabolomics contribute to the study of immune responses.

3.1 Terpenes, the largest class of specialized metabolites with multiple roles in plant defenses

Terpenes, or terpenoids, which constitute the largest class of SM, include C5 isoterpene units and derive from acetyl-CoA or glycolytic intermediates (Singh & Sharma, 2015). A well-known example of a defense terpene is the stress hormone abscisic acid (ABA), a sesquiterpene acting as a master regulator of stress signaling pathways (Pieterse, Leon-Reyes, Van der Ent, & Van Wees, 2009). Methods for profiling hormones, like ABA, have received considerable attention during the development of metabolomics technologies (Glauser, Vallat, & Balmer, 2014).

As defensive metabolites, terpenes not only have toxic and antimicrobial properties but also repellent roles, some owing to their volatile nature and strong scent to deter herbivores and pests (Hammerbacher, Coutinho, & Gershenzon, 2019). In *Mentha aquatica* for instance, a GC–MS study reported modulation of terpene synthesis toward the production of deterrent molecules against the herbivore *Chrysolina herbacea* (Atsbaha Zebelo et al., 2011). In other species, di- and sesquiterpenes act as antimicrobial phytoalexins like in rice against the fungus *Magnaporthe grisea* (Prisic, Xu, Wilderman, & Peters, 2004). In Arabidopsis, genetic engineering of herbivory-induced volatiles produced plants emitting terpenoids that attract carnivorous predatory mites thereby helping defense mechanisms (Kappers et al., 2005).

Furthermore, terpenes bear antioxidant activities and are able to process free radicals and thus mitigate the burst of ROS generated by stress (Graßmann, 2005). Carotenoids, in particular, have received special attention as pigments preventing photooxidation (Felemban, Braguy, Zurbriggen, & Al-Babili, 2019). Untargeted metabolomics has further highlighted the possible role of carotenoids in response to various biotic infections (Wilkinson, Pastor, Paplauskas, Pétriacq, & Luna, 2018; Williams, Pétriacq, Schwarzenbacher, Beerling, & Ton, 2018). Besides carotenoids, a recent untargeted metabolomics study on cotton reported the accumulation of terpenes, among other SM and carbohydrates, against *Aspergillus* leaf spot (Khizar et al., 2020). Similarly, untargeted metabolomics unraveled diterpene glycosides as potential chemical defenses against thrips in pepper (Macel et al., 2019). One such diterpene glycoside, capsianosides, displayed toxicity to insects and exhibited strong antimicrobial activities. Overall, metabolomics studies indicate various roles for terpenes in defense responses to biotic infections.

3.2 Phenolics, hallmarks of immune responses and mitigators of oxidative stress

Another diverse group of SM is the class of phenolics, also called polyphenols. Structurally, these metabolites consist of at least one aromatic ring bonded to one or more hydroxyl groups and other substituents (e.g., methoxyl or carboxyl moieties) which increase the molecule polarity thus allowing water solubility (López-Fernández et al., 2020). Phenolics derive from amino acid (e.g., phenylalanine) or the shikimate pathways and are divided into five main groups depending on their chemical complexity, namely phenolic acids, stilbenes, flavonoids, lignans, and others; flavonoids being the most abundant class (López-Fernández et al., 2020).

Many metabolomics studies, both LC-MS-based targeted and untargeted approaches, have reported the involvement of phenolics in plant-pathogen interactions (Castro-Moretti et al., 2020; López-Fernández et al., 2020). The most studied defense-responsive phenolic compound is the phytohormone SA which is synthesized from chorismate or phenylalanine in response to stress (Lefevere, Bauters, & Gheysen, 2020). SA accumulation (which can be quantified using reverse phase chromatography coupled to MS) and systemic transport has been demonstrated upon challenge with bacterial, fungal and herbivore pests, and in response to other stress-related events, such as defense priming, described below. Other specific phenolics such as coumarins, flavonoids, lignins and tannins have also been reported to be part of the plant defense system against pests and diseases (Kulbat, 2016). For instance, foliar phenolics, including flavonoids and coumarins, are implicated in constitutive resistance in *Fraxinus* species to Emerald Ash Borer (Qazi, Lombardo, & Abou-Zaid, 2018). As antioxidants, phenolics can also mitigate the redox alterations in response to biotic challenges like in vines in addition to their role in colors and flavors (Gabaston et al., 2017, 2018; Goufo, Singh, & Cortez, 2020).

Untargeted metabolomics approaches were more scarce but revealed accumulated pools of putative phenolics (e.g., flavonoids, anthocyanins, coumaric acids) in direct response to pathogenic infections (Camañes, Scalschi, Vicedo, González-Bosch, & García-Agustín, 2015; Mareya et al., 2019; Pétriacq, Stassen, & Ton, 2016; Tugizimana, Djami-Tchatchou, Steenkamp, Piater, & Dubery, 2018; Wilkinson et al., 2018), after external perception of different microbial molecular associated patterns (PAMP) and damage molecular associated patterns (DAMP) such as bacterial flagellin, fungal chitin or redox metabolites (Pétriacq, Ton, Patrit, Tcherkez, & Gakière, 2016), or in interaction with other environmental pressures like climate change or below-ground interactions (Pétriacq et al., 2017; Williams et al., 2018; Ziegler et al., 2016). Although phenolics are important SM involved in stressful conditions as activators of immune signaling cascades (e.g., SA, detailed below) or mitigators of the oxidative stress that results from (a)biotic stress, further investigations are required to determine the exact roles of these compounds in plant immune responses to pathogenic infections.

3.3 N- and S-containing metabolites

The possibility that plants can produce N- and S-containing metabolites adds another twist to the complex tale of SM. Plant nitrogenous compounds include nonprotein amino acids and specialized indoles, alkaloids and cyanogenic glucosides, many of which derive from amino acids (Singh, 2018). Benzoxazinoids are important indolic SM mostly found in Cereals and involved in biotic stress responses (Zhou, Richter, & Jander, 2018). They protect plants from insects and fungi (Handrick et al., 2016; Święcicka et al., 2020) and impact root-microbes interactions (Ahmad et al., 2011; Cotton et al., 2019). N-containing alkaloids have also received significant attention owing to their toxicity to humans and antiherbivore nature. A wealth of phytochemistry studies exemplified the defensive aspects of alkaloids as effective antiviral, antifungal and antimicrobial compounds

(Zaynab et al., 2018). Among other SM (cinnamic acids, benzenoids and flavonoids), steroidal glycoalkaloids accumulated as an immune response against curly stunt virus in tomato (Rossouw et al., 2019). Furthermore, metabolomics provided evidence that cyanogenic metabolism was partly involved in SA synthesis in peach (Diaz-Vivancos, Bernal-Vicente, Cantabella, Petri, & Hernández, 2017). Other untargeted studies reported accumulated pools of alkaloids together with terpenes and phenolics in tomato and Arabidopsis responses to different biotic foes (Wilkinson et al., 2018; Williams et al., 2018). This suggests that plants might potentiate their metabolic counterattack by involving several stress-responsive pathways. Although energetically costly, production of multiple SM likely strengthens plant immunity against destructive microbes.

Plant sulfurous SM include the thio-glucosides glucosinolates and indole-type phytoalexins like camalexin (Bednarek, 2012; Bednarek & Osbourn, 2009), their defensive roles identified thanks to (un)targeted metabolomics (Bednarek, 2012; Castro-Moretti et al., 2020). Glucosinolates, and their toxic hydrolysis products isothiocyanates, are particularly important in fungal interactions (Bednarek et al., 2009; Chen et al., 2020; Jogawat, Meena, Kundu, Varma, & Vadassery, 2020; Pétriacq, Stassen, & Ton, 2016; Su et al., 2018) and herbivory (Almuziny, Decker, Wang, Gerard, & Tharayil, 2017; Papazian et al., 2019). In Arabidopsis, SM relative quantitations showed that cell wall-mediated resistance underpinned changes in glucosinolate and camalexin levels that were associated with aphid challenge (Kloth et al., 2019). The same study reported compromised absolute levels of stress hormones when a defense-related pectine-modifying function was impaired. Moreover, in addition to microbial toxicity in leaves, camalexin potentially modulates the functionality of Arabidopsis root-associated microbes through chemical communication (Jacoby, Chen, Schwier, Koprivova, & Kopriva, 2020; Koprivova et al., 2019). This suggests that sulfurous SM can have both defensive and signaling roles depending on the interaction.

3.4 The case of redox compounds

Redox metabolites are often measured as important actors of plant–pathogen interactions, including many SM that directly process free radicals (mostly terpenes and phenolics) and other compounds that indirectly participate in the regulation of redox metabolism like ascorbate, glutathione and pyridine nucleotides (Decros et al., 2019; Noctor et al., 2011). These redox compounds are difficult to assay accurately due to their high reactivity and stability in plant matrices (Liu et al., 2020). However, they appear manifold in plant responses to biotic threats (Chaouch, Queval, & Noctor, 2012; Pétriacq et al., 2012; Pétriacq, Ton, et al., 2016; Zhang & Mou, 2009), highlighting their immune potential and emphasizing the need for their investigation in future work.

4. Plant metabolomics decodes priming responses

Many of the aforementioned pathways show a fast accumulation or primed production when adverse biotic conditions are likely. Immune priming is a physiological state of the plant providing induced resistance against a broad range of (a)biotic stresses (Pieterse et al., 2009). The capability to overcome a microbial attack depends, largely, on the speed of the response. Basal defenses are often slow and weak to reduce energy costs. However, plants can enhance their rapidity of response to resist microbial infections. This phenomenon, known as priming, involves intra- and extracellular communication carried out by phytohormones and mobile signals, some of them identified by targeted and untargeted metabolomics (Martínez-Medina et al., 2017). Plants can reach a primed state in a multitude of ways; from chemical induction, such as β-aminobutyric acid (BABA) treatments, to biological induction, such as *Trichoderma* or mycorrhiza-induced resistance. In each case, detection and identification of phytohormones and other specialized metabolites by LC-MS techniques have been instrumental in understanding the biological mechanisms behind these processes (Glauser et al., 2014).

Plants possess the ability to synthesize mobile signals to prime systemic responses in distal tissue, in a phenomenon commonly known as Systemic Acquired Resistance (SAR). Many of the SAR signals such as salicylic acid (SA), methyl salicylate (MeSA), azelaic acid (AzA), glycerol-3-phosphate (G3P) and pipecolic acid (PIP) were identified as a result of HPLC-MS targeted methodology (Chanda et al., 2011; Dempsey, Vlot, Wildermuth, & Klessig, 2011; Jung, Tschaplinski, Wang, Glazebrook, & Greenberg, 2009; Návarová et al., 2012; Park, Kaimoyo, Kumar, Mosher, & Klessig, 2007).

Additionally, untargeted metabolomics has shown that different SM such as phenylpropanoids, flavonoids, alkaloids and terpenes also have a significant function in SAR (Aliferis & Jabaji, 2012; Choi et al., 2006; López-Gresa et al., 2012; Mhlongo, Steenkamp, Piater, Madala, & Dubery, 2016). Mutants of the FMO1 gene, which synthesizes N-hydroxypipecolic acid (N-OH-PIP), are

SAR negative with azelaic acid, pipecolic acid and dehydroabietinal (Chen et al., 2018). A comparative metabolomic analysis between wild-type and *fmo1* mutant plants demonstrated the absence of N-OH-Pip accumulation in *fmo1* mutant lines after bacterial infection, but that supplementation with N-OH-Pip led to the reestablishment of SAR (Chen et al., 2018). Phytohormones are the master regulators of the SAR response; targeted metabolomics methods have demonstrated that SAR signaling induces the biosynthesis of SA, and, that the SA and JA cross-talk controls the resistance against pathogens with different infection strategies (Vlot, Dempsey, & Klessig, 2009).

Unlike SAR, Induced Systemic Resistance (ISR) is generally controlled by JA and ethylene signaling. ISR requires the interaction between plants roots and beneficial microorganisms, conferring benefits in growth and stimulating the plant immune system. A well-studied example is *Trichoderma*, a beneficial fungus used as a biocontrol agent in agricultural settings (Martínez-Medina et al., 2017). A targeted hormonal quantification demonstrated that *Trichoderma*-ISR in Arabidopsis infected with *Botrytis cinerea* involves changes in phytohormones, such as JA, SA and other metabolites (Contreras-Cornejo, Macías-Rodríguez, Beltrán-Peña, Herrera-Estrella, & López-Bucio, 2011; Martínez-Medina et al., 2017; Martínez-Medina, Del Mar Alguacil, Pascual, & Van Wees, 2014). Untargeted metabolite profiling in onion, bean and tomato inoculated with *Trichoderma* and infected with different soil-borne pathogens, showed that polyphenols, flavonoids and terpenes play a significant role in *Trichoderma*-ISR (Abdelrahman et al., 2016; Manganiello et al., 2018; Mayo-Prieto et al., 2019). Another well-established system to induce priming responses in plants is the mycorrhiza-induced resistance. Arbuscular mycorrhizal fungi (AMF) establish mutualistic associations with plant roots affecting the host physiology in terms of nutrition and protection. This interaction is finely regulated by plant phytohormones, depending on the developmental stage of the plant. Hormonal analysis by HPLC-MS and GC–MS in tomato roots revealed significant changes such as a reduction in ethylene and an increase in the JA-derivatives, jasmonate-isoleucine and oxo-phytodienoic acid (OPDA; (López-Ráez et al., 2010). AMF colonization also has a significant impact on the plant metabolome in roots and shoots. For instance, *Rhizophagus irregularis* and *Funneliformis mosseae* impacted the levels of primary and secondary metabolites in tomato roots, particularly sugars, amino acids, lignins and polyphenol derivatives (Rivero, Gamir, Aroca, Pozo, & Flors, 2015). Furthermore, leaves of plants inoculated with *Glomus intraradices* and *Rhizophagus irregularis* experienced an accumulation of amino acids, carbohydrates and products of TCA cycle metabolism (Gerlach et al., 2015). Together, this evidence shows that primary and secondary metabolites directly contribute to ISR, but the role and mechanistic details of these metabolomic shifts during immune priming require further research.

Plant priming can also be induced via chemical application. There are multiple natural or synthetic molecules capable of inducing a prime state in different plant species (Gamir, Sánchez-Bel, & Flors, 2014). Among these, the molecular mechanisms of the BABA-induced resistance (BABA-IR) have been the most studied for chemical priming (Balmer, Pastor, Gamir, Flors, & Mauch-Mani, 2015). An untargeted metabolomic study in BABA-treated plants revealed rapid mobilization of the primary metabolism during the priming phase, in the absence of the infection (Pastor, Balmer, Gamir, Flors, & Mauch-Mani, 2014). A combined metabolomic analysis in Arabidopsis plants infected with the fungus *Plectosphaerella (P.) cucumerina*, using BABA as a chemical inducer, found polyphenols, alkaloids and sugars among the metabolic profile of the primed state along with galacturonic acid, indole-3-carboxylic acid and hypoxanthine (Gamir, Pastor, Kaever, Cerezo, & Flors, 2014). These latter three metabolites were all shown to induce resistance in Arabidopsis against *P. cucumerina* when applied topically. In tomato fruit, BABA priming triggered metabolic signatures depending on the pathogens and involving SM, while soluble sugars acted as good predictors of fruit resistance (Luna et al., 2020; Wilkinson et al., 2018). Similar to BABA, hexanoic acid participated in the plasticity of priming responses in tomato plants infected with the fungus *B. cinerea* or the bacteria *Pst* (Camañes et al., 2015). Primed tomato plants infected with *B. cinerea,* accumulated more phenolic derivatives and alkaloids, in contrast, tomato plants treated with hexanoic acid and infected by *Pst* accumulated primary metabolites and fatty acids. Another comparative metabolomics study of different priming inducers, acibenzolar-*S*-methyl, riboflavin and azelaic acid, showed the accumulation of 20 cinnamic acid conjugates in primed plants, indicating the activation of the polyphenol signaling pathway as a common priming response in plants (Mhlongo et al., 2016). Overall, targeted and untargeted metabolomics profiling has demonstrated to be a powerful approach to untangle priming responses in primary and secondary metabolism. Understanding the molecular stress responses under priming state is the first step toward the development of environmentally friendly bioprotectors.

5. Metabolomics of resistance during environmental change

The plant metabolome is multifaceted and exhibits plasticity in response to the (a)biotic conditions the plant is experiencing. As such, *de novo* metabolite production is essential for routine and dynamic orchestration of response to wildly changeable environmental stress. Seasonal fluctuations in temperature, water availability and nutrient access are common across agriculturally important regions (Osborne, Rose, & Wheeler, 2013). However, the global climate is shifting toward extremes; drought events, heatwaves and flooding are more frequent and severe, fuelled by rising anthropogenic concentrations of CO_2 (IPCC, 2018). Additionally, rises in tropospheric levels of phytotoxic pollutants are having increasingly detrimental impacts on plant metabolic function. In light of these heightened abiotic pressures, the ability to balance the production of defense metabolites and response to environmental stress will likely be crucial for plant survival (Tenllado & Canto, 2020).

It stands to reason that abiotic stresses will exacerbate existing pathogenic stress as similar metabolic pathways that control immune defense also govern response to adverse environmental pressures. For instance, drought affects abscisic acid (ABA), a well-known regulatory hormone for immune responses; elevated CO_2 impacts the antagonistic defense-hormones salicylic acid (SA) and jasmonic acid (JA; Kazan, 2018); and elevated ozone (O_3) alters the antioxidant pathway which has an important role in biotrophic pathogen resistance (Iriti & Faoro, 2009). There is evidence that plant responses to myriad (a)biotic stresses and immune metabolism are intricately linked, albeit detailed metabolomics studies are still relatively novel (Table 1). Thus, the following paragraphs summarize the state of knowledge regarding immune metabolomics during major abiotic stresses as well as highlight avenues of research where knowledge is lacking. Specifically, we will focus on three major abiotic stresses; drought/warming, CO_2, and toxic O_3. Other, related, examples will not be discussed here but would be prudent to consider in the context of holistic climate change impacts (including water-logging, soil salinity, nutrient deficits, biodiversity/habitat loss, erosion).

Drought is the most devastating abiotic stress to plants, especially in crop plants. As plants rely on water for primary metabolism, it is unsurprising that drought has negative and unpredictable effects on the plant immune system

Table 1 Immune metabolic response to various abiotic stresses.

Stress	Impact on resistance	Metabolomics change	Reference
CO_2	Resistance to various diseases impacted. Negative or positive interactions are context-dependent	Increase in SA-related immunity, increase antioxidant machinery. Increased ROS production. Decrease in JA related resistance	Mhamdi and Noctor (2016), Zhou et al. (2020)
Drought	Potentially increased resistance to biotrophs, but decreased resistance has also been illustrated	Increases in primary metabolites proline, glutamate and arginine change oxidative stress response	Prasch and Sonnewald (2013)
Heat/ warming	Pathogen ranges and "season" expected to increase, plant hosts expected to become less resistant and more prone to pathogen infection	Increased sucrose and glucose accumulation, as well as histidine and tyrosine	Prasch and Sonnewald (2013), Elad and Pertot (2014)
Ozone	Postulated increased resistance to biotrophs and decreased resistance to necrotrophs	Increased production of immune-related secondary metabolites. Increased redox signaling	Iriti and Faoro (2009)

SA is salicylic acid; JA is jasmonic acid; ROS is reactive oxygen species.

(Nejat & Mantri, 2017). Generally, pathogen damage is much higher during drought, for example droughted chickpea is more susceptible to fungal pathogens *Rhizoctonia bataticola* and *Fusarium solani* (Sinha, Irulappan, Mohan-Raju, Suganthi, & Senthil-Kumar, 2019). As a result of combined pathogenic and abiotic stress the immune metabolome changes drastically (Prasch & Sonnewald, 2013). In droughted Arabidopsis, infected with the Turnip Mosaic Virus, much greater changes in metabolite and transcriptional profiles were observed under combined drought and pathogen stress than in individual stress. Over 170 unique transcript features were identified during combined stress (including PR genes involved in pathogen resistance); which are related to changes in levels of proline, glutamate and arginine; the primary metabolites with osmoprotective properties. Similarly, drought-resistant grape varieties display an increased accumulation of primary metabolites related to polyamide homeostasis (including arginine, proline and GABA). Polyamides are redox regulators that are likely involved in the reduction of oxidative stress that occurs during drought (Hatmi et al., 2015). Interestingly, the increased levels of these compounds in grapes correspond with resistance against the necrotrophic fungus *Botrytis cinerea*, as well as drought. These observations illustrate the utility of metabolomic investigation to provide a real-world application to manage both abiotic and biotic stresses in an otherwise sensitive crop. However, an unfortunate aspect of drought is that it often occurs alongside severe warm periods. Warming generally increases stomatal opening to allow transpiration, which presents an avenue for pathogen infection (Prasch & Sonnewald, 2013), with drastic effects on plant metabolic profiles especially in combination with viral infection. This is worsened during drought, leading to a breakdown in a plant's ability to function, and may render any drought-infection management strategies redundant.

CO_2 is an anthropogenic pollutant that has doubled in three generations and the current scientific consensus places elevated CO_2 (eCO_2) as the primary cause of global warming (IPCC, 2018). For a plant, rising levels of CO_2 can be directly beneficial, resulting in greater biomass, enhanced physiology and potentially higher yield (Ainsworth & Long, 2005). However, the current research offers little consensus on whether eCO_2 is beneficial or detrimental to plant defense (Kazan, 2018). For example, Williams et al. (2018) demonstrate an increased immune response against a biotrophic *Hyaloperonospora arabidopsidis* infection but increased susceptibility to the necrotrophic *P. cucumerina* under eCO_2. Although nuanced, many studies report increased resistance to biotic pathogens with eCO_2, which is associated with an increase in SA production and activity, as measured through mass spectrometry and qPCR transcripts of related gene expression (Mhamdi & Noctor, 2016; Williams et al., 2018; Zhou et al., 2020). Both primary (Mhamdi & Noctor, 2016) and secondary (Williams et al., 2018) metabolism have been investigated in Arabidopsis during eCO_2 and pathogen stress. Indeed, more in-depth metabolomics work captures a clear role of the antioxidant machinery, increased H_2O_2 production, presence of defense-related metabolites, changes in amino acid and carbohydrate profiles and activity of oxidase enzymes (Mhamdi & Noctor, 2016; Williams et al., 2018). Interestingly, Williams et al. also illustrate that secondary metabolism in plants grown in sub-ambient CO_2 ($saCO_2$) is connected to increased resistance to *H. arabidopsidis* through changes in redox status. This was highlighted by oxidation of amino acids, such as glycine, accumulation of defense metabolites, like glucosinolates, and changes in metabolite pools associated with peroxisomal ROS scavenging machinery. Specifically, *sa*CO_2 alters primary metabolism—photorespiration—resulting in greater H_2O_2 accumulation and altered outcome for disease. Increased resistance in *e*CO_2 is also postulated to result from changes in primary metabolism; sugar production, changes in sugar signaling and the knock-on impacts seen in antioxidant status were implicit (Mhamdi & Noctor, 2016; Williams et al., 2018). Latterly, some headway has been made using molecular tools to discern the role of the carbonic anhydrases, metalloenzymes involved in CO_2 homeostasis, in *e*CO_2 induced resistance against *Pseudomonas syringae*. This resistance is associated with altered peroxisomal reactive oxygen species (ROS) production (Zhou, Van Leeuwen, Pieterse, Bakker, & Van Wees, 2019).

However, it is unclear whether crop plants adopt a similar response as the work discussed thus far has largely been performed in model organisms such as Arabidopsis. There is compelling evidence that in maize *e*CO_2 results in a reduction in both SA and JA abundance, with detrimental impacts on resistance against necrotrophic *Fusarium verticillioides*. Furthermore, the production of the phytoalexins kauralexin and zealexin decreased through attenuation of JA signaling (Vaughan et al., 2014). Transcriptomics analysis of maize grown in *e*CO_2 has subsequently revealed alterations in immune-related gene expression of MPKs, WRKYs and MYC/JAZ as well as pathways associated with oxidation/reduction (Ge et al., 2018), but direct studies of redox-related metabolites are lacking. It is worth mentioning that increased ROS signaling is a characteristic of necroptosis, so in *e*CO_2 there is potential to increase hypersensitive response and programmed cell death during infection, as that is its secondary role (Coll, Epple, & Dangl, 2011). For a necrotrophic pathogen, metabolizing dead cellular content, this presents an avenue for infection and increased virulence in *e*CO_2 conditions. Since many crop plants show increased susceptibility to necrotrophs (Kazan, 2018) it is tempting to speculate that aberrations in redox homeostasis driven by *e*CO_2 and its impacts on primary metabolism are responsible (as suggested by Mhamdi & Noctor, 2016).

Anthropogenic sources of air pollution, besides CO_2, are also increasing globally (IPCC, 2018). Of particular concern is tropospheric (i.e., the atmospheric zone where terrestrial life exists) O_3, a highly toxic substance formed through photochemical reactions of sunlight with NOx, CO or volatile organic carbons. O_3 toxicity in plants leads to oxidative stress (Iriti & Faoro, 2009) and as such causing large metabolic shifts involving reprogramming in both primary and secondary metabolism, with increases of many immune-related secondary metabolites including polyphenols, isoprenoids and alkaloids (Iriti & Faoro, 2009). These alterations in defense-related secondary metabolites reduce the gravity and progression of pathogenic infections. Hence, it is postulated that pathogens become less virulent as metabolic responses to O_3 kick-in (Sandermann, 2008), but general effects on disease outcomes are still elusive. Leaf rust was largely inhibited by increased O_3 in wheat (von Tiedemann & Firsching, 2000) but infection of broad bean by *Uromyces viciae-fabae* resulted in reduced symptoms of toxicity on subsequent O_3 treatment (Lorenzini, Medeghini Bonatti, Nali, & Baroni Fornasiero, 1994). Peroxidase activity was primed in the latter study, potentially providing the protective effect, but as specific molecules (such as ROS or protective anthocyanins) were not measured this remains speculative. Unfortunately, there are very few studies that investigate both O_3 toxicity and pathogen resistance, let alone the metabolic underpinnings of these interactions. Given the increasing concentrations, and risk from O_3 migrating into agriculturally important zones (IPCC, 2018), these interactions certainly deserve further consideration.

It is clear that the field of interactive abiotic and biotic stresses is still young and further research is vital. Plant metabolic pathways determining multifactorial stress responses are both ecologically and agronomically important—crops are even more sensitive than ever to (a)biotic stress as their genetic diversity is typically lower than wild counterparts, reducing their potential to adapt (Roy, Tucker, & Tester, 2011). However, the study of crop plant, or cultivar, responses to interacting stresses are not prominent in the literature. As the expression of defense-related metabolites may differ across various plant species or cultivars, the knowledge we glean from model organisms provides key foundations but is limited. Many of the environmental challenges discussed in this section are further worsened through a loss of biodiversity. This includes both species diversity, genetic diversity in our domesticated cultivars and, subsequently, diversity of rare, unique or agriculturally relevant metabolites that may aid in positive responses to climate change. To safeguard our natural and managed ecosystems from the challenges we face, real attention must be paid to the diversity and role of novel defense metabolites. Developing a clear understanding of the plant defensome during pathogen challenge across different environments will help identify areas that are particularly sensitive to stress and allow specific mitigation strategies to be developed.

6. Concluding remarks and perspectives

Metabolomics provides a valuable molecular and phenotypic readout to encapsulate the chemical diversity of plants and their metabolic responses to biotic challenges. The many examples pointed out in this chapter unambiguously testify that metabolites, most particularly SM, are central components of the defensome in phytopathological systems; with various direct or indirect roles for pathogen resistance. Beyond metabolite profiling, the future trend would undoubtedly foster more global, comprehensive strategies based on large-scale multiomics strategies that are crucial to providing important insights at the systems level (Zander et al., 2020). Metabolomics will also be key to ecological biochemistry (i.e., ecometabolomics reviewed in chapter "A practical guide to implementing metabolomics in plant ecology and biodiversity research" by Uthe et al.) uncovering a plethora of metabolites with ecological and defensive roles (Peters et al., 2018). In this broader environmental context, metabolomics will allow plant biologists to untangle the complex molecular interplay between metabolic responses to biotic challenges and abiotic stresses.

One major challenge in phytopathological metabolomics lies in the difficulty to distinguish the plant, host metabolome vs the microbial metabolites (Allwood, Clarke, Goodacre, & Mur, 2010; Allwood, Heald, Lloyd, Goodacre, & Mur, 2011). Most compounds are ubiquitous to different organisms or exchanged between the host and pathogens as signals or metabolic responses. Furthermore, in most phytopathological studies, plants and microbes are typically incubated together and removing pathogens before the metabolite extraction is rarely performed. While this seems irrelevant for transcriptomics and proteomics that exploit species-specific databases, metabolomics suffers from cross-contamination of plant and microbial metabolites. One study addressed this pressing concern by implementing a stable isotope labeling of a bacterial pathogen metabolome prior to incubation of labeled bacteria with Arabidopsis cells (Pang et al., 2018). A washing step allowed removing excessive bacterial cells, and plant metabolites were analyzed by targeted LC-MS, highlighting an infection reprogramming of plant signaling and primary metabolic pathways including amino and organic acids, carbohydrates and nucleotides. This elegant study paved the way to labeling approach for differentiating host and pathogen metabolomes.

Metabolomics is a multidisciplinary research field including analytical chemistry, biology, computer science and statistics. In contrast to other "-omics" approaches that are focused on families of compounds with similar physio-chemical properties such as RNA for transcriptomics, proteins for proteomics, plant metabolites can vary from small secondary metabolites with high polarity to long apolar chain fatty acids. The chemical diversity of the plant specialized metabolites makes the development of the "all-in-one" extraction procedure impossible to achieve (Mushtaq, Choi, Verpoorte, & Wilson, 2014). For these reasons, the use of different combinations of solvents must be applied depending on the nature of the desired metabolites. The mixture of water and methanol is the most popular combination of nontargeted metabolomics analysis. This combination has a proven efficiency for the extraction of primary and secondary metabolites such as sugars, amino acids, organic acids, alkaloids and phenolic compounds (De Vos et al., 2007). Nevertheless, there is still a need to improve extraction procedures for greater coverage of the metabolome under consideration.

Tandem mass spectrometry techniques can detect thousands of signals from one biological sample. However, compound identification procedure remains a challenge and bottleneck in the metabolomics workflow. Bioinformatic tools have improved the methodology to curate metabolomics data even though thousands of signals (measured from a metabolomics study) still remain unidentified. Hence, improving compound identification is one of the biggest challenges in MS-based metabolomics. Currently, exact mass detectors display sub-ppm mass accuracy, allowing a higher degree of identification. As a result, online databases are expanding their MS library, helping the scientific community to increase the number of annotated compounds. (Lazar, Romanciuc, Socaciu, & Socaciu, 2015). Standardization of the protocols and data reproducibility is becoming an obvious need and several initiatives for public repositories, such as MetaboLights at EMBL-EBI, have been compiled to aide this (Haug et al., 2019). Future bioinformatic development for metabolomics will move us toward reduced human involvement and higher throughput data analysis (Meier, Ruttkies, Treutler, & Neumann, 2017). Hence, despite these challenges, the gap in metabolomics capabilities is being filled with innovative tools allowing more diverse applications and sophisticated study designs for a better understanding of biological processes (Singh, 2020).

References

Abdelrahman, M., Abdel-Motaal, F., El-Sayed, M., Jogaiah, S., Shigyo, M., Ito, S., et al. (2016). Dissection of Trichoderma longibrachiatum-induced defense in onion (Allium cepa L.) against Fusarium oxysporum f. Sp. Cepa by target metabolite profiling. *Plant Science*, *246*, 128–138. https://doi.org/10.1016/j.plantsci.2016.02.008.

Ahmad, S., Veyrat, N., Gordon-Weeks, R., Zhang, Y., Martin, J., Smart, L., et al. (2011). Benzoxazinoid metabolites regulate innate immunity against aphids and fungi in maize. *Plant Physiology*, *157*(1), 317–327. https://doi.org/10.1104/pp.111.180224.

Ainsworth, E. A., & Long, S. P. (2005). What have we learned from 15 years of free-air CO_2 enrichment (FACE)? A meta-analytic review of the responses of photosynthesis, canopy properties and plant production to rising CO_2. *New Phytologist*, *165*(2), 351–372. https://doi.org/10.1111/j.1469-8137.2004.01224.x.

Aliferis, K. A., & Jabaji, S. (2012). FT-ICR/MS and GC-EI/MS metabolomics networking unravels global potato Sprout's responses to Rhizoctonia solani infection. *PLoS One*, 7(8). https://doi.org/10.1371/journal.pone.0042576.

Allwood, J. W., Clarke, A., Goodacre, R., & Mur, L. A. J. (2010). Dual metabolomics: A novel approach to understanding plant–pathogen interactions. *Phytochemistry*, *71*(5–6), 590–597. https://doi.org/10.1016/j.phytochem.2010.01.006.

Allwood, J. W., Heald, J., Lloyd, A. J., Goodacre, R., & Mur, L. A. J. (2011). Separating the inseparable: The metabolomic analysis of plant–pathogen interactions. In N. W. Hardy & R. D. Hall (Éds.), Plant metabolomics (p. 31–49). Totowa, NJ: Humana Press. https://doi.org/10.1007/978-1-61779-594-7_3.

Almuziny, M., Decker, C., Wang, D., Gerard, P., & Tharayil, N. (2017). Nutrient supply and simulated Herbivory differentially alter the metabolite pools and the efficacy of the glucosinolate-based defense system in Brassica species. *Journal of Chemical Ecology*, *43*(2), 129–142. https://doi.org/10.1007/s10886-016-0811-y.

Atsbaha Zebelo, S., Bertea, C. M., Bossi, S., Occhipinti, A., Gnavi, G., & Maffei, M. E. (2011). Chrysolina herbacea modulates terpenoid biosynthesis of Mentha aquatica L. *PLoS One*, *6*(3). https://doi.org/10.1371/journal.pone.0017195.

Balmer, A., Pastor, V., Gamir, J., Flors, V., & Mauch-Mani, B. (2015). The 'prime-ome': Towards a holistic approach to priming. *Trends in Plant Science*, *20*(7), 443–452. https://doi.org/10.1016/j.tplants.2015.04.002.

Bednarek, P. (2012). Sulfur-containing secondary metabolites from Arabidopsis thaliana and other Brassicaceae with function in plant immunity. *Chembiochem: A European Journal of Chemical Biology*, *13*(13), 1846–1859. https://doi.org/10.1002/cbic.201200086.

Bednarek, P., & Osbourn, A. (2009). Plant-microbe interactions: Chemical diversity in plant defense. *Science*, *324*(5928), 746–748. https://doi.org/10.1126/science.1171661.

Bednarek, P., Pislewska-Bednarek, M., Svatos, A., Schneider, B., Doubsky, J., Mansurova, M., et al. (2009). A glucosinolate metabolism pathway in living plant cells mediates broad-spectrum antifungal defense. *Science*, *323*(5910), 101–106. https://doi.org/10.1126/science.1163732.

Berger, S., Sinha, A. K., & Roitsch, T. (2007). Plant physiology meets phytopathology: Plant primary metabolism and plant pathogen interactions. *Journal of Experimental Botany*, *58*(15–16), 4019–4026. https://doi.org/10.1093/jxb/erm298.

Bezrutczyk, M., Yang, J., Eom, J.-S., Prior, M., Sosso, D., Hartwig, T., et al. (2018). Sugar flux and signaling in plant-microbe interactions. *The Plant Journal*, *93*(4), 675–685. https://doi.org/10.1111/tpj.13775.

Bolton, M. D. (2009). Primary metabolism and plant defense—Fuel for the fire. *Molecular Plant-Microbe Interactions®*, *22*(5), 487–497. https://doi.org/10.1094/MPMI-22-5-0487.

Camañes, G., Scalschi, L., Vicedo, B., González-Bosch, C., & García-Agustín, P. (2015). An untargeted global metabolomic analysis reveals the biochemical changes underlying basal resistance and priming in Solanum lycopersicum, and identifies 1-methyltryptophan as a metabolite involved in plant responses to Botrytis cinerea and Pseudomonas sy. *The Plant Journal*, *84*(1), 125–139. https://doi.org/10.1111/tpj.12964.

Castro-Moretti, F. R., Gentzel, I. N., Mackey, D., & Alonso, A. P. (2020). Metabolomics as an emerging tool for the study of plant–pathogen interactions. *Metabolites*, *10*(2), 52. https://doi.org/10.3390/metabo10020052.

Chanda, B., Xia, Y., Mandal, M. K., Yu, K., Sekine, K., Gao, Q., et al. (2011). Glycerol-3-phosphate is a critical mobile inducer of systemic immunity in plants. *Nature Genetics*, *43*(5), 421–427. https://doi.org/10.1038/ng.798.

Chaouch, S., Queval, G., & Noctor, G. (2012). AtRbohF is a crucial modulator of defence-associated metabolism and a key actor in the interplay between intracellular oxidative stress and pathogenesis responses in Arabidopsis. *The Plant Journal for Cell and Molecular Biology*, *69*(4), 613–627. https://doi.org/10.1111/j.1365-313X.2011.04816.x.

Chen, Y.-C., Holmes, E. C., Rajniak, J., Kim, J.-G., Tang, S., Fischer, C. R., et al. (2018). *N*-hydroxy-pipecolic acid is a mobile metabolite that induces systemic disease resistance in *Arabidopsis*. *Proceedings of the National Academy of Sciences of the United States of America*, *115*(21), E4920–E4929. https://doi.org/10.1073/pnas.1805291115.

Chen, L.-Q., Hou, B.-H., Lalonde, S., Takanaga, H., Hartung, M. L., Qu, X.-Q., et al. (2010). Sugar transporters for intercellular exchange and nutrition of pathogens. *Nature*, *468*(7323), 527–532. https://doi.org/10.1038/nature09606.

Chen, J., Ullah, C., Reichelt, M., Beran, F., Yang, Z.-L., Gershenzon, J., et al. (2020). The phytopathogenic fungus Sclerotinia sclerotiorum detoxifies plant glucosinolate hydrolysis products via an isothiocyanate hydrolase. *Nature Communications*, *11*(1), 3090. https://doi.org/10.1038/s41467-020-16921-2.

Choi, Y. H., Kim, H. K., Linthorst, H. J. M., Hollander, J. G., Lefeber, A. W. M., Erkelens, C., et al. (2006). NMR metabolomics to revisit the tobacco mosaic virus infection in *Nicotiana t abacum* leaves. *Journal of Natural Products*, *69*(5), 742–748. https://doi.org/10.1021/np050535b.

Coll, N. S., Epple, P., & Dangl, J. L. (2011). Programmed cell death in the plant immune system. *Cell Death & Differentiation*, *18*(8), 1247–1256. https://doi.org/10.1038/cdd.2011.37.

Contreras-Cornejo, H. A., Macías-Rodríguez, L., Beltrán-Peña, E., Herrera-Estrella, A., & López-Bucio, J. (2011). *Trichoderma*-induced plant immunity likely involves both hormonal- and camalexin-dependent mechanisms in *Arabidopsis thaliana* and confers resistance against necrotrophic fungi *Botrytis cinerea*. *Plant Signaling & Behavior*, *6*(10), 1554–1563. https://doi.org/10.4161/psb.6.10.17443.

Copley, T. R., Aliferis, K. A., Kliebenstein, D. J., & Jabaji, S. H. (2017). An integrated RNAseq-1H NMR metabolomics approach to understand soybean primary metabolism regulation in response to Rhizoctonia foliar blight disease. *BMC Plant Biology*, *17*(1), 84. https://doi.org/10.1186/s12870-017-1020-8.

Cotton, T. E. A., Pétriacq, P., Cameron, D. D., Meselmani, M. A., Schwarzenbacher, R., Rolfe, S. A., et al. (2019). Metabolic regulation of the maize rhizobiome by benzoxazinoids. *The ISME Journal*, *13*(7), 1647–1658. https://doi.org/10.1038/s41396-019-0375-2.

De Vos, R. C., Moco, S., Lommen, A., Keurentjes, J. J., Bino, R. J., & Hall, R. D. (2007). Untargeted large-scale plant metabolomics using liquid chromatography coupled to mass spectrometry. *Nature Protocols*, *2*(4), 778–791. https://doi.org/10.1038/nprot.2007.95.

Deborde, C., Moing, A., Roch, L., Jacob, D., Rolin, D., & Giraudeau, P. (2017). Plant metabolism as studied by NMR spectroscopy. *Progress in Nuclear Magnetic Resonance Spectroscopy*, *102-103*, 61–97. https://doi.org/10.1016/j.pnmrs.2017.05.001.

Decros, G., Baldet, P., Beauvoit, B., Stevens, R., Flandin, A., Colombié, S., et al. (2019). Get the balance right: ROS homeostasis and redox signalling in fruit. *Frontiers in Plant Science*, *10*, 1091. https://doi.org/10.3389/fpls.2019.01091.

Deeken, R., Engelmann, J. C., Efetova, M., Czirjak, T., Müller, T., Kaiser, W. M., et al. (2006). An integrated view of gene expression and solute profiles of *Arabidopsis* tumors: A genome-wide approach. *The Plant Cell*, *18*(12), 3617–3634. https://doi.org/10.1105/tpc.106.044743.

Dempsey, D. A., Vlot, A. C., Wildermuth, M. C., & Klessig, D. F. (2011). Salicylic acid biosynthesis and metabolism. *The Arabidopsis Book*, *9*. https://doi.org/10.1199/tab. 0156.

Diaz-Vivancos, P., Bernal-Vicente, A., Cantabella, D., Petri, C., & Hernández, J. A. (2017). Metabolomics and biochemical approaches link salicylic acid biosynthesis to Cyanogenesis in peach plants. *Plant & Cell Physiology*, *58*(12), 2057–2066. https://doi.org/10.1093/pcp/pcx135.

Doehlemann, G., Wahl, R., Horst, R. J., Voll, L. M., Usadel, B., Poree, F., et al. (2008). Reprogramming a maize plant: Transcriptional and metabolic changes induced by the fungal biotroph *Ustilago maydis*: *Plant response to* U. maydis *infection*. *The Plant Journal*, *56*(2), 181–195. https://doi.org/10.1111/j.1365-313X.2008.03590.x.

Duplessis, S., Cuomo, C. A., Lin, Y.-C., Aerts, A., Tisserant, E., Veneault-Fourrey, C., et al. (2011). Obligate biotrophy features unraveled by the genomic analysis of rust fungi. *Proceedings of the National Academy of Sciences of the United States of America*, *108*(22), 9166–9171. https://doi.org/10.1073/pnas.1019315108.

Elad, Y., & Pertot, I. (2014). Climate change impacts on plant pathogens and plant diseases. *Journal of Crop Improvement*, *28*(1), 99–139. https://doi.org/10.1080/15427528.2014.865412.

Fabro, G., Kovács, I., Pavet, V., Szabados, L., & Alvarez, M. E. (2004). Proline accumulation and *AtP5CS2* gene activation are induced by plant-pathogen incompatible interactions in *Arabidopsis*. *Molecular Plant-Microbe Interactions*, *17*(4), 343–350. https://doi.org/10.1094/MPMI.2004.17.4.343.

Fabro, G., Rizzi, Y. S., & Alvarez, M. E. (2016). *Arabidopsis* proline dehydrogenase contributes to flagellin-mediated PAMP-triggered immunity by affecting RBOHD. *Molecular Plant-Microbe Interactions*, *29*(8), 620–628. https://doi.org/10.1094/MPMI-01-16-0003-R.

Felemban, A., Braguy, J., Zurbriggen, M. D., & Al-Babili, S. (2019). Apocarotenoids involved in plant development and stress response. *Frontiers in Plant Science*, *10*, 1168. https://doi.org/10.3389/fpls.2019.01168.

Fernandez, O., Urrutia, M., Bernillon, S., Giauffret, C., Tardieu, F., Le Gouis, J., et al. (2016). Fortune telling: Metabolic markers of plant performance. *Metabolomics*, *12*(10). https://doi.org/10.1007/s11306-016-1099-1.

Gabaston, J., Cantos-Villar, E., Biais, B., Waffo-Teguo, P., Renouf, E., Corio-Costet, M.-F., et al. (2017). Stilbenes from Vitis vinifera L. waste: A sustainable tool for controlling Plasmopara Viticola. *Journal of Agricultural and Food Chemistry*, *65*(13), 2711–2718. https://doi.org/10.1021/acs.jafc.7b00241.

Gabaston, J., El Khawand, T., Waffo-Teguo, P., Decendit, A., Richard, T., Mérillon, J.-M., et al. (2018). Stilbenes from grapevine root: A promising natural insecticide against Leptinotarsa decemlineata. *Journal of Pest Science*, *91*(2), 897–906. https://doi.org/10.1007/s10340-018-0956-2.

Gagneul, D., Aïnouche, A., Duhazé, C., Lugan, R., Larher, F. R., & Bouchereau, A. (2007). A reassessment of the function of the so-called compatible solutes in the halophytic Plumbaginaceae *Limonium latifolium*. *Plant Physiology*, *144*(3), 1598–1611. https://doi.org/10.1104/pp.107.099820.

Gamir, J., Pastor, V., Kaever, A., Cerezo, M., & Flors, V. (2014). Targeting novel chemical and constitutive primed metabolites against Plectosphaerella cucumerina. *The Plant Journal*, *78*(2), 227–240. https://doi.org/10.1111/tpj.12465.

Gamir, J., Sánchez-Bel, P., & Flors, V. (2014). Molecular and physiological stages of priming: How plants prepare for environmental challenges. *Plant Cell Reports*, *33*(12), 1935–1949. https://doi.org/10.1007/s00299-014-1665-9.

Ge, Y., Guo, B., Cai, Y., Zhang, H., Luo, S., & Gallois, P. (2018). Transcriptome analysis identifies differentially expressed genes in maize leaf tissues in response to elevated atmospheric [CO_2]. *Journal of Plant Interactions*, *13*(1), 373–379. https://doi.org/10.1080/17429145.2018.1483035.

Gebauer, P., Korn, M., Engelsdorf, T., Sonnewald, U., Koch, C., & Voll, L. M. (2017). Sugar accumulation in leaves of Arabidopsis sweet11/sweet12 double mutants enhances priming of the salicylic acid-mediated defense response. *Frontiers in Plant Science*, *8*, 1378. https://doi.org/10.3389/fpls.2017.01378.

Geier, B., Sogin, E. M., Michellod, D., Janda, M., Kompauer, M., Spengler, B., et al. (2020). Spatial metabolomics of in situ host–microbe interactions at the micrometre scale. *Nature Microbiology*, *5*(3), 498–510. https://doi.org/10.1038/s41564-019-0664-6.

Gerlach, N., Schmitz, J., Polatajko, A., Schlüter, U., Fahnenstich, H., Witt, S., et al. (2015). An integrated functional approach to dissect systemic responses in maize to arbuscular mycorrhizal symbiosis: Systemic responses in mycorrhizal maize. *Plant, Cell & Environment*, *38*(8), 1591–1612. https://doi.org/10.1111/pce.12508.

Glauser, G., Vallat, A., & Balmer, D. (2014). Hormone profiling. *Methods in Molecular Biology*, *1062*, 597–608. https://doi.org/10.1007/978-1-62703-580-4_31.

Gonzalez-Mula, A., Lachat, J., Mathias, L., Naquin, D., Lamouche, F., Mergaert, P., et al. (2019). The biotroph *Agrobacterium tumefaciens* thrives in tumors by exploiting a wide spectrum of plant host metabolites. *New Phytologist*, *222*(1), 455–467. https://doi.org/10.1111/nph.15598.

Goufo, P., Singh, R. K., & Cortez, I. (2020). A reference list of phenolic compounds (including Stilbenes) in grapevine (Vitis vinifera L.) roots, woods, canes, stems, and leaves. *Antioxidants (Basel, Switzerland)*, *9*(5). https://doi.org/10.3390/antiox9050398.

Graßmann, J. (2005). Terpenoids as plant antioxidants. In *Vol. 72. Vitamins & hormones* (pp. 505–535). Elsevier. https://doi.org/10.1016/S0083-6729(05)72015-X.

Guerriero, G., Berni, R., Muñoz-Sanchez, J., Apone, F., Abdel-Salam, E., Qahtan, A., et al. (2018). Production of plant secondary metabolites: Examples, tips and suggestions for biotechnologists. *Genes*, *9*(6), 309. https://doi.org/10.3390/genes9060309.

Hall, R. D. (2011). Plant metabolomics in a nutshell: Potential and future challenges. In R. D. Hall (Éd.), Annual plant reviews Vol. 43 (p. 1–24). Oxford, UK: Wiley-Blackwell. https://doi.org/10.1002/9781444339956.ch1.

Hammerbacher, A., Coutinho, T. A., & Gershenzon, J. (2019). Roles of plant volatiles in defence against microbial pathogens and microbial exploitation of volatiles. *Plant, Cell & Environment*, *42*(10), 2827–2843. https://doi.org/10.1111/pce.13602.

Handrick, V., Robert, C. A. M., Ahern, K. R., Zhou, S., Machado, R. A. R., Maag, D., et al. (2016). Biosynthesis of 8-O-methylated benzoxazinoid defense compounds in maize. *The Plant Cell*, *28*(7), 1682–1700. https://doi.org/10.1105/tpc.16.00065.

Hatmi, S., Gruau, C., Trotel-Aziz, P., Villaume, S., Rabenoelina, F., Baillieul, F., et al. (2015). Drought stress tolerance in grapevine involves activation of polyamine oxidation contributing to improved immune response and low susceptibility to Botrytis cinerea. *Journal of Experimental Botany*, *66*(3), 775–787. https://doi.org/10.1093/jxb/eru436.

Haug, K., Cochrane, K., Nainala, V. C., Williams, M., Chang, J., Jayaseelan, K. V., et al. (2019). MetaboLights: A resource evolving in response to the needs of its scientific community. *Nucleic Acids Research*. https://doi.org/10.1093/nar/gkz1019. gkz1019.

Hong, J., Yang, L., Zhang, D., & Shi, J. (2016). Plant metabolomics: An indispensable system biology tool for plant science. *International Journal of Molecular Sciences*, *17*(6). https://doi.org/10.3390/ijms17060767.

Hossain, M. S., Persicke, M., ElSayed, A. I., Kalinowski, J., & Dietz, K.-J. (2017). Metabolite profiling at the cellular and subcellular level reveals metabolites associated with salinity tolerance in sugar beet. *Journal of Experimental Botany*, *68*(21 – 22), 5961–5976. https://doi.org/10.1093/jxb/erx388.

Huai, B., Yang, Q., Qian, Y., Qian, W., Kang, Z., & Liu, J. (2019). ABA-induced sugar transporter TaSTP6 promotes wheat susceptibility to stripe rust. *Plant Physiology*, *181*(3), 1328–1343. https://doi.org/10.1104/pp.19.00632.

Huot, B., Yao, J., Montgomery, B. L., & He, S. Y. (2014). Growth–defense tradeoffs in plants: A balancing act to optimize fitness. *Molecular Plant*, 7(8), 1267–1287. https://doi.org/10.1093/mp/ssu049.

Intergovernmental Panel on Climate Change (IPCC). (2018). An IPCC special report on the impacts of global warming of 1.5°C above pre-industrial levels and related global greenhouse gas emission pathways, in the context of strengthening the global response to the threat of climate change, sustainable development, and efforts to eradicate poverty. In V. Masson-Delmotte, P. Zhai, H. O. Pörtner, D. Roberts, J. Skea, P. R. Shukla, & T. Waterfield (Eds.), *Global warming of 1.5°C*. Cambridge, UK: Cambridge University Press. /z-wcorg/.

Iriti, M., & Faoro, F. (2009). Chemical diversity and defence metabolism: How plants cope with pathogens and ozone pollution. *International Journal of Molecular Sciences*, *10*(8), 3371–3399. https://doi.org/10.3390/ijms10083371.

Jacoby, R. P., Chen, L., Schwier, M., Koprivova, A., & Kopriva, S. (2020). Recent advances in the role of plant metabolites in shaping the root microbiome. *F1000Research*, *9*. https://doi.org/10.12688/f1000research.21796.1.

Jogawat, A., Meena, M. K., Kundu, A., Varma, M., & Vadassery, J. (2020). Calcium channel CNGC19 mediates basal defense signaling to regulate colonization by Piriformospora indica in Arabidopsis roots. *Journal of Experimental Botany*, *71*(9), 2752–2768. https://doi.org/10.1093/jxb/eraa028.

Jung, H. W., Tschaplinski, T. J., Wang, L., Glazebrook, J., & Greenberg, J. T. (2009). Priming in systemic plant immunity. *Science*, *324*(5923), 89–91. https://doi.org/10.1126/science.1170025.

Kanwar, P., & Jha, G. (2019). Alterations in plant sugar metabolism: Signatory of pathogen attack. *Planta*, *249*(2), 305–318. https://doi.org/10.1007/s00425-018-3018-3.

Kappers, I. F., Aharoni, A., van Herpen, T. W. J. M., Luckerhoff, L. L. P., Dicke, M., & Bouwmeester, H. J. (2005). Genetic engineering of terpenoid metabolism attracts bodyguards to Arabidopsis. *Science (New York, N.Y.)*, *309*(5743), 2070–2072. https://doi.org/10.1126/science.1116232.

Kazan, K. (2018). Plant-biotic interactions under elevated CO_2: A molecular perspective. *Environmental and Experimental Botany*, *153*, 249–261. https://doi.org/10.1016/j.envexpbot.2018.06.005.

Kessler, A., & Kalske, A. (2018). Plant secondary metabolite diversity and species interactions. *Annual Review of Ecology, Evolution, and Systematics*, *49*(1), 115–138. https://doi.org/10.1146/annurev-ecolsys-110617-062406.

Khizar, M., Shi, J., Saleem, S., Liaquat, F., Ashraf, M., Latif, S., et al. (2020). Resistance associated metabolite profiling of Aspergillus leaf spot in cotton through non-targeted metabolomics. *PLoS One*, *15*(2). https://doi.org/10.1371/journal.pone.0228675.

Kloth, K. J., Abreu, I. N., Delhomme, N., Petřík, I., Villard, C., Ström, C., et al. (2019). PECTIN ACETYLESTERASE9 affects the transcriptome and metabolome and delays aphid feeding1 [OPEN]. *Plant Physiology*, *181*(4), 1704–1720. https://doi.org/10.1104/pp.19.00635.

Koprivova, A., Schuck, S., Jacoby, R. P., Klinkhammer, I., Welter, B., Leson, L., et al. (2019). Root-specific camalexin biosynthesis controls the plant growth-promoting effects of multiple bacterial strains. *Proceedings of the National Academy of Sciences of the United States of America, 116*(31), 15735–15744. https://doi.org/10.1073/pnas.1818604116.

Kulbat, K. (2016). The role of phenolic compounds in plant resistance. *Biotechnology and Food Science, 80*(2). Consulté à l'adresse http://yadda.icm.edu.pl/yadda/element/bwmeta1.element.baztech-9ee7afbc-1b02-46b3-a9fa-dceaf3af6199.

Lazar, A. G., Romanciuc, F., Socaciu, M. A., & Socaciu, C. (2015). Bioinformatics tools for Metabolomic data processing and analysis using untargeted liquid chromatography coupled with mass spectrometry. *Bulletin of University of Agricultural Sciences and Veterinary Medicine Cluj-Napoca. Animal Science and Biotechnologies, 72*(2), 103–115. https://doi.org/10.15835/buasvmcn-asb:11536.

Lefevere, H., Bauters, L., & Gheysen, G. (2020). Salicylic acid biosynthesis in plants. *Frontiers in Plant Science, 11*, 338. https://doi.org/10.3389/fpls.2020.00338.

Lemonnier, P., Gaillard, C., Veillet, F., Verbeke, J., Lemoine, R., Coutos-Thévenot, P., et al. (2014). Expression of Arabidopsis sugar transport protein STP13 differentially affects glucose transport activity and basal resistance to Botrytis cinerea. *Plant Molecular Biology, 85*(4–5), 473–484. https://doi.org/10.1007/s11103-014-0198-5.

Liu, C., Atanasov, K. E., Tiburcio, A. F., & Alcázar, R. (2019). The polyamine Putrescine contributes to H_2O_2 and RbohD/F-dependent positive feedback loop in Arabidopsis PAMP-triggered immunity. *Frontiers in Plant Science, 10*, 894. https://doi.org/10.3389/fpls.2019.00894.

Liu, H., Xu, F., Gao, Y., Pang, Y., Xie, C., & Jiang, C. (2020). An integrated LC-MS/MS strategy for quantifying the oxidative-redox metabolome in multiple biological samples. *Analytical Chemistry, 92*(13), 8810–8818. https://doi.org/10.1021/acs.analchem.0c00242.

Long, D. E., Fung, A. K., McGEE, E. E. M., Cooke, R. C., & Lewis, D. H. (1975). The activity of invertase and its relevance to the accumulation of storage polysaccharides in leaves infected by biotrophic fungi. *New Phytologist, 74*(2), 173–182. https://doi.org/10.1111/j.1469-8137.1975.tb02603.x.

López-Fernández, O., Domínguez, R., Pateiro, M., Munekata, P. E. S., Rocchetti, G., & Lorenzo, J. M. (2020). Determination of polyphenols using liquid chromatography-tandem mass spectrometry technique (LC-MS/MS): A review. *Antioxidants (Basel, Switzerland), 9*(6). https://doi.org/10.3390/antiox9060479.

López-Gresa, M. P., Lisón, P., Kim, H. K., Choi, Y. H., Verpoorte, R., Rodrigo, I., et al. (2012). Metabolic fingerprinting of tomato mosaic virus infected Solanum lycopersicum. *Journal of Plant Physiology, 169*(16), 1586–1596. https://doi.org/10.1016/j.jplph.2012.05.021.

López-Ráez, J. A., Verhage, A., Fernández, I., García, J. M., Azcón-Aguilar, C., Flors, V., et al. (2010). Hormonal and transcriptional profiles highlight common and differential host responses to arbuscular mycorrhizal fungi and the regulation of the oxylipin pathway. *Journal of Experimental Botany, 61*(10), 2589–2601. https://doi.org/10.1093/jxb/erq089.

Lorenzini, G., Medeghini Bonatti, P., Nali, C., & Baroni Fornasiero, R. (1994). The protective effect of rust infection against ozone, sulphur dioxide and paraquat toxicity symptoms in broad bean. *Physiological and Molecular Plant Pathology, 45*(4), 263–279. https://doi.org/10.1016/S0885-5765(05)80058-X.

Lou, Y.-R., Bor, M., Yan, J., Preuss, A. S., & Jander, G. (2016). Arabidopsis NATA1 acetylates putrescine and decreases defense-related hydrogen peroxide accumulation. *Plant Physiology*, 00446.2016. https://doi.org/10.1104/pp.16.00446.

Luna, E., Flandin, A., Cassan, C., Prigent, S., Chevanne, C., Kadiri, C. F., et al. (2020). Metabolomics to exploit the primed immune system of tomato fruit. *Metabolites, 10*(3), 96. https://doi.org/10.3390/metabo10030096.

Macel, M., Visschers, I. G. S., Peters, J. L., Kappers, I. F., de Vos, R. C. H., & van Dam, N. M. (2019). Metabolomics of Thrips resistance in pepper (Capsicum spp.) reveals monomer and dimer acyclic diterpene glycosides as potential chemical defenses. *Journal of Chemical Ecology*, *45*(5–6), 490–501. https://doi.org/10.1007/s10886-019-01074-4.

Manganiello, G., Sacco, A., Ercolano, M. R., Vinale, F., Lanzuise, S., Pascale, A., et al. (2018). Modulation of tomato response to Rhizoctonia solani by Trichoderma harzianum and its secondary metabolite Harzianic acid. *Frontiers in Microbiology*, *9*, 1966. https://doi.org/10.3389/fmicb.2018.01966.

Mareya, C. R., Tugizimana, F., Piater, L. A., Madala, N. E., Steenkamp, P. A., & Dubery, I. A. (2019). Untargeted metabolomics reveal defensome-related metabolic reprogramming in Sorghum bicolor against infection by Burkholderia andropogonis. *Metabolites*, *9*(1). https://doi.org/10.3390/metabo9010008.

Martínez-Medina, A., Del Mar Alguacil, M., Pascual, J. A., & Van Wees, S. C. M. (2014). Phytohormone profiles induced by Trichoderma isolates correspond with their biocontrol and plant growth-promoting activity on melon plants. *Journal of Chemical Ecology*, *40*(7), 804–815. https://doi.org/10.1007/s10886-014-0478-1.

Martínez-Medina, A., Fernandez, I., Lok, G. B., Pozo, M. J., Pieterse, C. M. J., & Van Wees, S. C. M. (2017). Shifting from priming of salicylic acid- to jasmonic acid-regulated defences by *Trichoderma* protects tomato against the root knot nematode *Meloidogyne incognita*. *New Phytologist*, *213*(3), 1363–1377. https://doi.org/10.1111/nph.14251.

Mayo-Prieto, S., Marra, R., Vinale, F., Rodríguez-González, Á., Woo, S., Lorito, M., et al. (2019). Effect of Trichoderma velutinum and Rhizoctonia solani on the Metabolome of Bean Plants (Phaseolus vulgaris L.). *International Journal of Molecular Sciences*, *20*(3), 549. https://doi.org/10.3390/ijms20030549.

Meier, R., Ruttkies, C., Treutler, H., & Neumann, S. (2017). Bioinformatics can boost metabolomics research. *Journal of Biotechnology*, *261*, 137–141. https://doi.org/10.1016/j.jbiotec.2017.05.018.

Meteier, E., La Camera, S., Goddard, M.-L., Laloue, H., Mestre, P., & Chong, J. (2019). Overexpression of the VvSWEET4 transporter in grapevine hairy roots increases sugar transport and contents and enhances resistance to Pythium irregulare, a Soilborne pathogen. *Frontiers in Plant Science*, *10*, 884. https://doi.org/10.3389/fpls.2019.00884.

Mhamdi, A., & Noctor, G. (2016). High CO_2 primes plant biotic stress defences through redox-linked pathways. *Plant Physiology*, *172*(2), 929–942. https://doi.org/10.1104/pp.16.01129.

Mhlongo, M. I., Steenkamp, P. A., Piater, L. A., Madala, N. E., & Dubery, I. A. (2016). Profiling of altered metabolomic states in *Nicotiana tabacum* cells induced by priming agents. *Frontiers in Plant Science*, 7. https://doi.org/10.3389/fpls.2016.01527.

Mushtaq, M. Y., Choi, Y. H., Verpoorte, R., & Wilson, E. G. (2014). Extraction for metabolomics: Access to the metabolome: Extraction for metabolomics: Access to the metabolome. *Phytochemical Analysis*, *25*(4), 291–306. https://doi.org/10.1002/pca.2505.

Návarová, H., Bernsdorff, F., Döring, A.-C., & Zeier, J. (2012). Pipecolic acid, an endogenous mediator of defense amplification and priming, is a critical regulator of inducible plant immunity. *The Plant Cell*, *24*(12), 5123–5141. https://doi.org/10.1105/tpc.112.103564.

Nejat, N., & Mantri, N. (2017). Plant immune system: Crosstalk between responses to biotic and abiotic stresses the missing link in understanding plant defence. *Current Issues in Molecular Biology*, *23*, 1–16. https://doi.org/10.21775/cimb.023.001.

Noctor, G., Mhamdi, A., Chaouch, S., Han, Y., Neukermans, J., Marquez-Garcia, B., et al. (2011). Glutathione in plants: An integrated overview. *Plant, Cell & Environment*. https://doi.org/10.1111/j.1365-3040.2011.02400.x.

O'Leary, B. M., Neale, H. C., Geilfus, C.-M., Jackson, R. W., Arnold, D. L., & Preston, G. M. (2016). Early changes in apoplast composition associated with defence and disease in interactions between *Phaseolus vulgaris* and the halo blight pathogen *Pseudomonas syringae* Pv. phaseolicola : Leaf apoplast composition during *P. syringae* infection. *Plant, Cell & Environment, 39*(10), 2172–2184. https://doi.org/10.1111/pce.12770.

Osborne, T., Rose, G., & Wheeler, T. (2013). Variation in the global-scale impacts of climate change on crop productivity due to climate model uncertainty and adaptation. *Agricultural and Forest Meteorology, 170*, 183–194. https://doi.org/10.1016/j.agrformet.2012.07.006.

Pang, Q., Zhang, T., Wang, Y., Kong, W., Guan, Q., Yan, X., et al. (2018). Metabolomics of early stage plant cell–microbe interaction using stable isotope labeling. *Frontiers in Plant Science, 9*. https://doi.org/10.3389/fpls.2018.00760.

Papazian, S., Girdwood, T., Wessels, B. A., Poelman, E. H., Dicke, M., Moritz, T., et al. (2019). Leaf metabolic signatures induced by real and simulated herbivory in black mustard (Brassica nigra). *Metabolomics: Official Journal of the Metabolomic Society, 15*(10), 130. https://doi.org/10.1007/s11306-019-1592-4.

Park, S.-W., Kaimoyo, E., Kumar, D., Mosher, S., & Klessig, D. F. (2007). Methyl salicylate is a critical mobile signal for plant systemic acquired resistance. *Science, 318*(5847), 113–116. https://doi.org/10.1126/science.1147113.

Pastor, V., Balmer, A., Gamir, J., Flors, V., & Mauch-Mani, B. (2014). Preparing to fight back: Generation and storage of priming compounds. *Frontiers in Plant Science, 5*. https://doi.org/10.3389/fpls.2014.00295.

Perez de Souza, L., Alseekh, S., Naake, T., & Fernie, A. (2019). Mass spectrometry-based untargeted plant metabolomics. *Current Protocols in Plant Biology, 4*(4). https://doi.org/10.1002/cppb.20100.

Peters, K., Worrich, A., Weinhold, A., Alka, O., Balcke, G., Birkemeyer, C., et al. (2018). Current challenges in plant eco-metabolomics. *International Journal of Molecular Sciences, 19*(5). https://doi.org/10.3390/ijms19051385.

Pétriacq, P., de Bont, L., Hager, J., Didierlaurent, L., Mauve, C., Guérard, F., et al. (2012). Inducible NAD overproduction in Arabidopsis alters metabolic pools and gene expression correlated with increased salicylate content and resistance to Pst-AvrRpm1. *The Plant Journal, 70*(4), 650–665. https://doi.org/10.1111/j.1365-313X.2012.04920.x.

Pétriacq, P., Stassen, J. H., & Ton, J. (2016). Spore density determines infection strategy by the plant-pathogenic fungus Plectosphaerella cucumerina. *Plant Physiology, 170*, 2325–2339. https://doi.org/10.1104/pp.15.00551.

Pétriacq, P., Ton, J., Patrit, O., Tcherkez, G., & Gakière, B. (2016). NAD acts as an integral regulator of multiple defense layers. *Plant Physiology, 172*(3), 1465–1479. https://doi.org/10.1104/pp.16.00780.

Pétriacq, P., Williams, A., Cotton, A., McFarlane, A. E., Rolfe, S. A., & Ton, J. (2017). Metabolite profiling of non-sterile rhizosphere soil. *The Plant Journal for Cell and Molecular Biology, 92*(1), 147–162. https://doi.org/10.1111/tpj.13639.

Pieterse, C. M. J., Leon-Reyes, A., Van der Ent, S., & Van Wees, S. C. M. (2009). Networking by small-molecule hormones in plant immunity. *Nature Chemical Biology, 5*(5), 308–316. https://doi.org/10.1038/nchembio.164.

Prasch, C. M., & Sonnewald, U. (2013). Simultaneous application of heat, drought, and virus to Arabidopsis plants reveals significant shifts in Signaling networks. *Plant Physiology, 162*(4), 1849–1866. https://doi.org/10.1104/pp.113.221044.

Prisic, S., Xu, M., Wilderman, P. R., & Peters, R. J. (2004). Rice contains two disparate ent-copalyl diphosphate synthases with distinct metabolic functions. *Plant Physiology, 136*(4), 4228–4236. https://doi.org/10.1104/pp.104.050567.

Qazi, S. S., Lombardo, D. A., & Abou-Zaid, M. M. (2018). A Metabolomic and HPLC-MS/MS analysis of the foliar phenolics, flavonoids and coumarins of the fraxinus species resistant and susceptible to emerald ash borer. *Molecules (Basel, Switzerland), 23*(11). https://doi.org/10.3390/molecules23112734.

Rivero, J., Gamir, J., Aroca, R., Pozo, M. J., & Flors, V. (2015). Metabolic transition in mycorrhizal tomato roots. *Frontiers in Microbiology, 6*. https://doi.org/10.3389/fmicb.2015.00598.

Rolin, D. (Éd.). (2013). Metabolomics coming of age with its technological diversity (1st ed.). Amsterdam: Elsevier.

Rossouw, L. T., Madala, N. E., Tugizimana, F., Steenkamp, P. A., Esterhuizen, L. L., & Dubery, I. A. (2019). Deciphering the resistance mechanism of tomato plants against whitefly-mediated tomato curly stunt virus infection through ultra-high-performance liquid chromatography coupled to mass spectrometry (UHPLC-MS)-based metabolomics approaches. *Metabolites, 9*(4). https://doi.org/10.3390/metabo9040060.

Roy, S. J., Tucker, E. J., & Tester, M. (2011). Genetic analysis of abiotic stress tolerance in crops. *Current Opinion in Plant Biology, 14*(3), 232–239. https://doi.org/10.1016/j.pbi.2011.03.002.

Rudd, J. J., Kanyuka, K., Hassani-Pak, K., Derbyshire, M., Andongabo, A., Devonshire, J., et al. (2015). Transcriptome and metabolite profiling of the infection cycle of *Zymoseptoria tritici* on wheat reveals a biphasic interaction with plant immunity involving differential pathogen chromosomal contributions and a variation on the hemibiotrophic lifestyle definition. *Plant Physiology, 167*(3), 1158–1185. https://doi.org/10.1104/pp.114.255927.

Sandermann, H. (2008). Ecotoxicology of ozone: Bioactivation of extracellular ascorbate. *Biochemical and Biophysical Research Communications, 366*(2), 271–274. https://doi.org/10.1016/j.bbrc.2007.12.018.

Singh, S. K. (2018). Explorations of plant's chemodiversity: Role of nitrogen-containing secondary metabolites in plant defense. In Archana Singh & I. K. Singh (Éds.), Molecular aspects of plant-pathogen interaction (p. 309–332). Singapore: Springer. https://doi.org/10.1007/978-981-10-7371-7_14.

Singh, A. (2020). Tools for metabolomics. *Nature Methods, 17*(1), 24. https://doi.org/10.1038/s41592-019-0710-6.

Singh, B., & Sharma, R. A. (2015). Plant terpenes: Defense responses, phylogenetic analysis, regulation and clinical applications. *3 Biotech, 5*(2), 129–151. https://doi.org/10.1007/s13205-014-0220-2.

Sinha, R., Irulappan, V., Mohan-Raju, B., Suganthi, A., & Senthil-Kumar, M. (2019). Impact of drought stress on simultaneously occurring pathogen infection in field-grown chickpea. *Scientific Reports, 9*(1), 5577. https://doi.org/10.1038/s41598-019-41463-z.

Smith, D., Muscatine, L., & Lewis, D. (1969). Carbohydrate movement from autotrophs to heterotrophs in parasitic and mutualistic symbiosis. *Biological Reviews, 44*(1), 17–85. https://doi.org/10.1111/j.1469-185X.1969.tb00821.x.

Su, X., Lu, G., Guo, H., Zhang, K., Li, X., & Cheng, H. (2018). The dynamic transcriptome and metabolomics profiling in Verticillium dahliae inoculated Arabidopsis thaliana. *Scientific Reports, 8*(1), 15404. https://doi.org/10.1038/s41598-018-33743-x.

Święcicka, M., Dmochowska-Boguta, M., Orczyk, W., Grądzielewska, A., Stochmal, A., Kowalczyk, M., et al. (2020). Changes in benzoxazinoid contents and the expression of the associated genes in rye (Secale cereale L.) due to brown rust and the inoculation procedure. *PloS One, 15*(5). https://doi.org/10.1371/journal.pone.0233807.

Tauzin, A. S., & Giardina, T. (2014). Sucrose and invertases, a part of the plant defense response to the biotic stresses. *Frontiers in Plant Science, 5*. https://doi.org/10.3389/fpls.2014.00293.

Tenllado, F., & Canto, T. (2020). Effects of a changing environment on the defenses of plants to viruses. *Current Opinion in Virology, 42*, 40–46. https://doi.org/10.1016/j.coviro.2020.04.007.

Toyota, M., Spencer, D., Sawai-Toyota, S., Jiaqi, W., Zhang, T., Koo, A. J., et al. (2018). Glutamate triggers long-distance, calcium-based plant defense signaling. *Science*, *361*(6407), 1112–1115. https://doi.org/10.1126/science.aat7744.

Tugizimana, F., Djami-Tchatchou, A. T., Steenkamp, P. A., Piater, L. A., & Dubery, I. A. (2018). Metabolomic analysis of defense-related reprogramming in Sorghum bicolor in response to Colletotrichum sublineolum infection reveals a functional metabolic web of phenylpropanoid and flavonoid pathways. *Frontiers in Plant Science*, *9*, 1840. https://doi.org/10.3389/fpls.2018.01840.

Tugizimana, F., Mhlongo, M. I., Piater, L. A., & Dubery, I. A. (2018). Metabolomics in plant priming research: The way forward? *International Journal of Molecular Sciences*, *19*(6). https://doi.org/10.3390/ijms19061759.

Vaughan, M. M., Huffaker, A., Schmelz, E. A., Dafoe, N. J., Christensen, S., Sims, J., et al. (2014). Effects of elevated [CO_2] on maize defence against mycotoxigenic Fusarium verticillioides. *Plant, Cell & Environment*, *37*(12), 2691–2706. https://doi.org/10.1111/pce.12337.

Vlot, A. C., Dempsey, D. A., & Klessig, D. F. (2009). Salicylic acid, a multifaceted hormone to combat disease. *Annual Review of Phytopathology*, *47*, 177–206. https://doi.org/10.1146/annurev.phyto.050908.135202.

von Tiedemann, A., & Firsching, K. H. (2000). Interactive effects of elevated ozone and carbon dioxide on growth and yield of leaf rust-infected versus non-infected wheat. *Environmental Pollution (Barking, Essex: 1987)*, *108*(3), 357–363. https://doi.org/10.1016/s0269-7491(99)00214-6.

Wagner, G., Laperche, A., Lariagon, C., Marnet, N., Renault, D., Guitton, Y., et al. (2019). Resolution of quantitative resistance to clubroot into QTL-specific metabolic modules. *Journal of Experimental Botany*, *70*(19), 5375–5390. https://doi.org/10.1093/jxb/erz265.

Walerowski, P., Gündel, A., Yahaya, N., Truman, W., Sobczak, M., Olszak, M., et al. (2018). Clubroot disease stimulates early steps of phloem differentiation and recruits SWEET sucrose transporters within developing galls. *The Plant Cell*, *30*(12), 3058–3073. https://doi.org/10.1105/tpc.18.00283.

Wilkinson, S. W., Pastor, V., Paplauskas, S., Pétriacq, P., & Luna, E. (2018). Long-lasting β-aminobutyric acid-induced resistance protects tomato fruit against *Botrytis cinerea*. *Plant Pathology*, *67*(1), 30–41. https://doi.org/10.1111/ppa.12725.

Williams, A., Pétriacq, P., Schwarzenbacher, R. E., Beerling, D. J., & Ton, J. (2018). Mechanisms of glacial-to-future atmospheric CO_2 effects on plant immunity. *The New Phytologist*, *218*(2), 752–761. https://doi.org/10.1111/nph.15018.

Yamada, K., Saijo, Y., Nakagami, H., & Takano, Y. (2016). Regulation of sugar transporter activity for antibacterial defense in *Arabidopsis*. *Science*, *354*(6318), 1427–1430. https://doi.org/10.1126/science.aah5692.

Zander, M., Lewsey, M. G., Clark, N. M., Yin, L., Bartlett, A., Guzmán, J. P. S., et al. (2020). Integrated multi-omics framework of the plant response to jasmonic acid. *Nature Plants*, *6*(3), 290–302. https://doi.org/10.1038/s41477-020-0605-7.

Zaynab, M., Fatima, M., Abbas, S., Sharif, Y., Umair, M., Zafar, M. H., et al. (2018). Role of secondary metabolites in plant defense against pathogens. *Microbial Pathogenesis*, *124*, 198–202. https://doi.org/10.1016/j.micpath.2018.08.034.

Zhang, X., & Mou, Z. (2009). Extracellular pyridine nucleotides induce PR gene expression and disease resistance in Arabidopsis. *The Plant Journal*, *57*(2), 302–312. https://doi.org/10.1111/j.1365-313X.2008.03687.x.

Zhou, S., Richter, A., & Jander, G. (2018). Beyond defense: Multiple functions of benzoxazinoids in maize metabolism. *Plant and Cell Physiology*, *59*(8), 1528–1537. https://doi.org/10.1093/pcp/pcy064.

Zhou, Y., Van Leeuwen, S. K., Pieterse, C. M. J., Bakker, P. A. H. M., & Van Wees, S. C. M. (2019). Effect of atmospheric CO_2 on plant defense against leaf and root pathogens of

Arabidopsis. *European Journal of Plant Pathology, 154*(1), 31–42. https://doi.org/10.1007/s10658-019-01706-1.

Zhou, Y., Vroegop-Vos, I. A., Van Dijken, A. J. H., Van Der Does, D., Zipfel, C., Pieterse, C. M. J., et al. (2020). *Carbonic anhydrases CA1 and CA4 function in atmospheric CO2-modulated disease resistance*. https://doi.org/10.5167/UZH-188176.

Ziegler, J., Schmidt, S., Chutia, R., Müller, J., Böttcher, C., Strehmel, N., et al. (2016). Non-targeted profiling of semi-polar metabolites in Arabidopsis root exudates uncovers a role for coumarin secretion and lignification during the local response to phosphate limitation. *Journal of Experimental Botany, 67*(5), 1421–1432. https://doi.org/10.1093/jxb/erv539.

Further reading

Fagard, M., Launay, A., Clement, G., Courtial, J., Dellagi, A., Farjad, M., et al. (2014). Nitrogen metabolism meets phytopathology. *Journal of Experimental Botany, 65*(19), 5643–5656. https://doi.org/10.1093/jxb/eru323.

Rojas, C. M., Senthil-Kumar, M., Tzin, V., & Mysore, K. S. (2014). Regulation of primary plant metabolism during plant-pathogen interactions and its contribution to plant defense. *Frontiers in Plant Science, 5*. https://doi.org/10.3389/fpls.2014.00017.

CHAPTER FOUR

Plant metabolomics to the benefit of crop protection and growth stimulation

Cédric Bertrand[a,b,*], Azucena Gonzalez-Coloma[c,*], and Claire Prigent-Combaret[d]
[a]PSL Research University: EPHE-UPVD-CNRS, USR 3278. CRIOBE, Université de Perpignan, Perpignan Cedex, France
[b]Société Akinao, Perpignan Cedex, France
[c]Instituto de Ciencias Agrarias, CSIC, Madrid, Spain
[d]Université Lyon, Université Claude Bernard Lyon 1, CNRS, INRA, VetAgro Sup, UMR5557 Ecologie Microbienne, Villeurbanne, France
*Corresponding authors: e-mail address: cedric.bertrand@univ-perp.fr; azu@ica.csic.es

Contents

Advances in Botanical Research, Volume 98
ISSN 0065-2296
https://doi.org/10.1016/bs.abr.2020.11.002

Abstract

Biotic and abiotic stresses strongly affect plant growth, quality of production and crop yield. Respectively biocontrol products and biostimulants have been proposed as agronomic tools to counteract those stresses. Research and development on biocontrol and biostimulant products require the study and characterization of the biotic and abiotic interactions involved in these processes. Plant metabolomics approach plays a key role in the identification of chemical mediators involved in these two types of interactions. Metabolomics is a useful tool for the characterization of volatile organic compounds (VOCs) involved in plant-insect interactions, or for the bioguided identification of active compounds from new raw materials. On the other hand, metabolomics approach can be used for the characterization of metabolic pathways affected by plant priming, biostimulant or elicitor application and/or for mechanistic study of Plant-PGPR (Plant Growth Promoting Rhizobacteria) interaction through biomarker specific metabolic fingerprint study.

1. Introduction

The success of modern-day intensive farming practices is often attributed to the use of agrochemicals. However, the development of these chemicals has led to many negative environmental and health consequences. Moreover, with the emergence of an increasingly frequent phenomenon of resistance among conventional pesticides and in a regulatory context for a use of pesticides better in line with sustainable development (EU pesticide package), the interest of manipulating metabolomics technologies in order to identify the plant's ability to interact metabolically/chemically with its environment and to identify levers that make it possible to implement solutions of protection and optimization of growth may lead to alternative solutions.

Plants interact with other organisms and particularity with other plants, insects and a large diversity of microorganisms that can be defined as the plant microbiota.

In one hand, some crude ethanolic or aqueous plant extracts are traditionally used by organic or biodynamic farmers for pests or microbial disease management, but relatively few studies exist regarding the effectiveness, standardization and composition of these extracts. On the other hand, the use of products from natural origin seems a sustainable option due to their perceived eco-friendly profile over environmentally harmful chemical pesticides. In this context, biocontrol products (Bps) are booming. Bps mostly include the use of microorganisms (viruses, bacteria and fungi) and botanical substances (essential oils, plant extracts, minerals, etc.) for biocontrol operations (Bertrand & Prigent-Combaret, 2012). Botanical substances include pure natural compounds, plant extracts and botanical preparations. Except for pure compounds, one of the main intrinsic qualities of plant bio-based products used for the development of biocontrol is the complexity of the preparation, as the mixture of metabolites could allow a high synergy of activity and avoid the development of pest resistance. In this context, traditional botanical preparations are getting increasing attention from the stakeholders (OECD, ENV/JM/M87MONO(2012)36). Nevertheless, the botanical ingredients being usually complex mixtures composed of several active ingredients, their composition needs to be characterized (Andreu et al., 2018), and metabolomics methods are well-adapted to best identify the composition of natural products.

Nowadays, plant metabolomics is increasingly useful in agrochemistry and in particular in biocontrol and development of alternative biobased solutions for crop protection or biostimulation. Metabolomics is also a useful tool for understanding plant-microorganism interactions and for the identification of bioactive compounds involved in those interactions that could lead to innovative biocontrol products.

Metabolomic tools are thus helpful for the development of biological control mainly to achieve the following different objectives: characterization of biomass; identification of active substances; characterization of complex mixture and characterization of biomarkers and mechanisms. This chapter deals with the subject of crop protection and biostimulation and the interest of manipulating metabolomic technologies in order to identify the plant's ability to chemically interact with its environment and to identify levers to implement solutions of protection and optimization of growth. It takes up concepts on plant metabolomics, chemical ecology, abiotic and biotic stress, and phytochemicals resources.

2. Plant attraction/repulsion/defense processes are associated to chemical strategies: A playground for metabolomics

All tissues of the plants (roots, stems, leaves and fruits) are associated to distinct microbial communities either at their surface (epiphyte) or inside the tissue (endophyte), the whole constituting the plant holobiont (Rosenberg & Zilber-Rosenberg, 2016). Among the plant microbiota, some bacteria are able to promote plant development, using plant growth-promotion and plant protection properties. A large part of these microorganisms is associated to the plant roots that exude a wide variety of molecules within the rhizosphere. Exudates may represent between 20% and 50% of photosynthetic carbon produced by the plant and are able to attract and sustain the growth of a large diversity of microbial species (Haichar, Santaella, & Heulin, 2014). In return, plant-beneficial microorganisms (also known as *Plant Growth-Promoting Rhizobacteria*, PGPR) are capable to promote the plant growth and protect plants against pathogens or abiotic stresses, thanks to several indirect or direct mechanisms (Vacheron et al., 2013). Among them, certain microorganisms will favor the growth of the plant using plant-beneficial functions such as nitrogen fixation, phosphate solubilization or even by modulating the plant hormonal pathways (via auxin production, deamination of 1-aminocyclopropane-1- carboxylate) (Drogue, Doré, Borland, Wisniewski-Dyé, & Prigent-Combaret, 2012). Others have rather a plant protective activity and protect the plant against pathogens through three main mechanisms: competition, production of antimicrobial compounds, or induction of systemic resistance in the plant (Besset-Manzoni, Rieusset, Joly, Comte, & Prigent-Combaret, 2018).

Regarding the use of these microorganisms in agriculture, plant-beneficial microorganisms or compounds they secrete can be used commercially to inoculate plants in order to stimulate their growth or protect them against diseases. They are in that case known as biostimulants and biocontrol products, respectively.

The European Biostimulants Industry Council (EBIC) have defined plant biostimulants as products "containing substance(s) and/or microorganisms whose function when applied to plants or the rhizosphere is to stimulate natural processes to enhance/benefit nutrient uptake, nutrient efficiency, tolerance to abiotic stress, and crop quality." (European Biostimulants Industry Council, 2012). This category regroups a wide diversity of biological and organic products including living microorganisms, algae products, animal or plant fermentation products, protein hydrolysates, composts and manures, food wastes, etc. (Brown & Saa, 2015) but excludes mineral fertilizers. Biostimulants act by improving the plant nutrition and growth either directly by interacting with plant signaling cascades or indirectly through the stimulation of plant-beneficial microorganisms to produce molecules that will improve the plant nutrition and growth. Thus, biostimulant products are often containing high amounts of simple or complex carbohydrates that may act as a source of energy for plant-beneficial microorganisms or that may interact with plant signaling cascades. Bacteria of the genus *Azospirillum* is one of the most emblematic biostimulants of bacterial origin used in agriculture, particularly in South America, where members of this genus are often applied in fields to improve plant growth, with more than 100 products already in the market (Cassán et al., 2020). Many studies have shown that the application of microbe-based biostimulants benefits crops by a biostimulation process but it can also protect plants against biotic stresses, in which case the product can be used as a biocontrol product.

Biocontrol corresponds to a set of methods for protecting crops using living organisms or natural substances and natural mechanisms in the context of integrated pest management. Biological control is the most successful, most cost-effective and environmentally safest way of pest management. (IOBC Internet Book of Biological Control Version 6, Spring 2012 Copyright IOBC 1 IOBC Internet Book of Biological Control, version 6). Biological control or biocontrol and Biocontrol products development are also based on the knowledge of biotic interaction, and the characterization of the mechanisms of this interaction (Fig. 1). All those interactions are based on exchanges of metabolite (including volatiles) sometimes named allelochemical compounds or biomediators. It is now known that this allelochemicals exhibit high antifungal, anti-herbivory and antimicrobial properties or can have phytotoxic effects.

Bps are divided into four categories: macro-organisms, microorganisms, chemical mediators, and natural substances of animal or plant origins (Fig. 2). Macroorganisms include invertebrate organisms (insects, nematodes, mites) that have direct or indirect activity on pests. A distinction is made between predatory macro-organisms for which the pest is a prey (i.e., ladybugs against aphids) and parasitic macro-organisms for which the pest is an essential host for its development (trichogramma). Microorganisms (bacteria, fungi, viruses) have various activities (anti-insect, anthelmintic, antifungal, anti-bacterial, etc.). One hundred of bacterial strains have been registered as biocontrol agents around the world. In order of importance, they belong to *Bacillus, Pseudomonas* and *Streptomyces* genera (van Lenteren, Bolckmans, Köhl, Ravensberg, & Urbaneja, 2018). Chemical mediators include insect pheromones and kairomones. Pheromones are signal molecules involved in the communication between insects of the same species, whereas kairomones are semio-chemical molecules released into the environment by producing-organisms to receptor organisms belonging to other species. Chemical mediators are used in two ways: biological control by trapping techniques or by mating disruption methods. Natural substances include all products of mineral, vegetable, animal or microbial origins. Their mechanisms of action are diverse. They are used against insects, bacteria, and fungi. This category is largely

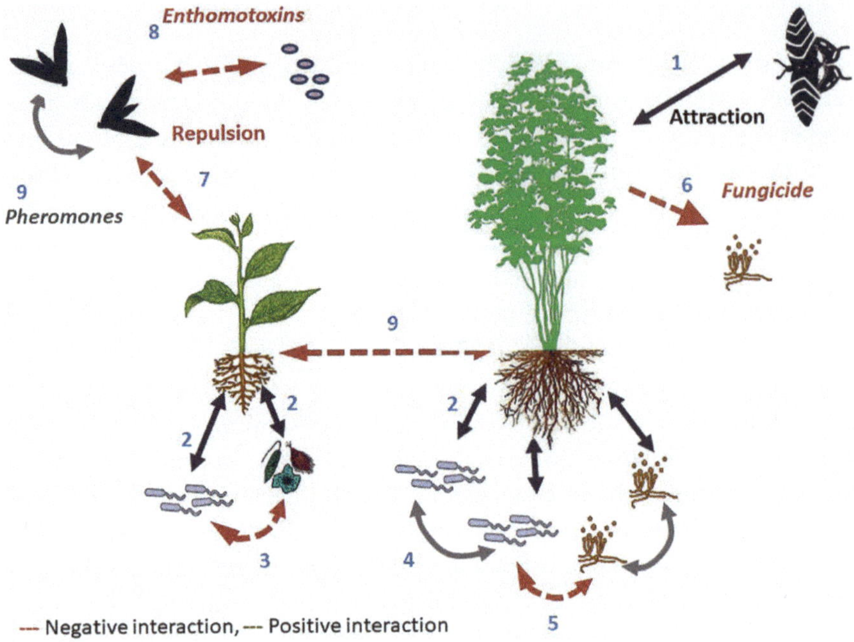

Fig. 1 Biotic interactions between organisms in plant systems: 1—the plant emits large quantities of volatile compounds to attract pollinators but these compounds can also guide phytophagous insects. 2—Compounds emitted by the plant attract symbiotic but also phytopathogenic microorganisms. 3 and 5 Antibiosis: compounds emitted by certain microorganisms can inhibit other microorganisms. 4—In the context of a phenomenon called quorum sensing, certain compounds can be perceived within a population and induce a phase change, leading to the modification of the bacteria metabolism and, for example, to the production of virulence factors. 6—The plant is protected by the production of phytoalexins, antifungal or repellent compounds with respect to phytophagous organisms. 7—Compounds are emitted by the plant to repel biting sucking phytoagritic insects, it may be, for example, citronellol. 8—Entomopathogenic bacteria (e.g., *Bacillus thuringiensis*). 9—Phytotoxic compounds can also be produced by the plant to limit the growth of other plants, this is the phenomenon of allelopathy.

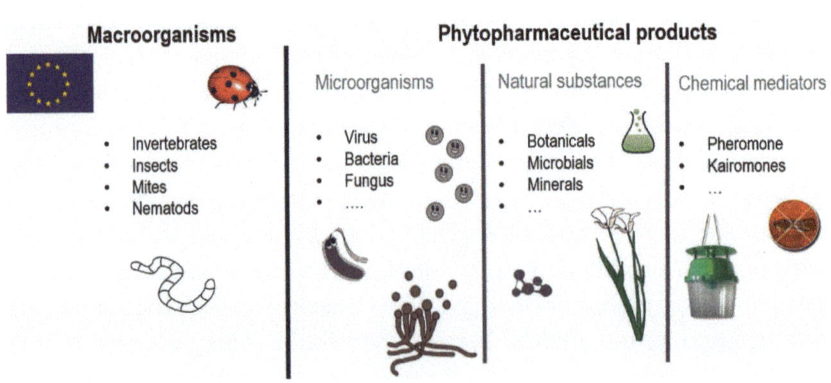

Fig. 2 Biocontrol products are agents and products that use natural mechanisms within the framework of integrated pest management.

dominated by sulfur products and plant products. In addition to these product categories, there is also biocontrol by conservation, which aims to promote the action of auxiliary organisms already present in the environment, and biocontrol by acclimatization, which aims to obtain a permanent establishment of an exotic auxiliary organism to regulate the populations of an invasive species.

Plant metabolomics is increasingly useful in agrochemistry and in particular in biocontrol and development of alternative biobased solutions for crop protection or biostimulation. Metabolomics enables to better understand plant-microorganisms' interactions and makes possible the identification of bioactive compounds involved in those interactions.

3. Volatilomics in plant-insect interactions, plant-plant dialogues and crop protection

The plant volatilome includes volatile essential oil components and volatile organic compounds (VOCs) (Kantsa, Sotiropoulou, Vaitis, & Petanidou, 2015). VOCs are biosynthesized from the mevalonic acid and methylerythritol phosphate pathways (isoprenoids), the lipoxygenase pathway (fatty acid derivatives), the shikimic acid pathway (benzenoids and phenylpropanoids), the amino acid derivatives pathway (Baldwin, 2010; Conchou et al., 2019) and other metabolic routes giving alkenes and small oxygenated compounds (ethylene acetaldehyde, acetone, methanol) (Jardine et al., 2017). The release of these substances is frequently associated with biotic and abiotic stresses that include herbivory by insects (Arimura & Pearse, 2017).

VOC emissions vary within species (Vivaldo, Masi, Taiti, Caldarelli, & Mancuso, 2017), with environmental conditions (including above or below ground interactions) (Delory, Delaplace, Fauconnier, & du Jardin, 2016), and within individual plants depending on the organs. For example, pinene emissions from maritime pine branches or stems are composed of different enantiomers (Staudt, Byron, Piquemal, & Williams, 2019). Plant-associated microorganisms such as epiphytic (Helletsburger, Dötterl, Ruprecht, & Junker, 2017) or endophytic bacteria (Maggini et al., 2019) can significantly affect the VOC composition released by a plant organ.

3.1 Plant-insect interactions and VOCs

Plants release VOCs as a response to insect attack (Holopainen & Gershenzon, 2010). Herbivore-induced plant volatiles consist mainly of terpenes, green leaf volatiles, phenylpropanoids and plant signals (Fig. 3).

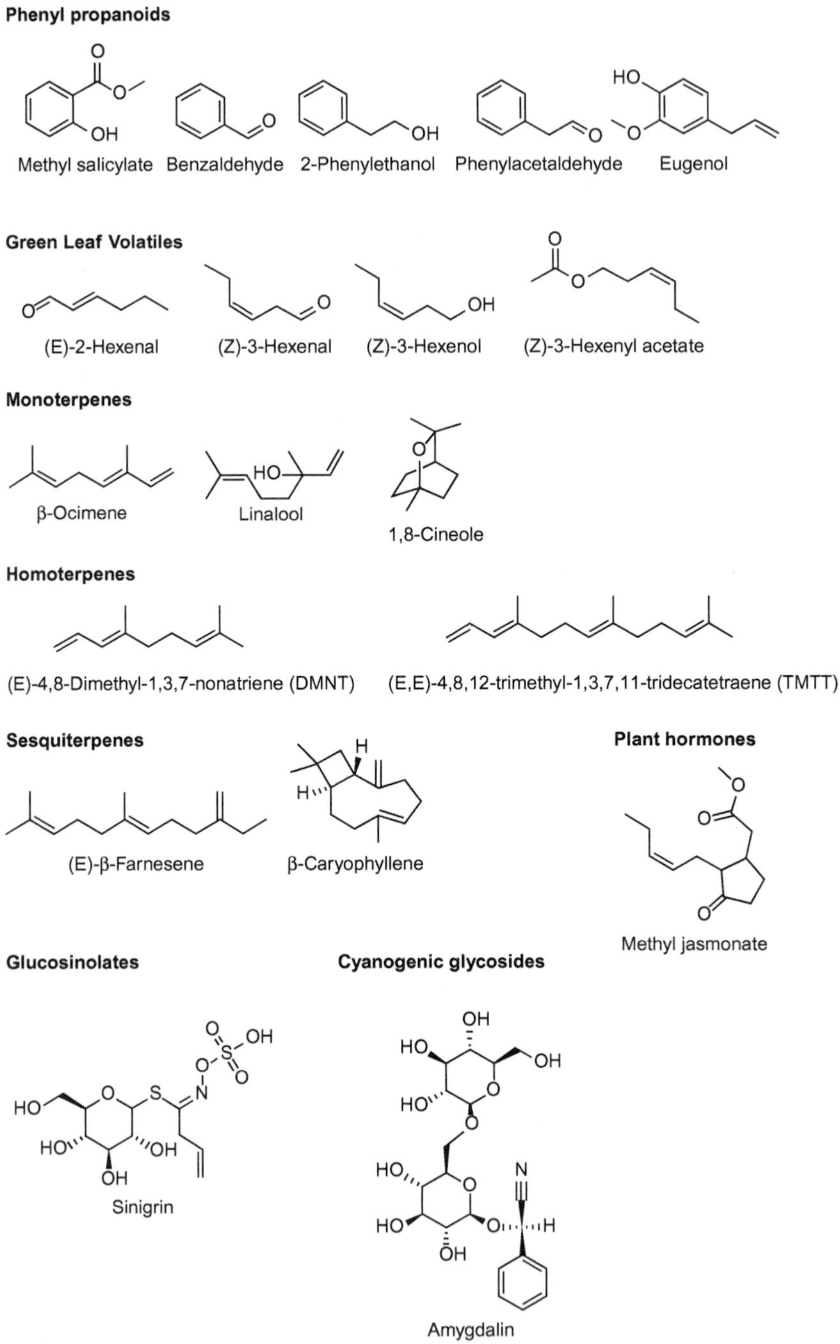

Fig. 3 Main volatile organic compounds involved in plant-insect interactions.

Jasmonic acid (JA) is a herbivore-related plant defense signal. JA regulates the emission of herbivore-induced VOCs, mostly terpenes (Paudel Timilsena, Seidl-Adams, & Tumlinson, 2020). Other signaling molecules, such as ethylene and salicylic acid, act synergistically with JAs to regulate the biosynthesis of VOCs and other herbivore-induced plant defenses (Okada, Abe, & Arimura, 2015). These responses are highly specific to the herbivore species and ecology. *Spodoptera exigua, Mythimna separata,* or *Tetranychus urticae* induced different mixtures by feeding on lima bean plants (Arimura & Pearse, 2017). The specialist *Manduca sexta* induced the production of jasmonic acid and ethylene when feeding on *Nicotiana attenuata,* while the generalist *Spodoptera litura* induced salicylic acid (Paudel Timilsena et al., 2020). JAs do not fully explain how plants induce herbivore-specific VOC mixtures. Elicitor molecules from the herbivore saliva could regulate the JAs signaling pathway (Arimura & Pearse, 2017).

Terpenes are the most structurally diverse chemical group, originated by the terpene synthase family of enzymes. The structural diversity of aromatic compounds and phenylpropanoids comes from the action of methyl and acetyltransferases and includes C6/C1 (methyl salicylate, benzaldehyde), C6/C2 (2-phenylethanol, phenylacetaldehyde) and C6/C3 (eugenol, chavicol) subclasses. Green leaf volatiles originate from lipid oxidation through the lipoxygenase pathway and are the major volatiles in terrestrial plants, mostly composed of (*Z*)-3-hexenal, (*E*)-2-hexenal, (*Z*)-3-hexen-1-ol and (*Z*)-3- hexen-1-yl acetate (Arimura & Pearse, 2017).

Additionally, essential oils in aromatic plants, glucosinolates in Brassicaceae, cyanogenic glycosides in Rosaceae and oleoresins in conifers, are released after herbivore attacks due to physical damage of the storage tissues (Arimura & Pearse, 2017). Essential oils and oleoresins are mainly composed of terpenes (mono- and sesquiterpenes) and play a role in attracting natural herbivore enemies, protection against pathogens, plant to plant signaling and attracting pollinators among others (Sharifi-Rad et al., 2017). Glucosinolates are glycosylated secondary metabolites present in Brassicaceae. They contain a β-D-glucose moiety linked to a (*Z*)-*N*-hydroximinosulfate ester with an amino acid-derived side chain (Jeschke, Gershenzon, & Vassão, 2016; Kumar, 2017). The enzymes myrosinases transform glucosinolates to volatile toxic isothiocyanates after plant damage and are stored in myrosin cells in the phloem parenchyma and stomata guard cells (Kumar, 2017). Cyanogenic glycosides are glycosidic derivatives of α-hydroxynitriles stored in vacuoles separated from the β-glucosidases that hydrolyze them after tissue damage, releasing HCN, a potent toxin (Kumar, 2017; Zagrobelny, de Castro, Møller, & Bak, 2018).

3.2 Plant-plant communication and VOCs

VOCs can induce systemic VOCs by de novo synthesis in damaged or undamaged tissues (Holopainen & Gershenzon, 2010). This type of signaling has been described in hybrid poplar *(Populus deltoides* x nigra) damaged by gypsi moth (*Lymantria dispar*) larvae, in lima bean leaves damage by beetles (*Cerotoma ruficornis* and *Gynandrobrotica guerreroensis*) inducing the production of extrafloral nectar in adjacent leaves and in sagebrush and willow which VOCs induced defenses in clonal plants (Karban, Shiojiri, & Ishizaki, 2010; Pearse, Hughes, Shiojiri, Ishizaki, & Karban, 2013). Inter-branch communication has been shown in blueberry plants attacked by gypsy moth (Rodriguez-Saona, Rodriguez-Saona, & Frost, 2009). Insect herbivore feeding also modified the concentrations and the distribution of total and individual glucosinolates in the plant (Kumar, 2017).

VOCs can also prime plants, allowing for greater responses after a subsequent herbivory attack (Ali, Sugimoto, Ramadan, & Arimura, 2013; Holopainen & Gershenzon, 2010). Herbivore-induced indole primed mono and homoterpenes and the production of stress hormones (jasmonate-isoleucine conjugate and abscisic acid) in neighboring maize plants (Erb et al., 2015). There is no clear evidence on how species-specific is the priming effect. The generalist *Heliothis virescens* or the specialist *Manduca sexta* induced VOCs that primed the production of defensive volatiles and JA in *Nicotiana benthamiana* by a conspecific regurgitant (Paudel Timilsena et al., 2020). Other reports showed herbivore-specific induction of defense priming. The reproductive rate of a specialist or generalist aphid (*Uroleucon macolai, Aphis gossypii*) *in Baccharis salicifolia* plants was significantly reduced when primed plants were damaged by the same aphid species (Moreira, Nell, Katsanis, Rasmann, & Mooney, 2018). Furthermore, whitefly modified the composition of VOC emissions by tomato plants by mimicking a pathogen attack that increased the quality as hosts for the insect in neighboring plants (Zhang et al., 2019).

Other VOCs such as essential oils can also prime plant defenses. Thyme essential oil primed defense responses in apple fruit against the fungal pathogen *Botryitis cinerea* through the PR-8 gene of apple (Banani et al., 2018), however little is known on their priming for insect defenses.

3.3 Direct effects on insects

Many reports on herbivore–plant interactions mediated by induced VOCs showed attraction. (*E*)-4,8-Dimethyl-1,3,7-nonatriene was identified as an attractant to fall armyworm (*Spodoptera frugiperda*) in VOCs released by cowpea plants after insect infestation (Takabayashi & Shiojiri, 2019).

Volatile terpenes or green leaf volatiles deter oviposition of insect species (Holopainen & Gershenzon, 2010; Papanastasiou, Ioannou, & Papadopoulos, 2020). Aphids and thrips are repelled by volatile terpenes (Avellaneda, Díaz, Coy-Barrera, Rodríguez, & Osorio, 2019; Li et al., 2019) and green leaf volatiles (Ahmed et al., 2018). Furthermore, there are many examples in the literature on the direct repellent, antifeedant and insecticidal effects of essential oils and their components (Ikbal & Pavela, 2019; Pavela & Benelli, 2016).

Glucosinolates hydrolyzed by myrosinases after tissue damage give unstable aglucones that degrade to isothiocyanates, which are direct defenses against herbivores (Jeschke et al., 2016) and stimulate larval feeding, oviposition and host location by specialist feeders (Kumar, 2017). Plant cyanogenic glycosides stored in the vacuole are hydrolyzed upon tissue damage by specific cyanogenic β-glucosidases to the corresponding α-hydroxynitriles which release toxic HCN (Morant et al., 2008).

3.4 Indirect effects on insects

VOCs attract herbivore predators and parasitoids after above and below ground herbivore oviposition and/or feeding. Tests with VOCs and transgenic plants allowed for the identification of specific terpenes and green leaf volatiles involved in enemy attraction (Holopainen & Gershenzon, 2010). The parasitoid wasp *of Pieris rapae* larvae (*Cotesia glomerata*) is attracted to (*E*)-3-hexenal and (Z)-3-hexenyl acetate. The parasitoid of *Plutella xylostella* larvae (*C. vestalis*) is attracted to a mixture of four volatiles [(Z)-3-hexenyl acetate, α-pinene, sabinene, and n-heptanal]. VOCs induced by *Tetranychus urticae* attract their predatory mites *Phytoseiulus persimilis*. Lady beetles use different willow VOC mixtures to find prey patches (Takabayashi & Shiojiri, 2019).

Glucosinolates and their breakdown products have distinct roles in multitrophic interactions. Parasitoids of Brassicaceae specialist herbivores find their hosts using isothiocyanates (Kumar, 2017). Myrosinase activity has been detected in insects that can sequester glucosinolates from their food plants (the aphids *Lipaphis erysimi*, *Brevicoryne brassicae* and in the flea beetle *Phyllotreta striolata*). Upon attack by a predator, isothiocyanates are released, deter predators and are also synergists of the aphid alarm pheromone (*E*)-β-farnesene (Beran, Köllner, Gershenzon, & Tholl, 2019).

Cyanogenic glycosides are also a defense system against enemies. Larvae of the burnet moth (*Zygaena filipendulae*) sequester bitter linamarin and lotaustralin and the release of toxic HCN after their breakdown deter enemies (Beran et al., 2019).

3.5 Volatilomics and plant protection

The importance of the plant volatilome on plant-insect interactions has practical implications in plant protection. The introduction of VOC releasing plant species in a crop can reduce pest damage (Conchou et al., 2019). As previously mentioned VOCs can reduce oviposition, including non-host volatiles. Methylsalicylate emitted by birch trees affects mating of the processionary moth affecting nearby pines (Jactel, Birgersson, Andersson, & Schlyter, 2011). This can be used in push-pull strategies (Conchou et al., 2019).

The study of plant volatilomics requires gas chromatography (GC) coupled with mass spectrometry (GC–MS) (Bicchi & Maei, 2012). Hydrodistillation is a limited extraction technique used for essential oils from aromatic plants. The most common method for efficient extraction of VOCs is the use of Solid Phase Microextraction (SPME), based on the absorption-adsorption of the head-space volatiles into a coated fiber (Boiteux et al., 2018). This technique allows for the study of biogenic VOCs (Wong, Yan, Shellie, Sciarrone, & Marriott, 2019).

4. Plant metabolomics applied to biostimulation and biocontrol products development

Throughout the globe approximately 2 million tonnes of pesticides are utilized, out of which 47.5% are herbicides, 29.5% are insecticides, 17.5% are fungicides and 5.5% are other pesticides (Sharma et al., 2019). In this context research of alternative to synthetic herbicides is a priority. Among the options for developing alternatives to synthetic herbicides, exploring the diversity of allelochemical compounds and simultaneously identifying sources of biomass supply or proposing a green chemical synthesis of the natural pure compounds could be a valuable option.

Securing high quality biomass is the basic condition for any development plants-based biocontrol products. As an example, citral would be a good candidate for bioherbicide development. Citral is a monoterpene commonly found as volatile component in many different aromatic plants. It has shown a high herbicidal effect on weeds (Graña, Sotelo, Díaz-Tielas, Reigosa, & Sánchez-Moreiras, 2013). *Elionurus muticus* is a grass from Pampa biome, Brazil known as lemongrass and is considered as an alternative source of citral. This species has been considered to have a large economically important potential among the main essential oils that are produced from grass (Füller, Bertrand, Simon, de Barros, & Neto, 2014). At the very beginning, chemical characterization of wild populations of *E. muticus* was needed. Several plants samples four different from four different populations were collected in southern Brazil. Metabolic profiling was performed using gas chromatography and mass spectrometry. The results showed that the studied populations have chemical variability based on their geographic origin, which indicates the possibility of chemotypes (Füller et al., 2017).

Based on the results of this first metabolomic approach, the cluster constituted with the plants collected from Morro Santa and Sao Borja would be selected, because of a high quantity of citral and stable quality of the essential oil. This "chemotype" could be a good candidate for the development of a bioherbicide.

Some traditional plants used for plant protection are nowadays commercialized within the framework of the EU regulation as Basic substances. Basic substances are substances that are not predominantly used for plant protection purposes but are useful in plant protection. Further criteria according to article 23 of Regulation (EC) No 1107/2009 are that they are not substances of concern, that they do not have the capacity to cause endocrine-disrupting, neurotoxic or immunotoxic effects, and that they are not placed on the market as plant protection products. Some crude aqueous extracts of those basics' substances are traditionally used by organic or biodynamic farmers for pests or microbial disease management, but relatively few studies exist, on the effectiveness of these extracts (Andreu et al., 2018). For example, white willow (*Salix alba*) bark, horsetail (*Equisetum arvense*), *Urtica* spp. are used in organic farming for various claimed activities. These plants are commonly used within aqueous decoction or maceration, and the observed variability in activity in the field has been widely described. One of the supposed causes of this is variability is non-standardized preparation workflows or poor quality of the plant. For the first point EU regulation proposes a "Mode of use" (e.g., Basic Substance, *Equisetum arvense* L. SANCO/12386/2013–rev. 7 20 July 2017) for each approved basic substance. Regarding plant quality assessment metabolic profiling could be used for the identification and characterization of the potential mixtures of bioactives metabolites and/or quality biomarkers. *Salix alba* cortex has been shown to have antifungal activity against *Plasmopara viticola* (Andreu et al., 2018; Deniau et al., 2019). In vitro assays have shown willow bark extracts have a potential anti-germinative effect on spore of *Plasmopara viticola*. This anti-germinative activity of *Salix alba* cortex extract would be related with the high amount and the diversity of the polyphenol characterized by HPLC-DAD-MS/MS-based metabolomic profiling (Andreu et al., 2018).

We have shown that *M. gale* fruits and leaves extracts exhibit phytotoxic activities against different plant species, including against the invasive species *Fallopia* (Popovici et al., 2011). The phytotoxicity is due to the presence of 3-(1-oxo-3-phenylpropyl)-1,1,5-trimethylcyclohexane-2,4,6-trione (myrigalone A), a dihydrochalcone present in both the fruits and leaves. We hypothesize that *M. gale* could be used in the future to being a potential source of environmentally friendly herbicides. Identification of the active metabolite (Myrigalone A) was reached within a classical bioguided fractionation workflow following by dereplication procedure combining uses of spectral properties of targeted metabolite (Popovici et al., 2011; Popovici, Bertrand, & Comte, 2009). This last method is now judged to be inadequate in terms of time and cost, particularly in case of screening of new raw materials as micro-mycetes, bacteria or micro-algae. For this purpose, new metabolomic workflows are emerging usually: a rapid micro-fractionation of crude extracts is carried out followed by multivariate data analysis of their UPLC-HRMS/MS profiles. In parallel bioactivity of crude extracts and micro-fractionation of each active crude extract are evaluated. For example, herbicide potential can be evaluated based on the inhibition of plants specifics enzymes using microtiter plate screening systems. Mass spectral similarity networking is used to highlight clusters involved in the evaluated activity.

5. Metabolomics to understand and improve the mechanisms of perception and responses to the environment for health benefits

During their growth, plants interact with a huge diversity of microorganisms. These biotic interactions are mediated by a very wide variety of secondary metabolites produced either by PGPR, other members of the soil microbial community and by the plant (Moco et al., 2007; Venturi & Keel, 2016). Indeed, both plant and microbial metabolomics studies are key methods to better understand the molecular dialogue established when plants are interacting with biocontrol or biostimulant PGPR. After the perception of PGPR through their cell surface components (i.e., flagella, lipopolysaccharides; Jourdan, Ongena, & Thonart, 2008; Ramamoorthy, Viswanathan, Raguchander, Prakasam, & Samiyappan, 2001), plants respond, among others, by modulating their metabolic pathways. Several studies have reported that primary and secondary metabolite profiles in roots, shoots and xylem sap were modified upon inoculation of PGPR (Chamam et al., 2013; Mishra, Singh, Jaiswal, Kumar, & Maurya, 2006; Osbourn, Qi, Townsend, & Qin, 2003; Rozier et al., 2016; Walker et al., 2011).

5.1 Impact of PGPR on primary metabolism of plants

Several studies evidence that PGPR are able to modify the primary metabolism of plants and in particular the pool of carbon compounds in plants at both above and below ground (Lade et al., 2018). This results in the alteration of root exudation patterns upon *Azospirillum* inoculation with an increase of organic compounds and sugars potentially exudated by roots (Rozier et al., 2016). *Pseudomonas* strains are also able to modify plant root exudates (Kuzmicheva et al., 2017; Matilla et al., 2010).

Metabolic contents of xylem sap of maize inoculated by the biostimulant *A. lipoferum* CRT1 was analyzed (Rozier et al., 2016). Xylem sap contains a wide range of small, water-soluble, organic substances such as primary metabolites (sugars, amino acids, organic acids, etc.). *A. lipoferum* triggers decrease of primary metabolite contents in the xylem sap, and release of metabolites outside of the roots as root exudates (Rozier et al., 2016). It is therefore hypothesized that PGPR increases the flux of metabolites from the interior to the exterior of roots to favor the PGPR growth. Shikimic acid (lignin precursor) and quinic acid content also decrease in xylem sap after *Azospirillum* inoculation. This may be due to the potential sequestration of phenylproponoids in roots and the modification of the lignin content in maize (Chamam et al., 2013; El Zemrany et al., 2006).

A. brasilense Sp7 was recently shown to interact differently with tomato and maize, used as models of C_3- and C_4-host plants, respectively (Lade et al., 2018). In tomato, the *Azospirillum* strain induces higher stress responses with the production of reactive oxygen species (ROS) and higher damages of photosynthetic proteins than in maize, where the photosynthetic apparatus appears much more protected. This leads to a better interaction of the PGPR and plant growth with inoculated maize compared to inoculated-tomato.

5.2 Impact of bioproducts on the secondary metabolite profiles of plants

The inoculation of cereals (maize, rice, wheat) by *Azospirillum lipoferum* and *A. brasilense* strains leads to variable modifications of the secondary metabolism of the plant partner, depending on bacterial strain-plant species or plant cultivar combinations, suggesting strain-dependent responses (Walker et al., 2011).

In rice, secondary metabolite profiling revealed that *Azospirillum* inoculation significantly affected hydroxycinnamic acid (HCA), hydroxycinnamic acid amides (HCAA), resorcinol derivatives and flavonoids content in a strain-dependent manner (Chamam et al., 2013; Valette, Rey, Gerin, Comte, & Wisniewski-Dyé, 2020). In maize, the metabolites, whose contents are modified in response to the inoculation of *Azospirillum*, are mainly benzoxazoninones and benzoxazolinones (Walker et al., 2011). To date, these compounds have been identified in many Poaceae such as corn, wheat and rye. It is important to note that major secondary metabolism changes could arise without morphological effects on root architecture, indicating that PGPR can affect plant secondary metabolism without influencing the physiological functions linked to plant growth (Chamam et al., 2013; Rozier et al., 2016; Walker et al., 2011).

To analyze better the specificity of PGPR-plant interaction, the effects of two strains of *Azospirillum* sp. B510 and *A. lipoferum* 4B on the secondary metabolism of two rice cultivars were compared (Chamam et al., 2013). Modulation of the metabolic profiles of the two cultivars in the presence of the strain B510, and only one cultivar in the presence of the strain 4B. In this case, the cultivar corresponds to the cultivar from which the 4B strain was originally isolated. Moreover, the strongest plant phytostimulatory responses were observed when each strain was inoculated onto its host cultivar (Chamam et al., 2013). Furthermore, the positive plant growth responses are combined to the reduced expression of genes involved in stress responses and plant defense (Drogue et al., 2014). This result suggests that co-evolution between a host plant and a PGPR may result in attenuation of plant defense reactions.

The *Azospirillum* sp. B510 is able to colonize the interior of the tissues of the two rice cultivars thanks to its ability to produce a set of pectinolytic and cellulolytic enzymes promoting the degradation of plant walls and its entry into plant tissues (Boyer, Bally, Perrotto, Chaintreuil, & Wisniewski-Dyé, 2008). Although it is a plant-beneficial bacteria, during the first stages of the interaction, the plant response to this PGPR has strong similarities to that induced by the pathogen *Burkholderia glumae* (Chamam, Wisniewski-Dyé, Comte, Bertrand, & Prigent-Combaret, 2015). Among the discriminant compounds, *p*-coumaric acid, feruloyl-quinic acid and feruloyl hexoses have been identified in rice (mainly the cultivar Cigalon) as being induced in the presence of the pathogen and the endophytic PGPR B510. *C*-glycosylated flavones are other discriminating phenolic compounds induced in the presence of the pathogen and the endophytic PGPR. All these compounds have been described as having antioxidant or antimicrobial activities (Brazier-Hicks et al., 2009; Qiu, Liu, & Beta, 2009), or as being intermediates in the biosynthesis of lignin for feruloyl-quinic acid and feruloyl hexoses (Vanholme, De Meester, Ralph, & Boerjan, 2019).

In a wider study where the Nipponbare cultivar was inoculated with several *Azospirillum* strains initially isolated from rice or other plants and other PGPR not belonging to the *Azospirillum* genus, like *Paraburkholderia phytofirmans* and *Herbaspirillum seropedicae*, it was shown that some common plant metabolic signature could be however observed between all the studied strains (Valette et al., 2020). Three compounds that exhibit the greatest accumulation in response to all PGPR have been identified as HCA derivatives: feruloyl quinic acid, *N*-*p*-coumaroylputrescine and *N*-feruloylputrescine, the two latter being HCA amides (HCAA). Contrarily, when the rice was inoculated with a pathogen *B. glumae* the N-feruloylputrescine content appeared to be 11-fold reduced compared to the control condition. However, the main role of HCAA reported in the literature concerns the protection of the plant against abiotic and biotic stresses, some HCAA being associated with antimicrobial properties (Bassard, Ullmann, Bernier, & Werck-Reichhart, 2010). Six compounds exhibited a lower accumulation in response to inoculation by all the PGPR. Among them, three have been identified as resorcinol derivatives (Valette et al., 2020).

Among biocontrol mechanisms controlling crop diseases, using induced resistance is a real challenge for pesticide-free agriculture development. Biological inducer (i.e., PGPR, fungi or plant and algal extracts) are widely reported to induce resistance against a broad spectrum of pathogens in many plant species. An important consideration when using induced resistance is the crop plant itself. Different varieties of a crop plant can exhibit marked differences in the expression of induced resistance. It is possible that induced resistance might not be appropriate for use in some crops, especially if research shows the crop to be poor at responding to elicitor (Walters, Ratsep, & Havis, 2013). It is necessary to evaluate the efficiency of the elicitor according to plant species and stress conditions. In this context, elucidation of the plant metabolic responses that occur after elicitation would help develop strategies based on natural defense stimulation. Study of elicitation within NMR technique applied on leaves of grape wine was found to be relevant as it covers a wide range of the plant metabolome (Burdziej et al., 2019).

LCMS profiling would be also suitable for elicitation characterization of a wide range of plant, white radish (Rakpenthai, Khaksar, Burow, Olsen, & Sirikantaramas, 2019), *Medicago truncatula* (Broeckling et al., 2005), wheat (Gawlik-Dziki et al., 2016). Plant metabolomics would be an appropriate tool for evaluating elicitation efficiency depending on plant species, variety, and environmental stress.

5.3 Plant and bacterial metabolomics to help the development of bioinoculants and multi-species consortia

Bioinoculant products have been for a long time composed of only one microorganism. Several studies had stressed the positive impact of multi-species consortia on plant growth, plant health and plant physiology (Besset-Manzoni et al., 2018). By combining microorganisms with different plant-beneficial functions one can expect better effects on the plant due to synergistic effects of the microorganisms rather than when one plant-beneficial microorganism is inoculated.

The impact of multi-species bioinoculants on plant growth was studied as early as 1995 with *Azospirillum* and *Arthrobacter* strains used as nitrogen fixers, and *Agrobacterium* as a phosphate-solubilizing bacterium (Belimov, Kojemiakov, & Chuvarliyeva, 1995). Indeed, inoculation of barley with the consortium allowed increased nitrogen and phosphate plant uptakes, compared to single inoculations. In other studies, the inoculation of mycorrhizal fungi with bacteria was shown to promote plant growth (Kim et al., 2010; Wu, Cao, Li, Cheung, & Wong, 2005). Consortia were also shown to better protect plants against abiotic stress or pathogens than individual strains (Sarma, Yadav, Singh, & Singh, 2015; Vurukonda, Vardharajula, & Shrivastava, 2016).

The molecular mechanisms of interactions between consortium's microorganisms leading to additive or synergistic effects on the plants are often not investigated. The interaction of antagonistic biocontrol DAPG-producing *Pseudomonas* and biostimulant *Azospirillum brasilense* deserves particular attention because of the growth-inhibiting activity of *Pseudomonas* antimicrobial metabolites, such as 2,4-diacetylphloroglucinol against *Azospirillum* (Couillerot et al., 2011). However, synergistic effects might also exist since DAPG has been shown to increase, at low molecular concentrations the expression of genes encoding plant-beneficial -functions in *A. brasilense,* like the *ppdC* gene required for the biosynthesis of the auxin acid indole-3-acetic acid (Combes-Meynet, Pothier, Moënne-Loccoz, & Prigent-Combaret, 2011). In addition, the two strains were shown to grow better when they are co-cultivated in mixed biofilms (Pagnussat et al., 2016). Biofilms correspond to bacterial populations adhering to each other and/or to surfaces and enclosed in an extracellular polysaccharide matrix, and where the exchange of metabolites between bacteria is particularly favored. Thanks to an untargeted bacterial metabolomic analysis, Rieusset and collaborators evidenced that biocontrol *Pseudomonas* strains produced a higher and more diversified content of secondary metabolites in biofilms than in planktonic lifestyle (Rieusset et al., 2020). The increased production of secondary metabolites in biofilms will influence the biotic interactions of fluorescent *Pseudomonas* with their host plant and with other members of the rhizo-microbiota. Promoting biofilm formation when developing bioinoculants might thus help to enhance their biocontrol or biostimulant activities. Similar observations have also been made with plant-associated *Bacillus* strains that overproduce antimicrobial compounds in biofilms with the up-regulation at a transcriptional level of *fenA* involved in the synthesis of the antifungal cyclic lipopeptide, fengycin; *bmyA*, involved in the synthesis of the antifungal compound, bacillomycin D; and *dhbA*, involved in the synthesis of the siderophore, bacillibactin (Pandin et al., 2019). Furthermore, the production of *Bacillus* bioactive secondary metabolites and developmental processes like biofilm formation and sporulation are modified upon biotic interactions of bacilli PGPR with rhizosphere-inhabiting competitors or fungal pathogens (Andrić, Meyer, & Ongena, 2020).

Currently, there is great enthusiasm for the design and study of synthetic communities (SynCom) of rhizosphere microorganisms. Plant-inoculated SynCom should be comprised of multi-species populations possessing complementary modes of actions on plants (Liu, Qin, & Bai, 2019). Knowledge of SynCom's microbial functioning can help to define the best composition of multi-species bioinoculants for improving the growth and protection of crops. It could lead to developing new sustainable agricultural practices. Indeed, as plant species and plant genotypes are selecting specific microbial rhizo-microbiota thanks to root exudates (Bouffaud, Poirier, Muller, & Moënne-Loccoz, 2014; Guyonnet et al., 2018; Haichar et al., 2008; Pérez-Jaramillo, Mendes, & Raaijmakers, 2016), we can expect that the use of multi-species consortia or SynCom can improve the rhizosphere establishment and expression of plant-beneficial activities of at least a minimum of partners of the bio-inoculant in any environmental conditions and crop systems.

Plant metabolomics can help to better understand the mechanisms of action of PGPR consortia on plant and to identify those that best interact with plants. For instance, it was recently shown that a consortium of PGPR can alleviate drought stress and decrease damages to chlorophyll and photosystem (Khan et al., 2019). The consortium has the capacity to enhance the levels of soluble sugars and sugar derivatives that can act as osmoprotectants, and to reduce the lipid peroxidation and production of reactive oxygen species, by increasing the content of some amino acids to enhance the biosynthesis of proteins, vitamins like riboflavin and nicotinamide, and compounds like citrulline, glycerol, pipecolate, or *N*-acetylputrescine, etc. In other studies, consortium ability to modify the content of some metabolic plant biomarkers have been looked for. Consortia comprising of one *Pseudomonas* strain, a mycorrhizal inoculant mix composed of three *Glomus* strains, and one strain of *Azospirillum* (i.e., three different consortia containing three different *Azospirillum* strains were constituted) were inoculated on maize seeds. The effects on maize were measured notably using secondary metabolic profiling and biomass measurements. All the three-component consortia have similar phytostimulation effects, despite different secondary metabolic responses (Couillerot et al., 2013; Walker et al., 2012). Maize secondary metabolite profiles varied between treatments, and especially when plants were inoculated by single PGPR and by consortia, pointing that consortium effects could not be predicted based on knowledge of single inoculation effects. In addition, maize elaborated specific metabolic patterns according to the *Azospirillum* strains present within the three-component consortia (Couillerot et al., 2013). Discriminant compounds involved in the separation of secondary metabolite profiles from consortium-inoculated maize plants are belonging to the family of benzoxazinoids and cinnamic acids that are involved in biotic interactions and key plant biomarkers (Zhou, Richter, & Jander, 2018).

6. Conclusion

Metabolomics is a useful tool for the characterization of volatile organic compounds (VOCs) involved in plant-insect interactions, or the bioguided identification of active compounds from new raw materials. On the other hand, metabolomics approach can be used for the characterization of metabolic pathways affected by plant priming, biostimulant or elicitor application and/or for mechanistic study of Plant/PGPR interaction through biomarker specific metabolic fingerprint study.

References

Ahmed, N., Darshanee, H. L. C., Fu, W. Y., Hu, X.-S., Fan, Y., & Liu, T. X. (2018). Resistance of seven cabbage cultivars to green peach aphid (Hemiptera: Aphididae). *Journal of Economic Entomology, 111*, 909–916.

Ali, M., Sugimoto, K., Ramadan, A., & Arimura, G. (2013). Memory of plant communications for priming anti-herbivore responses. *Science Reports, 3*, 1872.

Andreu, V., Levert, A., Amiot, A., Cousin, A., Aveline, N., & Bertrand, C. (2018). Chemical composition and antifungal activity of plant extracts traditionally used in organic and biodynamic farming. *Environmental Science and Pollution Research, 25*, 29971–29982.

Andrić, S., Meyer, T., & Ongena, M. (2020). *Bacillus* responses to plant-associated fungal and bacterial communities. *Frontiers in Microbiology, 11*, 1350. https://doi.org/10.3389/fmicb.2020.01350.

Arimura, G., & Pearse, I. S. (2017). From the lab bench to the Forest: Ecology and Defence mechanisms of volatile-mediated talking trees. *Advances in Botanical Research, 82*, 3–17.

Avellaneda, J., Díaz, M., Coy-Barrera, E., Rodríguez, D., & Osorio, C. (2019). Rose volatile compounds allow the design of new control strategies for the western flower thrips (Frankliniella occidentalis). *Journal of Pest Science*. https://doi.org/10.1007/s10340-019-01131-7.

Baldwin, I. T. (2010). Plant volatiles. *Current Bioloiy, 20*, R392–R397.

Banani, H., Olivieri, L., Santoro, K., Garibaldi, A., Gullino, M. L., & Spadaro, D. (2018). Thyme and savory essential oil efficacy and induction of resistance against Botrytis cinerea through priming of defense responses in apple. *Food, 2018*(7), 11.

Bassard, J. E., Ullmann, P., Bernier, F., & Werck-Reichhart, D. (2010). Phenolamides: Bridging polyamines to the phenolic metabolism. *Phytochemistry, 71*, 1808–1824.

Belimov, A. A., Kojemiakov, A. P., & Chuvarliyeva, C. V. (1995). Interaction between barley and mixed cultures of nitrogen fixing and phosphate-solubilizing bacteria. *Plant and Soil, 173*, 29–37.

Beran, F., Köllner, T. G., Gershenzon, J., & Tholl, D. (2019). Chemical convergence between plants and insects: Biosynthetic origins and functions of common secondary metabolites. *New Phytologist, 223*, 52–67.

Bertrand, C., & Prigent-Combaret, C. (2012). Alternatives to pesticides. *Biofutur, 330*, 39–40.

Besset-Manzoni, Y., Rieusset, L., Joly, P., Comte, G., & Prigent-Combaret, C. (2018). Exploiting rhizosphere microbial cooperation for developing sustainable agriculture strategies. *Environmental Science and Pollution Research, 25*, 1–18.

Bicchi, C., & Maei, M. (2012). The plant Volatilome: Methods of analysis. In J. Normanly (Ed.), *Vol. 918. High-throughput phenotyping in plants: Methods and protocols, methods in molecular biology* (pp. 289–310). Totowa, NJ, USA: Humana Press.

Boiteux, J., Monardez, C., Fernández, M.Á., Espino, M., Pizzuolo, P., & Silva, M. F. (2018). *Larrea divaricata* volatilome and antimicrobial activity against *Monilinia fructicola*. *Microchemical Journal, 142*, 1–8.

Bouffaud, M. L., Poirier, M. A., Muller, D., & Moënne-Loccoz, Y. (2014). Root microbiome relates to plant host evolution in maize and other Poaceae. *Environmental Microbiology, 16*, 2804–2814.

Boyer, M., Bally, R., Perrotto, S., Chaintreuil, C., & Wisniewski-Dyé, F. (2008). A quorum-quenching approach to identify quorum-sensing-regulated functions in *Azospirillum lipoferum*. *Research in Microbiology, 159*, 699–708.

Brazier-Hicks, M., Evans, K. M., Gershater, M. C., Puschmann, H., Steel, P. G., & Edwards, R. (2009). The C-glycosylation of flavonoids in cereals. *Journal of Biological Chemistry, 284*, 17926–17934.

Broeckling, C. D., Huhman, D. V., Farag, M. A., Smith, J. T., May, G. D., Mendes, P., et al. (2005). Metabolic profiling of *Medicago truncatula* cell cultures reveals the effects of biotic and abiotic elicitors on metabolism. *Journal of Experimental Botany*, *56*, 323–336. https://doi.org/10.1093/jxb/eri058.

Brown, P., & Saa, S. (2015). Biostimulants in agriculture. *Frontiers in Plant Science*, *6*, 671.

Burdziej, A., Da Costa, G., Gougeon, L., Le Mao, I., Bellée, A., Corio-Costet, et al. (2019). Impact of different elicitors on grapevine leaf metabolism monitored by 1H NMR spectroscopy. *Metabolomics*, *15*, 67. https://doi.org/10.1007/s11306-019-1530-5.

Cassán, F., Coniglio, A., López, G., Molina, R., Nievas, S., Le Noir de Carlan, C., et al. (2020). Everything you must know about *Azospirillum* and its impact on agriculture and beyond. *Biology and Fertility of Soils*, *56*, 461–479.

Chamam, A., Sanguin, H., Bellvert, F., Meiffren, G., Comte, G., Wisniewski-Dyé, F., et al. (2013). Plant secondary metabolite profiling evidences strain-dependent effect in the *Azospirillum-Oryza sativa* association. *Phytochemistry*, *87*, 65–77.

Chamam, A., Wisniewski-Dyé, F., Comte, G., Bertrand, C., & Prigent-Combaret, C. (2015). Differential responses of *Oryza sativa* secondary metabolism to biotic interactions with cooperative, commensal and phytopathogenic bacteria. *Planta*, *242*(6), 1439–1452.

Combes-Meynet, E., Pothier, J. F., Moënne-Loccoz, Y., & Prigent-Combaret, C. (2011). The *Pseudomonas* secondary metabolite 2,4-diacetylphloroglucinol is a signal inducing rhizoplane expression of *Azospirillum* genes involved in plant-growth promotion. *Molecular Plant Microbe Interaction*, *24*, 271–284. https://doi.org/10.1094/MPMI-07-10-0148.

Conchou, L., Lucas, P., Meslin, C., Proffit, M., Staudt, M., & Renou, M. (2019). Insect Odorscapes: From plant volatiles to natural olfactory scenes. *Frontiers in Physiology*, *10*, 972. https://doi.org/10.3389/fphys.2019.00972.

Couillerot, O., Combes-Meynet, E., Pothier, J. F., Bellvert, F., Challita, E., Poirier, M. A., et al. (2011). The role of the antimicrobial compound 2,4-diacetylphloroglucinol in the impact of biocontrol *Pseudomonas fluorescens* F113 on *Azospirillum brasilense* phytostimulators. *Microbiology*, *157*, 1694–1705. https://doi.org/10.1099/mic.0.043943-0.

Couillerot, O., Ramírez-Trujillo, A., Walker, V., von Felten, A., Jansa, J., Maurhofer, M., et al. (2013). Comparison of prominent *Azospirillum* strains in *Azospirillum-Pseudomonas-Glomus* consortia for promotion of maize growth. *Applied Microbiology and Biotechnology*, *97*, 4639–4649. https://doi.org/10.1007/s00253-012-4249-z.

Delory, B. M., Delaplace, P., Fauconnier, M. L., & du Jardin, P. (2016). Root-emitted volatile organic compounds: Can they mediate belowground plant-plant interactions? *Plant and Soil*, *402*, 1–26.

Deniau, M., Bonafos, R., Chovelon, M., Parvaud, C., Furet, A., Bertrand, C., et al. (2019). Willow extract (Salix cortex), a basic substance of agronomical interests. *International Journal of Bio-resource and Stress Management*, *10*, 408–418.

Drogue, B., Doré, H., Borland, S., Wisniewski-Dyé, F., & Prigent-Combaret, C. (2012). Which specificity in cooperation between phytostimulating rhizobacteria and plants? *Research in Microbiology*, *163*, 500–510.

Drogue, B., Sanguin, H., Chamam, A., Mozar, M., Llauro, C., Panaud, O., et al. (2014). Plant root transcriptome profiling reveals a strain-dependent response during *Azospirillum*-rice cooperation. *Frontiers in Plant Science*, *5*, 607.

El Zemrany, H., Cortet, J., Lutz, M. P., Chabert, A., Baudoin, E., Haurat, J., et al. (2006). Field survival of the phytostimulator *Azospirillum lipoferum* CRT1 and functional impact on maize crop, biodegradation of crop residues, and soil faunal indicators in a context of decreasing nitrogen fertilization. *Soil Biology and Biochemistry*, *38*, 1712–1726.

Erb, M., Veyrat, N., Robert, C. A., Xu, H., Frey, M., Ton, J., et al. (2015). Indole is an essential herbivore-induced volatile priming signal in maize. *Nature Communications*, *6*, 6273.

European Biostimulants Industry Council. (2012). Towards an optimal regulatory framework for biostimulants. *EBIC position paper*, 1–7.

Füller, T. N., Bertrand, C., Antonello, B., Wesp, C., de Barros, I. B. I., & Neto, J. F. B. (2017). Chemical diversity of wild populations of *Elionurus muticus* (Spreng.) and the allelopathic effect of its essential oil. *Journal of Essential Oil Research*, *29*, 499–506. https://doi.org/10.1080/10412905.2017.1313786.

Füller, T. N., Bertrand, C., Simon, A., de Barros, I. B. I., & Neto, J. F. B. (2014). *Elionurus muticus* as an alternative source of Citral from Pampa biome, Brazil. *Journal of Oleo Science*, *63*, 1109–1116.

Gawlik-Dziki, U., Dziki, D., Nowak, R., Świeca, M., Olech, M., & Pietrzak, W. (2016). Influence of sprouting and elicitation on phenolic acids profile and antioxidant activity of wheat seedlings. *Journal of Cereal Science*, *70*, 221–228. https://doi.org/10.1016/j.jcs.2016.06.011.

Graña, E., Sotelo, T., Díaz-Tielas, C., Reigosa, M., & Sánchez-Moreiras, A. (2013). The phytotoxic potential of the terpenoid citral on seedlings and adult plants. *Weed Science*, *61*, 469–481. https://doi.org/10.1614/WS-D-12-00159.1.

Guyonnet, J. P., Guillemet, M., Dubost, A., Simon, L., Ortet, P., Barakat, M., et al. (2018). Plant nutrient resource use strategies shape active rhizosphere microbiota through root exudation. *Frontiers in Plant Science*, *9*, 1662.

Haichar, F. Z., Marol, C., Berge, O., Rangel-Castro, J. I., Prosser, J. I., Balesdent, J., et al. (2008). Plant host habitat and root exudates shape soil bacterial community structure. *ISME Journal*, *2*, 1221–1230.

Haichar, Z., Santaella, C., & Heulin, T. (2014). Root exudates mediated interactions belowground. *Soil Biology and Biochemistry*, *77*, 69–80.

Helletsburger, C., Dötterl, S., Ruprecht, U., & Junker, R. R. (2017). Epiphytic bacteria alter floral scent emission. *Journal of Chemical Ecololy*, *43*, 1073–1077.

Holopainen, J. K., & Gershenzon, J. (2010). Multiple stress factors and the emission of plant VOCs. *Trends in Plant Science*, *15*, 176–184. https://doi.org/10.1016/j.tplants.2010.01.006.

Ikbal, C., & Pavela, R. (2019). Essential oils as active ingredients of botanical insecticides against aphids. *Journal of Pest Science*, *92*, 971–986.

Jactel, H., Birgersson, G., Andersson, S., & Schlyter, F. (2011). Non-host volatiles mediate associational resistance to the pine processionary moth. *Oecologia*, *166*, 703–711.

Jardine, K. J., Fernandes de Souza, F. V., Oikawa, P., Higuchi, N., Bill, M., Porrasghb, R., et al. (2017). Integration of C1 and C2 metabolism in trees. *International Journal of Molecular Sciences*, *18*, E2045.

Jeschke, V., Gershenzon, J., & Vassão, D. G. (2016). Insect detoxification of glucosinolates and their hydrolysis products. In S. Kopriva (Ed.), *Vol. 80. Advances in botanical research* (pp. 199–245). Amsterdam: Elsevier Ltd.

Jourdan, E., Ongena, M., & Thonart, P. (2008). Molecular characteristics of the rhizobacteria-triggered plant immunity. *Biotechnologie Agronomie Société Et Environnement*, *12*, 437–449.

Kantsa, A., Sotiropoulou, S., Vaitis, M., & Petanidou, T. (2015). Plant Volatilome in Greece: A review on the properties, prospects, and Chemogeography. *Chemistry & Biodiversity*, *12*, 1466–1480.

Karban, R., Shiojiri, K., & Ishizaki, S. (2010). An air transfer experiment confirms the role of volatile cues in communication between plants. *The American Naturalist*, *176*, 381e384.

Khan, N., Bano, A., Rahman, M. A., Guo, J., Kang, Z., & Babar, M. A. (2019). Comparative physiological and metabolic analysis reveals a complex mechanism involved in drought tolerance in chickpea (*Cicer arietinum* L.) induced by PGPR and PGRs. *Scientific Reports*, *9*, 2097. https://doi.org/10.1038/s41598-019-38702-8.

Kim, K., Yim, W., Trivedi, P., Madhaiyan, M., Boruah, H. P. D., Islam, M. R., et al. (2010). Synergistic effects of inoculating arbuscular mycorrhizal fungi and *Methylobacterium oryzae* strains on growth and nutrient uptake of red pepper (*Capsicum annuum* L.). *Plant and Soil, 327*, 429–440. https://doi.org/10.1007/s11104-009-0072-4.

Kumar, S. (2017). Plant secondary metabolites (PSMs) of Brassicaceae and their role in plant defense against insect herbivores – A review. *Journal of Applied and Natural Science, 9*, 508–519.

Kuzmicheva, Y. V., Shaposhnikov, A. I., Petrova, S. N., Makarova, N. M., Tychinskaya, I. L., Puhalsky, J. V., et al. (2017). Variety specific relationships between effects of rhizobacteria on root exudation, growth and nutrient uptake of soybean. *Plant and Soil, 419*, 83–96.

Lade, S. B., Román, C., Cueto-Ginzo, A. I., Serrano, L., Sin, E., Achón, M. A., et al. (2018). Host-specific proteomic and growth analysis of maize and tomato seedlings inoculated with *Azospirillum brasilense* Sp7. *Plant Physiology and Biochemistry, 129*, 381–393.

Li, J., Hu, H., Mao, J., Yu, L., Stoopen, G., Wang, M., et al. (2019). Defense of pyrethrum flowers: Repelling herbivores and recruiting carnivores by producing aphid alarm pheromone. *New Phytologist, 223*, 1607–1620.

Liu, Y. X., Qin, Y., & Bai, Y. (2019). Reductionist synthetic community approaches in root microbiome research. *Current Opinion in Microbiology, 49*, 97–102.

Maggini, V., Bandeira Reidel, R. V., De Leo, M., Mengoni, A., Gallo, E. R., Miceli, E., et al. (2019). Volatile profile of Echinacea purpurea plants after in vitro endophyte infection. *Natural Product Research*. https://doi.org/10.1080/14786419.2019.1579810.

Matilla, M. A., Ramos, J. L., Bakker, P. A. H. M., Doornbos, R., Badri, D. V., Vivanco, J., et al. (2010). *Pseudomonas putida* KT2440 causes induced systemic resistance and changes in *Arabidopsis* root exudation. *Environmental Microbiology Report, 2*, 381–388.

Mishra, R. P. N., Singh, R. K., Jaiswal, H. K., Kumar, V., & Maurya, S. (2006). *Rhizobium*-mediated induction of phenolics and plant growth promotion in rice (*Oryza sativa* L.). *Current Microbiology, 52*(5), 383–389.

Moco, S., Vervoort, J., Moco, S., Bino, R. J., De Vos, R. C. H., & Bino, R. (2007). Metabolomics technologies and metabolite identification. *Trends in Analytical Chemistry, 26*, 855–866.

Morant, A. V., Jørgensen, K., Jørgensen, C., Paquette, S. M., Sánchez-Pérez, R., & Møller, B. L. (2008). β-Glucosidases as detonators of plant chemical defense. *Phytochemistry, 69*, 1795–1813.

Moreira, X., Nell, C. S., Katsanis, A., Rasmann, S., & Mooney, K. A. (2018). Herbivore specificity and the chemical basis of plant—Plant communication in Baccharis salicifolia (Asteraceae). *New Phytologist, 220*, 703–713.

Okada, K., Abe, H., & Arimura, G. (2015). Jasmonates induce both defense responses and communication in monocotyledonous and dicotyledonous plants. *Plant & Cell Physiology, 56*, 16e27.

Osbourn, A. E., Qi, X., Townsend, B., & Qin, B. (2003). Dissecting plant secondary metabolism—Constitutive chemical defences in cereals. *New Phytologist, 159*(1), 101–108.

Pagnussat, L. A., Salcedo, F., Maroniche, G., Keel, C., Valverde, C., & Creus, C. M. (2016). Interspecific cooperation: Enhanced growth, attachment and strain-specific distribution in biofilms through *Azospirillum brasilense-Pseudomonas protegens* co-cultivation. *FEMS Microbiology Letters, 363*, fnw238. https://doi.org/10.1093/femsle/fnw238.

Pandin, C., Darsonval, M., Mayeur, C., Le Coq, D., Aymerich, S., & Briandet, R. (2019). Biofilm formation and synthesis of antimicrobial compounds by the biocontrol agent *Bacillus velezensis* QST713 in an *Agaricus bisporus* compost micromodel. *Applied in Environmental Microbiology, 85*. https://doi.org/10.1128/AEM.00327-19. e00327–19.

Papanastasiou, S. A., Ioannou, C. S., & Papadopoulos, N. T. (2020). Oviposition-deterrent effect of linalool—A compound of citrus essential oils—On female Mediterranean fruit flies, Ceratitis capitata (Diptera: Tephritidae). *Pest Management Science*. https://doi.org/10.1002/ps.5858.

Paudel Timilsena, B., Seidl-Adams, I., & Tumlinson, J. H. (2020). Herbivore-specific plant volatiles prime neighboring plants for nonspecific defense responses. *Plant, Cell & Environment, 43*, 787–800.

Pavela, R., & Benelli, G. (2016). Essential oils as ecofriendly biopesticides? Challenges and constraints. *Trends in Plant Science, 21*, 1000–1007.

Pearse, I. S., Hughes, K., Shiojiri, K., Ishizaki, S., & Karban, R. (2013). Interplant volatile signaling in willows: Revisiting the original talking trees. *Oecologia, 172*, 869e875.

Pérez-Jaramillo, J. E., Mendes, R., & Raaijmakers, J. M. (2016). Impact of plant domestication on rhizosphere microbiome assembly and functions. *Plant Molecular Biology, 90*, 635–644.

Popovici, J., Bertrand, C., Comte, G., (2009). Use of a plant myrica scale for the production of an agent weedkiller. Patent. WO2010012945A2.

Popovici, J., Bertrand, C., Jacquemoud, D., Bellvert, F., Fernandez, M. P., Comte, G., et al. (2011). An allelochemical from *Myrica gale* with strong phytotoxic activity against highly invasive *Fallopia x bohemica* taxa. *Molecules, 16*, 2323–2333.

Qiu, Y., Liu, Q., & Beta, T. (2009). Antioxidant activity of commercial wild rice and identification of flavonoid compounds in active fractions. *Journal of Agricultural and Food Chemistry, 57*, 7543–7551.

Rakpenthai, A., Khaksar, G., Burow, M., Olsen, C. E., & Sirikantaramas, S. (2019). Metabolic changes and increased levels of bioactive compounds in white radish (Raphanus sativus L. cv. 01) sprouts elicited by Oligochitosan. *Agronomy, 9*, 467.

Ramamoorthy, V., Viswanathan, R., Raguchander, T., Prakasam, V., & Samiyappan, R. (2001). Induction of systemic resistance by plant growth promoting rhizobacteria in crop plants against pests and diseases. *Crop Protection, 20*, 1–11.

Rieusset, L., Rey, M., Muller, D., Vacheron, J., Gerin, F., Dubost, A., et al. (2020). Secondary metabolites from plant-associated *Pseudomonas* are overproduced in biofilm. *Microbial Biotechnology, 0*, 1–19. https://doi.org/10.1111/1751-7915.13598.

Rodriguez-Saona, C. R., Rodriguez-Saona, L. E., & Frost, C. J. (2009). Herbivore-induced volatiles in the perennial shrub, Vaccinium corymbosum, and their role in inter-branch signaling. *Journal of Chemical Ecology, 35*, 163–175.

Rosenberg, E., & Zilber-Rosenberg, I. (2016). Microbes drive evolution of animals and plants: The hologenome concept. *MBio, 7*, 1–8.

Rozier, C., Erban, A., Hamzaoui, J., Prigent-Combaret, C., Comte, G., Kopka, J., et al. (2016). Xylem sap metabolite profile changes during phytostimulation of maize by the plant growth-promoting rhizobacterium, *Azospirillum lipoferum* CRT1. *Metabolomics (Los Angel), 6*, 3.

Sarma, B. K., Yadav, S. K., Singh, S., & Singh, H. B. (2015). Microbial consortium-mediated plant defense against phytopathogens: Readdressing for enhancing efficacy. *Soil Biology and Biochemistry, 87*, 25–33.

Sharifi-Rad, J., Sureda, A., Tenore, G. C., Daglia, M., Sharifi-Rad, M., Valussi, M., et al. (2017). Biological activities of essential oils: From plant Chemoecology to traditional healing systems. *Molecules, 22*, 70. https://doi.org/10.3390/molecules22010070.

Sharma, A., Kumar, V., Shahzad, B., Tanveer, M., Sidhu, G. P. S., Handa, N., et al. (2019). Worldwide pesticide usage and its impacts on ecosystem. *SN Applied Sciences, 1*, 1446.

Staudt, M., Byron, J., Piquemal, K., & Williams, J. (2019). Compartment specific chiral pinene emissions identified in a maritime pine forest. *Science of the Total Environment, 654*, 1158–1166.

Takabayashi, J., & Shiojiri, K. (2019). Multifunctionality of herbivory-induced plant volatiles in chemical communication in tritrophic interactions. *Current Opinion in Insect Science, 32*, 110–117.

Vacheron, J., Desbrosses, G., Bouffaud, M. L., Touraine, B., Moënne-Loccoz, Y., Muller, D., et al. (2013). Plant growth-promoting rhizobacteria and root system functioning. *Frontiers in Plant Science, 4*, 356.

Valette, M., Rey, M., Gerin, F., Comte, G., & Wisniewski-Dyé, F. (2020). A common metabolomic signature is observed upon inoculation of rice roots with various rhizobacteria. *Journal of Integrative Plant Biology*, *62*(2), 228–246.

van Lenteren, J. C., Bolckmans, K., Köhl, J., Ravensberg, W. J., & Urbaneja, A. (2018). Biological control using invertebrates and microorganisms: Plenty of new opportunities. *BioControl*, *63*, 39–59.

Vanholme, R., De Meester, B., Ralph, J., & Boerjan, W. (2019). Lignin biosynthesis and its integration into metabolism. *Current Opinion in Biotechnology*, *56*, 230–239.

Venturi, V., & Keel, C. (2016). Signaling in the rhizosphere. *Trends in Plant Science*, *21*, 187–198.

Vivaldo, G., Masi, E., Taiti, C., Caldarelli, G., & Mancuso, S. (2017). The network of plant volatile organic compounds. *Scientific Reports*, 7, 11050.

Vurukonda, S. S. K., Vardharajula, S., Shrivastava, M., & SkZ, A. (2016). Enhancement of drought stress tolerance in crops by plant growth promoting rhizobacteria. *Microbiological Research*, *184*(supplement C), 13–24.

Walker, V., Bertrand, C., Bellvert, F., Moënne-Loccoz, Y., Bally, R., & Comte, G. (2011). Host plant secondary metabolite profiling shows a complex, strain-dependent response of maize to plant growth-promoting rhizobacteria of the genus *Azospirillum*. *New Phytologist*, *189*(2), 494–506.

Walker, V., Couillerot, O., Von Felten, A., Bellvert, F., Jansa, J., Maurhofer, M., et al. (2012). Variation of secondary metabolite levels in maize seedling roots induced by inoculation with *Azospirillum*, *Pseudomonas* and *Glomus* consortium under field conditions. *Plant and Soil*, *356*, 151–163.

Walters, D. R., Ratsep, J., & Havis, N. D. (2013). Controlling crop diseases using induced resistance: Challenges for the future. *Journal of Experimental Botany*, *64*, 1263–1280. https://doi.org/10.1093/jxb/ert026.

Wong, Y. F., Yan, D. D., Shellie, R. A., Sciarrone, D., & Marriott, P. J. (2019). Rapid plant volatiles screening using headspace SPME and person-portable gas chromatography–mass spectrometry. *Chromatographia*, *82*, 297–305.

Wu, S. C., Cao, Z. H., Li, Z. G., Cheung, K. C., & Wong, M. H. (2005). Effects of biofertilizer containing N-fixer, P and K solubilizers and AM fungi on maize growth: A greenhouse trial. *Geoderma*, *125*, 155–166.

Zagrobelny, M., de Castro, E. C. P., Møller, B. L., & Bak, S. (2018). Cyanogenesis in arthropods: From chemical warfare to nuptial gifts. *Insects*, *9*, 51.

Zhang, P. J., Wei, J. N., Zhao, C., Zhang, Y. F., Li, C. Y., Liu, S. S., et al. (2019). Airborne host–plant manipulation by whiteflies via an inducible blend of plant volatiles. *Proceedings of the National Academy of Sciences of United States of America*, *116*, 7387–7396.

Zhou, S., Richter, A., & Jander, G. (2018). Beyond defense: Multiple functions of benzoxazinoids in maize metabolism. *Plant and Cell Physiology*, *59*, 1528–1533.

CHAPTER FIVE

Metabolomics in plant-microbe interactions in the roots

Li Chen, Melina Schwier, Jenna Krumbach, Stanislav Kopriva*, and Richard P. Jacoby

Institute for Plant Sciences, Cluster of Excellence on Plant Sciences (CEPLAS), University of Cologne, Cologne, Germany

*Corresponding author: e-mail address: skopriva@uni-koeln.de

Contents

Abstract

In the last decade the importance of microbiota for plant performance has been overwhelmingly demonstrated. Plant microbiomes, both in roots and shoots, fulfill a plethora of functions that can strongly influence various plant traits. Metabolites are the main tools that plants use to actively shape their microbiome. Mechanistically, plants exude a complex mix of primary and secondary metabolites from roots, which serve as growth substrates for certain microbial strains, exert toxic effects on others, or act as signals in mediating the plant microbe interactions. Flavonoids and strigolactones have long been known to play important roles in enabling microbial symbiosis, even though their full complexity may not yet be fully elucidated. Recently, several other plant metabolites have also been implicated as important players mediating the interaction with the root microbiome, including molecules such as coumarins, camalexin and triterpenes. Because plants use such a wide range of chemical classes to shape their microbiome, mechanistic studies of plant-microbe interactions are increasingly applying high-throughput metabolomics techniques. In this review, we describe how metabolomics approaches have profiled the chemical complexity of plant rhizodeposits, and discuss how metabolomics can be utilized to functionally characterize specific metabolites that influence plant-microbe interactions. We will also discuss the advanced metabolite

Advances in Botanical Research, Volume 98
ISSN 0065-2296
https://doi.org/10.1016/bs.abr.2020.09.018

analyses that are needed to disentangle the complex metabolic interactions between plant hosts and their associated microbial communities.

1. Introduction

Plants in their natural environments live surrounded by a vast number and variety of other organisms. Some of them use the plants as their source of food, some compete with plants for resources, and some help plants to acquire nutrients or to overcome the competition. Microorganisms, i.e., bacteria, fungi, and protists form the largest and most diverse group of plant neighbors. Often these microorganisms and their roles in the plant ecosystems are characterized as "the good, the bad, and the ugly" to represent beneficial microbiota, plant pathogens, and the soil-borne human pathogens (Mendes, Garbeva, & Raaijmakers, 2013). Indeed, some microbial species in rhizosphere soil are beneficial to plants, functioning as plant growth promoting rhizobacteria (PGPR) or symbionts to improve plant health and nutrition. Other microbes display either commensalism, by maintaining a neutral relationship with host plants or pathogenicity, resulting in infection symptoms, such as plant tissue degradation, rusts, or powdery mildews (Bulgarelli, Schlaeppi, Spaepen, Ver Loren van Themaat, & Schulze-Lefert, 2013; van der Heijden, de Bruin, Luckerhoff, van Logtestijn, & Schlaeppi, 2016). It is thus of great importance to uncover mechanisms behind these complex interactions and as they might underpin agricultural approaches to improving food production and sustainability.

As plants are sessile, they have developed a number of mechanisms to interact with surrounding microorganisms. These include defense mechanisms against pathogens and synthesis of attractants for symbiotic organisms. Although genetically encoded, the plant-microbe communications rely on metabolites: their synthesis, transport, and metabolism. For successful interaction with soil microorganisms, plants excrete metabolites into the soil in the form of root exudates (Sasse, Martinoia, & Northen, 2018). Metabolite analysis in the multi-organism systems, however, is far from trivial and therefore the knowledge on metabolic interactions lags behind the understanding of the genetics of the interactions, particularly with respect to plant immunity. Nevertheless, the last decade brought a significant progress in awareness of the role metabolites play in plant microbe interactions and in adapting metabolomics approaches to further dissect these interactions (Fig. 1). Here we describe some of the challenges of studying metabolic interactions

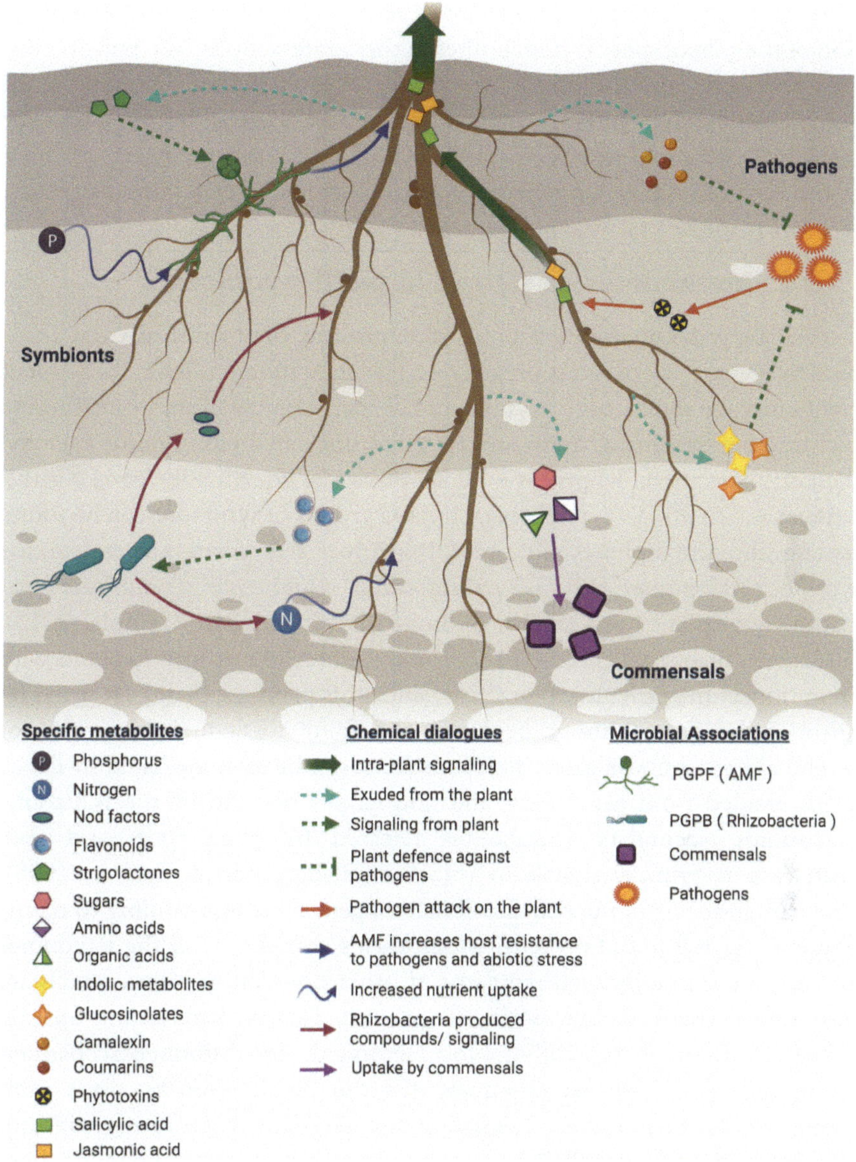

Fig. 1 Metabolic dialogues between plants and microbes. Plant-microbe interactions range from pathogenic to mutualistic, but a common theme is the important role played by metabolites. This figure highlights some of the key metabolic exchanges known to underpin plant-microbe interactions in the roots. In symbiotic interactions, plants exude a set of signaling molecules that recruit AMF and BNF symbionts. These microbes boost plant nutrition by mobilizing N and P. Pathogenic interactions between plants and microbes are characterized by a chemical warfare, whereby both parties either attack one another, manipulate or defend by releasing a set of bioactive secondary metabolites. In the commensal interactions, root exudation provides carbon substrates to fuel the proliferation of a diverse microbial community. Metabolomics provides a high-throughput platform to characterize these metabolic exchanges in greater depth. The figure was created with BioRender.com.

between plants and microbes and provide examples of successful identification of the role of specific metabolites in these interactions. We will also discuss the ways metabolomics approaches may help to disentangle the complex metabolic interactions between plant hosts and their associated microbial communities. We will focus this discussion to the interactions of the roots, as the area of the most significant increase in knowledge.

2. Metabolites as signals in plant symbiosis

Plant roots do not exist alone underground; their rhizosphere is occupied by a vast array of microorganisms. The rhizosphere is thus a hot spot of plant-microbe interactions as root exudates are the main source of carbon to fuel microorganisms' growth and to shape microbial community (Jacoby, Peukert, Succurro, Koprivova, & Kopriva, 2017; Mendes et al., 2013; Sasse et al., 2018). Two well characterized groups of symbiont can be found among the rhizosphere microbes, Rhizobia, that form nitrogen-fixing symbiosis in legumes, and mycorrhizal fungi, which enter a symbiotic association with plant roots (Smith & Smith, 2012; Udvardi & Poole, 2013). Biological nitrogen fixation is an important source of nitrogen to plants with great impact on agricultural practice, since this process potentially provides nitrogen fertilizer without depleting the finite fossil fuel resources nor producing greenhouse gases. This functional symbiosis is mediated by flavonoids exuded from plant roots and nod factors released by the Rhizobia. Flavonoids, secondary metabolites released by plant roots and seed coats, act as primary signals to rhizobia (Bladergroen & Spaink, 1998). They are inducers of the rhizobacterial *nod* genes through binding to membrane associated nodD protein, which forms complex with the flavonoids and act as a transcriptional regulator of other nodulation genes, which are responsible for rhizobia synthesis of nod factors (Schlaman, Spaink, Okker, & Lugtenberg, 1989). Nod factors are lipochitooligosaccharides, often sulfated, which act as signals that can be detected by plant roots (Long, 1996). They induce a signaling cascade that triggers changes in root hair structure and cell wall division. This aids to trap rhizobia and guide them inside the roots, resulting in formation of nodules, in which symbiosis and nitrogen fixation take place (Cullimore, Ranjeva, & Bono, 2001; Long, 1996; Oldroyd, Murray, Poole, & Downie, 2011). Nod factors are highly specific, ensuring a strict specificity in nodulation between a particular plant species and rhizobial strain (Oldroyd et al., 2011). However, flavonoids play a broader role in plant symbiosis with Rhizobia. They are essential for nodule initiation due to regulation of auxin transport (Wasson, Pellerone, & Mathesius, 2006). In addition, flavonoids and isoflavonoids are also important for the other type of nodulation, in actinorhizal symbiosis with *Frankia* (Abdel-Lateif et al., 2013; Auguy et al., 2011).

The main type of mycorrhiza of herbaceous plants is formed by arbuscular mycorrhizal fungi (AMF) of the phylum Glomeromycota, which colonize many economically important crops, including cereals, vegetables, and horticultural plants (Smith & Smith, 2012). Mycorrhiza improve crops yield, supplying nutrition and alleviating abiotic stress. Mycelium of AMF scavenges effectively for phosphorus (P), which increases P uptake of plants and then leads to positive host plant growth. AMF could even improve plant growth under P deficiency (Harrison, Dewbre, & Liu, 2002; Smith, Smith, & Jakobsen, 2003). AMF can also significantly improve host plant resistance to several abiotic stresses like salinity, drought, cold/heat, nutrient deficiency and metal toxic stress (Jung, Martinez-Medina, Lopez-Raez, & Pozo, 2012). The signaling leading to AMF symbiosis is similar to signaling in nodulation, a plant signal triggers fungal hyphae differentiation and synthesis of a "myc" factor that together with other signals initiates a symbiotic signaling cascade, that has many common elements with the signaling in nodulation (Schmitz & Harrison, 2014). The primary plant signals are the sesquiterpenes named strigolactones, phytohormones with function in control of shoot and root branching, inducers of germination of parasitic plants, such as *Striga* (Akiyama, Matsuzaki, & Hayashi, 2005; Waters, Gutjahr, Bennett, & Nelson, 2017). Strigolactones are exuded from plant roots particularly during phosphate limitation and are triggering extensive branching in the AMF (Schmitz & Harrison, 2014). Strigolactones are, however, not the only plant metabolic signals to enable the symbiosis, additionally, C16 aliphatic fatty acids associated with cutin are necessary for forming appressoria and penetration of plant cells (Wang et al., 2012) and flavonoids also promote spore germination (Tsai & Phillips, 1991). Also for the long elusive myc signal, two groups of metabolites have been found that exert the same effect on plant roots as mycorrhiza fungi, lipochitooligosaccharides and chitin oligomers (Genre et al., 2013; Maillet et al., 2011). While the myc-lipochitooligosaccharides have a similar structure as nod factors and are also often sulfated, the chitin oligomers are common with various fungal pathogens (Schmitz & Harrison, 2014). How these signals are distinguished is still an open question, as is the number and specificity of the various molecules in fungal exudates, due to a very low abundance. An intriguing recent development is the identification of hydroxy- and carboxyblumenol glucosides in leaves of mycorrhized plants, accumulation of which responds quantitatively to the level of colonization (Wang et al., 2018). The blumenol derivatives have been detected in number of plants species and were induced by diverse mycorrhiza fungi. They might prove to be a valuable tool to estimate rapidly and in high throughput the mycorrhiza colonization levels, e.g., in genetic screens (Wang et al., 2018).

3. Metabolites involved in plant pathogen interaction

A number of nematodes, fungi, oomycetes and bacteria are notorious soil-borne plant pathogens (Back, Haydock, & Jenkinson, 2002; Kamoun et al., 2015; Mansfield et al., 2012). Plants infected by soil-borne pathogens suffer from root rot, root blackening, wilt, stunting or seedling damping-off. For example, *Phellinus weirii*, a widespread indigenous fungal pathogen in the coniferous forests of western North America, can cause laminated root rot in several conifer species and even kill some vigorous conifer trees (Hansen & Goheen, 2000). *P. weirii* penetrates roots and utilizes nutrients in the phloem and cambium. So even though the canopy of the tree looks green, the roots are irreparably damaged. This fungus is considered as the most economically damaging pathogen in the conifer forest in the west of the Cascade Mountains (Hansen & Goheen, 2000). Another typical pathogen is the fungal pathogen of soybeans, cereals, tomatoes, and other crops, *Fusarium* spp., which colonizes roots and causes root rot symptoms and wilting of the plants (Arias, Leandro, & Munkvold, 2013; Gordon, 2017).

Chemical cues in the rhizosphere are important also for root pathogens to recognize and infect root tissues. For example, zoospores and hyphae of *Phytophthora sojae*, an oomycete pathogen infecting soybean root, recognize trace amount of isoflavones daidzein and genistein, chemicals secreted by soybean, and use them as chemotrophic signals to direct their growth (Morris, Bone, & Tyler, 1998). Similarly, *Aphanomyces cochlioides* causing root rot disease in spinach is attracted by several flavonoids from spinach root exudates, such as Cochliophilin A (Tahara, Ohkawa, Takayama, & Ogawa, 2001). Root exudates generally play an important role in attracting pathogens to the roots as well as in their deterring or in triggering microbiome functions that help plants to defend against the pathogens (Haney, Samuel, Bush, & Ausubel, 2015; Jacoby, Chen, Schwier, Koprivova, & Kopriva, 2020; Sasse et al., 2018) and see below).

Multiple metabolites are part of the plant immune response to protect plants against a variety of pests. Most of our knowledge on these metabolites comes from experiments with leaf pathogens. One of the best examples are the tryptophan derived defense compounds in Arabidopsis, indolic glucosinolates and camalexin. Glucosinolates are sulfur rich secondary compounds of the Caparales that form the basis of the "mustard oil bomb" (Bones & Rossiter, 1996). Upon tissue damage, glucosinolates stored in the vacuoles come into contact with the enzyme myrosinase which cleaves a glucose residue and the remaining instable intermediate rearranges into organic isothiocyanates, nitriles, or thiocyanates (Halkier & Gershenzon, 2006). These pungent volatiles are a deterrent to herbivores but act also in immunity against fungal and bacterial pathogens (Bednarek et al., 2009; Clay, Adio, Denoux, Jander, & Ausubel, 2009). Glucosinolates are synthesized from amino acids, the main contributors being methionine for aliphatic glucosinolates, phenylalanine for aromatic, and tryptophan for indolic ones (Halkier & Gershenzon, 2006). Tryptophan is a precursor for another defense compounds of the Brassicaceae, the phytoalexin camalexin. Camalexin was first isolated from a plant of the Brassicaceae family, camelina (*Camelina sativa*), but detected and studied primarily in Arabidopsis, where it is formed upon infection with fungal pathogens, some bacterial strains, and also abiotic factors (Glawischnig, 2007). Variation in induction of camalexin synthesis has been shown to largely control the resistance with necrotrophic pathogen *Botrytis cinerea* in leaves of Arabidopsis accessions (Rowe & Kliebenstein, 2008). Both compounds have only recently been found in infected roots and in root exudates with important roles in interaction with plant growth promoting microorganisms (Iven et al., 2012; Koprivova et al., 2019; Lahrmann et al., 2015). However, apart from a clear inhibition of the growth of root fungal pathogen *Verticillium longisporum* by camalexin, little is known about the extent of the importance of indolic phytoalexins in root pathogen defense. One challenge in understanding of the mechanisms of action of plant derived indolic compounds is the variety of secondary modifications these compounds undergo. A number of hydroxylation, methylation, and glycosylation reactions contribute to the large structural diversity of indolic glucosinolates and other indole compounds in plant roots (Fahey, Zalcmann, & Talalay, 2001; Stahl et al., 2016). Thus, one opportunity for high-throughput plant metabolomics involves the development of new analysis pipelines to quantify and exactly determine the structures of such compounds. Phytoalexins, however, are not limited to glucosinolates and indoles. For example, flavonoids were shown to play an important role in protection of number of plant species against *Fusarium* and other fungal pathogens (reviewed in Mierziak,

Kostyn, & Kulma, 2014). Accumulation of phenolic compounds was the major difference between hops cultivars resistant or susceptible to soilborne pathogen *Verticillium nonalfalfae* (Kunej, Mikulic-Petkovsek, Radisek, & Stajner, 2020). For more details on how metabolomics helps to dissect the metabolic networks in plant defense see Chapter "Untangling plant immune responses through metabolomics" by Petriacq.

An interesting aspect of plant immunity is priming, i.e., intensifying immune response by previous triggers. For example, β-aminobutyric acid (BABA), a priming agent in Arabidopsis, increases plants resistance to necrotrophic fungus *Plectosphaerella cucumerina* by stimulating callose-rich cell wall depositions (Ton & Mauch-Mani, 2004). BABA-induced resistance in Arabidopsis might be based on formation of stable hormone conjugates, which can be rapidly hydrolyzed into active signals to synthesize defensive compounds upon attack (Dietz, Sauter, Wichert, Messdaghi, & Hartung, 2000). Therefore, it is essential to identify such hormone conjugates, which function in plant immunity system. Salicylic acid, jasmonic acid, abscisic acid (ABA), and indole derivatives are good candidates for studying plant immunity and a targeted metabolomics workflow can help to identify which are responsible for mediating the priming response. Revealing accumulation of indole-3-carboxylic acid in BABA-primed Arabidopsis plants and its function in defense against *P. cucumerina* is a good example for further studies (Gamir, Pastor, Cerezo, & Flors, 2012).

An additional important task for metabolomic analysis in plant microbe interactions involves characterizing the full repertoire of bioactive compounds released by plant pathogens. These molecules can be non-protein effectors used to attack the plant (Pusztahelyi, Holb, & Pocsi, 2015), and furthermore some of these metabolites also exert toxicity upon humans (Adeniji, Babalola, & Loots, 2020). A chemically diverse variety of toxins are produced by pathogens, including aflatoxin, alternariol, toxoflavin, fumonisin, or coronatine, synthesized and released by pathogenic species of *Aspergillus*, *Alternaria*, *Burkholderia*, *Fusarium*, and *Pseudomonas*, respectively (Adeniji et al., 2020; Lee & Ryu, 2017). It is estimated that globally 25% of crops are contaminated with mycotoxins, with negative consequences for human health. Metabolomics, particularly LC-MS technologies are available to detect mycotoxins in plant material and food with a sensitivity of $0.03\,\mu g\,kg^{-1}$ (Eshelli, Qader, Jambi, Hursthouse, & Rateb, 2018). There is a clear opportunity for metabolomics to provide new insights into pathogen biology, but further development of the methodology is necessary in order to minimize matrix effects and reduce interference of other plant metabolites (Eshelli et al., 2018).

4. Metabolomics for identification of plant metabolites shaping root microbiome

As sessile organisms, plants have adapted to make use of their microbiome to support their development in terms of health and growth. The impact of the microbial assemblages ranges from detrimental effects of soil borne pathogens to mutualistic or beneficial interactions that ameliorate abiotic stress tolerance, host nutrient acquisition and resistance against pests, pathogens and diseases (Bulgarelli et al., 2013). A shift in balance of the microbiota can thus substantially affect these essential roles, having impact on crop production in agricultural ecosystems. This has been demonstrated in a recent study whereby both plant development and nitrogen fertilization fueled changes in the structure of root-associated microbiomes of wheat (Chen et al., 2019). Manipulating root microbiome has the potential to address the challenges of increased demand for food due to expanding human population, with the underlining issues of climate change, reduced agricultural productivity and cultivable land, and assist to optimize crop yield, agricultural sustainability and reduce losses by biotic and abiotic stresses (Hartman et al., 2018). Therefore, the unprecedented attention to understand the microbiome composition and dynamics to encourage plant-beneficial relationships has become increasingly relevant and attractive.

The literature on understanding the microbiome assembly has long been dominated by geneticists. The emphasis has been on describing the taxonomic composition of the microbiota and on the changes in this composition through different factors (Bai et al., 2015; Bergelson, Mittelstrass, & Horton, 2019; Bulgarelli et al., 2012; Bulgarelli et al., 2013; Lundberg et al., 2012; Muller, Vogel, Bai, & Vorholt, 2016). These studies demonstrated clearly that while soil is the major determinant of microbiome composition, plants significantly shape their microbiome. For example, Lundberg et al. (2012) showed that different Arabidopsis accessions have distinct microbiomes and that its composition also depends on the developmental stage; Bai et al. (2015) found similarities between leaf and root microbiomes; and Bergelson et al. (2019), found evidence that there is a stronger influence of host effects on plant-fungal communities than bacterial taxa. These findings made apparent the important roles of plant genotypes in determining the composition and function of the microbiome (Jacoby et al., 2020). The primary mechanism by which plants may affect the microbiota is through root exudation. Indeed, root exudates play an important role in mediating plant interactions with rhizosphere symbionts and pathogens modulating their contribution to plant performance, nutrient uptake, defense, and response to abiotic stress. In order to take advantages of these interactions in the field of agriculture and ecology and reduce fertilizer inputs, it is essential to elucidate qualitative and quantitative composition of root exudates and microorganisms' metabolome and functions of their components. Root exudates include a diversity of chemical compounds which can be divided into primary and secondary metabolites, such as sugars, amino acids, organic acids, phenolics, and phytohormones (Monchgesang et al., 2016; Sasse et al., 2018; Zhalnina et al., 2018). To better understand the rhizosphere chemical environment and plant microbe interactions, underground metabolites should be able to be traced and analyzed. Due to the huge diversity of root exudates, metabolomics must be applied to identify and annotate the rhizosphere metabolome. Correspondingly, and given the importance of metabolites in the communication between plants and microbes, the role of metabolomics in their dissection is getting increasing attention.

Metabolomics provides an unbiased assessment of a cell's physiology and metabolite distribution in response to its environment. Metabolomics approaches have been efficacious in elucidating defense mechanisms such as induced resistance (Schwachtje, Fischer, Erban, & Kopka, 2018), allelopathy (Kong et al., 2018), or nutrient deficiency (Ziegler et al., 2016). Metabolomics studies of the rhizosphere can be challenging due to the complexity and dynamic interactive microenvironment. The composition of this underground ecosystem is rich in primary and secondary metabolites as forms of chemical communication to surrounding soil organisms. The highly diverse species- and genotype-specific metabolites, particularly secondary ones, enable the respective organism to adapt to demanding conditions, by recruiting beneficial and growth promoting microorganisms or activating defense mechanisms against pathogens (Jacoby et al., 2020). Thus, there is a constant shift of microbial assemblages, manipulated by root exudates that respond to the biotic and abiotic challenges (Kong et al., 2018; Monchgesang et al., 2016; Sasse et al., 2018). The root exudates play a central role as a source of metabolites for complex signaling systems, meaning that characterizing the composition and dynamics of root exudates would enhance our understanding of the plant-microbe network that is established for the overall fitness of the plant. Unbiased analytical metabolomics tools such as HPLC, LC-MS and NMR spectroscopy allows researchers to identify and quantify the metabolites of root exudates in response to various treatments to further understand complex mechanisms of the plant and their network with microorganisms (Monchgesang et al., 2016; Yuan et al., 2018; Zhalnina et al., 2018; Ziegler et al., 2016).

LC-MS based metabolome approaches has been widely used for detection and characterization of biologically active chemical compounds. Feussner and Feussner (2019) originally developed a methodology for the rapid and sensitive quantification of the well-known plant hormone jasmonate and its biosynthetic precursors from minimum plant material. This method of metabolite fingerprinting however, also allowed applications on metabolites in a great variety of plant species and respective plant organs as well as mycelia of various fungi and fluids derived from both plant and fungi (Feussner & Feussner, 2019). The workflow allows a broad metabolite coverage via two-phase extraction, separate analysis of polar and non-polar compounds by Ultra-performance liquid chromatography (UPLC), which is then subjected to both positive and negative electrospray ionization (ESI) time-of-flight (TOF)-MS analysis. For data analysis and visualizations, the researchers recommend MarVis-toolbox, which also allows for implementing custom databases. This flexible methodology grants the ability to apply a broader database for comparison to uncover metabolites responsible for triggering crucial signaling mechanisms. This metabolic fingerprinting approach was successful, for example, in identifying the antimicrobial coumarin scopoletin to be dominantly produced and secreted from Arabidopsis roots (Stringlis et al., 2018). Scopoletin is produced under iron deficiency as its potent chelator to improve uptake in plants. This study identified a transcription factor MYB72 and a β-glucosidase BGLU42 as essential for scopoletin exudation. In addition, the report found that scopoletin affects the association of Arabidopsis with rhizosphere bacterium *Pseudomonas simiae* WCS417, and is thus involved in shaping of plant microbiome (Stringlis et al., 2018; Voges, Bai, Schulze-Lefert, & Sattely, 2019).

Along with coumarins, triterpenes have been recently found to be involved in shaping of the microbiome. Triterpenes are natural plant products involved in antimicrobial activity and defense against various pathogens and pests (Papadopoulou, Melton, Leggett, Daniels, & Osbourn, 1999; Thimmappa, Geisler, Louveau, O'Maille, & Osbourn, 2014). They are synthesized via the mevalonate pathway and can accumulate inside the plant tissue as triterpene glycosides. Very recently, Huang et al. (2019), discovered that the disruption of synthesis of three Arabidopsis triterpenes, derived from thalianol and arabidiol gene clusters and accumulating in roots, resulted in altered root microbiota. The results of metabolite analyses and microbiome profiling suggest that triterpenes contribute significantly to selective modulation of the microbial assemblage of Arabidopsis roots (Huang et al., 2019). The complex pattern of triterpene in root exudates makes it difficult to identify single active metabolite, requiring combination of several methods, including GC-MS, lC-MS, and NMR. Accordingly, quantitative H NMR has demonstrated its value as a powerful method for analyzing triterpene content in crude plant extracts of roots and rhizomes as well as aerial sections of *Actaea racemosa*, *A. podocarpa*, and *A. cordifolia* (Imai, Lankin, Godecke, Chen, & Pauli, 2020). This methodology does not require reference compounds for all targets for quantitative analysis, but the hydrogen signals allow the quantification of less common derivates of cycloartane directly in the extracts.

Benzoxazinoids are a class of bioactive indole metabolites produced by grasses, which have long been implicated as important for influencing plant-microbe interactions. Recent work by Cotton et al. (2019) has applied high-throughput metabolomics to investigate maize mutants impaired in benzoxazinoid synthesis. Data analysis revealed that mutant plants exhibit widespread alterations to root metabolite profiles, which could mechanistically explain why these mutants recruit an altered microbial community. Methodologically, this study reports a promising approach to link metabolite abundance with microbiome composition, by searching for correlative links between metabolite peaks acquired via LC-MS versus OTUs acquired via amplicon sequencing. This integrative approach has strong potential to bridge the gap between biochemical analyses of plant metabolite composition and taxonomic analyses of microbial community composition.

Interestingly, metabolites have been shown to mediate also epigenetic regulation of plant-microbe interactions (Vilchez et al., 2020). The observation that a hypermetylation Arabidopsis mutant *rdd*, deficient in the DNA demethylases REPRESSOR OF SILENCING 1 (ROS1), DEMETER-LIKE 2 (DML2), and DML3, loses the plant growth promotion effect of *Bacillus megaterium* YC4 led to identification of *myo*-inositol as the modulator of this interaction. *Myo*-inositol was highly depleted in the exudates of *rdd* mutant compared to Col-0 and also other mutants with lower accumulation of this metabolite showed altered interaction with *B. megaterium* YC4 (Vilchez et al., 2020). Thus, metabolomics analyses might be necessary also for dissection of genetic and epi-genetic mechanisms controlling plant microbe interactions.

It is a high priority to identify the biochemical changes induced in non-exposed tissues following microbial inoculation at a distal position on the plant. These molecules could mediate signaling processes, either to communicate immune responses within the plant, or alternatively to modulate the surrounding microbial community by exuding substrates that nourish commensals or toxins that kill pathogens. Untargeted metabolomics offers a high-throughput platform to deepen our knowledge of microbe-induced plant signaling processes, either by identifying novel chemical agents that transduce these signaling events, or by characterizing the metabolic alterations occurring in distal tissues. Recently, Korenblum et al. (2020) measured the exudate profiles of tomato roots that were distally positioned from the microbial inoculation, by using a split-root growth system that physically separated the two root sections. A major conclusion was that acetylsugars and various glycosides are strongly responsive to long-distance signals imparted by microbial inoculation. These plant metabolic responses are not uniform, because the distal patterns of root exudation observed in the sterile compartment were significantly altered by the microbial treatment applied to local roots. This work shows that the combination of split-root techniques with untargeted metabolomics provides an excellent platform to identify new aspects of plant systemic signaling.

5. Exometabolomics

A promising new application of metabolomics in the field of plant-microbe interactions is exometabolomics, which involves measuring the metabolic footprint that microbes imprint upon their chemical environment. This approach has strong potential to deliver new mechanistic insights into plant microbiome research, because microbial substrate uptake is a central mechanism shaping the assembly of the plant-associated microbial community (Sasse et al., 2018). However, there is an incomplete knowledge of which plant metabolites are consumed by which microbial strains. Exometabolomics has the potential to fill this knowledge gap, by characterizing the nutritional niches occupied by individual microbes. This information about microbial substrate preferences can then be used to mechanistically explain microbial community assembly, mediated via processes such as niche differentiation and competitive exclusion (Jacoby & Kopriva, 2019).

Generally, exometabolomic workflows deploy high-throughput techniques to measure the specific plant metabolites that are utilized by microbial strains as growth substrates (Fig. 2). The derived datasets can deliver new insights into the nutritional interdependence between plants and their associated microbiome, by defining the specific plant metabolites used as a

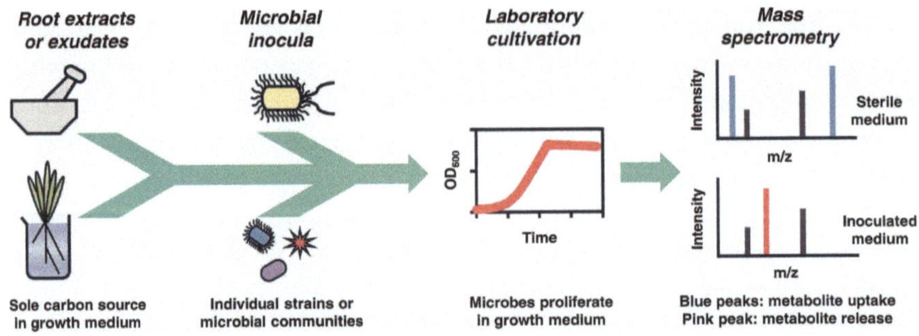

Fig. 2 Exometabolomics applied to study plant-microbe interactions in the roots. Exometabolomics is a methodological approach that uses high-throughput metabolomics techniques to measure microbial substrate preferences. In the field of plant-microbe interactions, exometabolomics has strong potential to deliver new information about the metabolic niches occupied by root-associated microbes. Here we present a schematic illustration showing how exometabolomics workflows can be applied to study plant-microbe interactions in the roots. First, root metabolites are provided as the sole carbon source in the growth medium, using either root extracts or root exudates. Next, microbes are inoculated into this medium, using either an individual strain or a microbial community. Following microbial proliferation, mass spectrometry is used to profile the metabolite composition of the culture supernatant. Data analysis involves identifying metabolite peaks with differential abundance between the inoculated medium versus the sterile medium. Here, the two blue peaks represent metabolites that were consumed as microbial growth substrates, which give information about the metabolic niche that this microbe occupies in the rhizosphere. Additionally, the pink peak represents a metabolite that was released by the microbe, which could play a role in cross-feeding, antibiosis or signaling.

carbon source for microbial proliferation. It has long recognized that the plant rhizosphere provides a rich habitat for microbial growth, because compared to the surrounding soil, the root zone is highly enriched with carbon substrates arising from root exudation and sloughed-off root cells (Jones, Nguyen, & Finlay, 2009). Biochemically, this substrate mixture is highly complex, containing both primary and secondary plant metabolites (Dennis, Miller, & Hirsch, 2010). Although it is widely accepted that the microbial consumption of these root metabolites plays a significant role in shaping the composition of the root microbiome, there is a relatively poor knowledge about the metabolic interdependencies that have coevolved between plants and microbes. For instance, it has been suggested that microbial specialization onto distinct plant hosts is mediated by specialized catabolic pathways to utilize distinct secondary metabolites only produced by that species (Chagas, Pessotti, Caraballo-Rodriguez, & Pupo, 2018). However, there have been relatively few studies that have characterized these nutrient exchanges in terms of individual metabolites. Meanwhile, it is often asserted that plants might modulate their exudate chemistries in order to recruit cooperative strains into their rhizosphere (Lambers, Mougel, Jaillard, & Hinsinger, 2009), but again there is relatively little information about the specific molecules involved. Here, we argue that exometabolomics is well suited to provide new information that will help to answer these outstanding questions.

The datasets derived from exometabolomics workflows provide much more detailed information about microbial substrate uptake compared to alternative techniques, such as minimal medium growth assays or phenotype microarrays. In the recent literature, multiple studies have deployed exometabolomics to study plant-associated microbes. Two recent studies have investigated the substrate uptake preferences of nitrogen-fixing symbionts cultivated on laboratory growth media, using LC-MS (Fei, diCenzo, Bowdish, McCarry, & Finan, 2016) and NMR (Montes-Grajales, Esturau-Escofet, Esquivel, & Martinez-Romero, 2019). One disadvantage of using laboratory growth media for exometabolomics studies is, however, that these media do not fully represent the biochemical environment of the plant host, particularly because they lack secondary metabolites specific to the target species. Therefore, we will focus on two recent studies that have applied exometabolomics to study the metabolic footprints of plant-associated microbes cultivated either on medium containing root exudates (Zhalnina et al., 2018) or root extracts (Jacoby, Martyn, & Kopriva, 2018). Here we will review these studies in detail, focusing on the methodological approaches and key biological insights.

Zhalnina et al. (2018) utilized untargeted metabolomics workflow to assess changes of root exudates during growth stages of an annual grass (*Avena barbata*), coupled with exometabolomics to reveal the relationship between root exudates and rhizosphere microbial community. This data was generated via HILIC-LC-MS/MS, by detecting metabolites present in root exudates at four different plant developmental stages, as well as measuring whether these metabolites were utilized as growth substrates by a panel of 16 bacterial strains when cultivated using root exudates as the sole carbon source. This set of bacterial isolates was chosen to encompass taxa that were either enriched or depleted in response to plant growth. Data analysis showed that bacterial strains enriched in the rhizosphere were significantly more likely to uptake aromatic organic acids and amino acids compared to strains that were depleted by plant growth. A broader analysis of root exudate consumption by isolated strains showed that the uptake of most root exudate components was generally quite similar among diverse bacterial isolates. However, a small set of metabolites including salicylic acid were more likely to be utilized via strains that successfully colonize the rhizosphere, which positions these metabolites as key mediators of microbial community assembly. Assessing the study, a key strength is that it combines the temporal analysis of plant root exudation traits with exometabolomic measurement of bacterial substrate uptake preferences. Integrating these two datasets, it can be confidently concluded that rhizosphere microbial assembly is shaped by root exudates, whereby younger plants recruit their microbiome by exuding the preferred metabolic substrates of strains that successfully colonize the rhizosphere. This work showcases the advantages of applying exometabolomics to study microbial community assembly, because the derived data pinpoints the adaptive metabolite exchanges that have coevolved between plants and microbes to confer a fitness benefit for both parties.

Jacoby et al. (2018) utilized untargeted metabolomics to investigate the substrate preferences of bacterial strains cultivated on Arabidopsis root extract, in order to compare the metabolic niches occupied by two bacterial strains isolated from field-grown Arabidopsis plants versus *E. coli* isolated from the human gut. Exometabolomic profiles were acquired by analyzing culture filtrates using C18-LC-MS, following cultivation on growth medium where Arabidopsis root extracts were provided as the sole carbon source. Data analysis showed that all three strains took up a common set of 62 MS-features, with m/z values mainly corresponding to primary metabolites such as amino acids. However, the two bacterial strains isolated from field-grown Arabidopsis roots were both capable of consuming a wide variety of MS-features matching to plant secondary metabolites such as scopoletin and xylobiose, none of which were catabolized by *E. coli*. This suggests that *E. coli* is poorly adapted to colonize the Arabidopsis rhizosphere, because it cannot establish a differential metabolic niche. Furthermore, there was relatively little overlap in the substrate preferences of the two plant-associated strains, implying that these two strains avoid direct competition by occupying complementary metabolic niches. Additionally, the study reports that all three strains release a significant number of novel MS-features into their growth media, which could potentially be important in rhizosphere ecology for microbe-microbe interactions such as cross-feeding. For the wider research community, the major takeaway from this work is the development of a relatively straightforward methodological approach for conducting rhizosphere exometabolomics in the model plant Arabidopsis. This workflow can be easily adapted to profile various Arabidopsis mutants or diverse microbial strains, with the major advantage that it is relatively straightforward to generate root extracts compared to root exudates. Potentially, this methodology will be a useful tool for pinpointing the molecular mechanisms that underpin root colonization and niche differentiation in the Arabidopsis microbiome.

Mass spectrometry imaging is a complementary technique to exometabolomics, because it provides information about the spatial distribution of metabolic substrates in the environment (Silva & Northen, 2015). Multiple recent studies have deployed either MALDI-imaging or LAESI-MS to investigate the in situ metabolomic changes that nitrogen-fixing bacteria induce upon legume root nodules (Gemperline, Jayaraman, Maeda, Ane, & Li, 2015; Stopka et al., 2017; Velickovic et al., 2018). Another study has recently applied LAESI-MS to analyze the in situ metabolic changes in *Setaria viridis* roots inoculated with growth-promoting bacteria (Agtuca et al., 2020). Compared to standard metabolomics approaches that involve sample homogenization, a key advantage of these spatially-resolved workflows is the ability to dissect how metabolite composition varies according to the plant's anatomy and histology. For research into plant-microbe interactions, spatially-resolved metabolomics approaches can give new insights into the metabolic alterations occurring at the distinct site of microbial colonization. There is an opportunity for future studies to couple mass spectrometry imaging with exometabolomics, because the imaging datasets provide spatial information about metabolite availability on the plant root, whereas the exometabolomic information defines the substrate preferences of microbial colonists. These two approaches can be integrated together, to give new information about the site-specific metabolic exchanges occurring between plants and microbes located at different anatomical regions along the root.

Assessing the literature applying exometabolomics to study plant-microbe interactions, it is clear that the field is still in its early stages of development, with publications only starting to emerge from 2016 onwards. However, we anticipate that the field will grow quickly over the coming years, because this methodological approach is the most appropriate tool for analyzing the microbial consumption of specific root metabolites. The derived data provide complementary information to genetic and taxonomic studies that dominate the literature, because these techniques are poorly suited to identifying the underlying mechanisms that shape microbial community assembly. Methodologically, exometabolomics is incredibly versatile, with various mass spectrometry platforms as well as NMR being successfully applied across different microbes (Pinu & Villas-Boas, 2017). In the context of plant root metabolites, HILIC chromatography provides better separation of the polar primary metabolites that presumably represent the bulk of nutrient exchange between plants and microbes (Zhalnina et al., 2018), whereas C18 chromatography offers sharp and reproducible elution profiles for hydrophobic secondary metabolites that could play key roles in microbial niche differentiation (Jacoby et al., 2018). Computationally, exometabolomics datasets can be analyzed using online repositories for natural product mass spectra (Wang et al., 2016) as well as microbial cross-feeding (Kosina et al., 2018), enabling researchers to integrate their primary data with results of pre-existing studies. In the longer term, it is tempting to speculate that exometabolomics could assist in strategies to rationally design the plant microbiome, by matching the substrate uptake profile of high-performing microbial inoculants versus the root exudation profile of new crop varieties.

6. Challenges of metabolomics of plant-microbe systems

6.1 Collection of root exudates

Studies of metabolite abundance and function in plant-microbe interactions pose several challenges beyond the classical use of plant metabolomics. The first problem is already the sampling of root exudates. Given the complexity of root architecture and soil structure, recovering root exudates from soil-grown plants unaltered by sorption and microbial activity is very difficult (Oburger & Jones, 2018). In addition, the exudates have to avoid contamination by metabolites released from soil organic matter and from dead root cells. Therefore, most of our knowledge on root exudates is derived from a hydroponics approach and collection of root exudation in nutrient solution or water (Carvalhais et al., 2013; Strehmel, Bottcher, Schmidt, & Scheel, 2014; Zhalnina et al., 2018; Ziegler et al., 2016). As this is obviously a limitation, Oburger and Jones (2018) questioned how ecologically relevant exudation results obtained under hydroponic conditions are in comparison to the natural conditions of soil. Unfortunately, so far this is a rhetorical question since even the best systems for exudate collection from soil grown plants are still only partly natural, e.g., through using a membrane or mesh separating the roots from the soil (Oburger et al., 2014) or sampling tubes absorbing specific compound classes (Weidenhamer, Mohney, Shihada, &

Rupasinghe, 2014). A growth system designed by Petriacq et al. (2017), in which plants grow in tubes filled with soil perforated at the bottom and exudates are extracted by a rapid flush of the tube with different concentrations of methanol, seems to offer a good solution to describe metabolic composition of the rhizosphere. However, to distinguish between plant exudates and soil organic compounds metabolized by the microbiota is not trivial and even isotopic labeling of plant metabolites might not solve this limitation due to rapid metabolic processes in the rhizosphere. Thus, unfortunately, a perfect universal extraction protocol does not yet exist and for most research questions the hydroponics approach with all its disadvantages still remains the method of choice.

Some questions on root exudation, particularly its dynamics, can be addressed by labeling techniques. For example, diverse metabolic sensors can be introduced into the rhizosphere to quantify the exuded metabolites. In a pioneering study Pini et al. (2017) developed biosensors based on *Rhizobium leguminosarum* bv *viciae* expressing luciferase under control of various chemically inducible promoters. These biosensors were able to detect amino acids, organic acids, sugars, and flavonoids and monitor the quantitative and spatial exudation pattern of these metabolites (Pini et al., 2017). Alternatively, sensors can be based on enzymes immobilized on roots producing colored or fluorescent products as successfully tested to image glucose (Voothuluru, Braun, & Boyer, 2018). Another possibility is to use isotope labeling, e.g., exposure of the leaves to $^{13}CO_2$ or $^{14}CO_2$, to quantify root exudation or for ^{14}C also image the spatial pattern of the exudation (Oburger & Jones, 2018). While the labeling approaches can bring some answers, they do not allow to assess the whole diversity of metabolites in the exudates.

6.2 Determining the origin of metabolites in plant-microbe interactions

When a metabolite is found in rhizosphere accumulating upon interaction of plant with specific microbes, it is often impossible to distinguish whether it was synthesized and exuded by the plant, the microbe, or both. This is particularly problematic during analyses of tissues with high proportion of microbes, e.g. nodules or highly arbusculated roots, or in experiments in which plant tissues are infiltrated with large titres of microbes (Allwood, Heald, Lloyd, Goodacre, & Mur, 2012). If the microbes are less abundant, it is possible to test for the presence and quantity of compounds not synthesized by plants and estimate the microbial contribution to the total metabolome. One example is ergosterol found almost exclusively in fungi and used as a marker for fungal biomass (Larsson & Saraf, 1997). Combined with physical separation of the microbes away from plant tissues, this would help to identify metabolites specifically enriched in one of the fractions, as such compounds would allow to estimate the purity of the fraction. The physical separation is similar to the pipeline allowing determination of bacterial transcriptome changes during plant immune response (Nobori et al., 2018). While possibly applicable for metabolomics studies of infiltrated bacteria in plant leaves, this approach has not been tested for the roots and is not suitable for metabolites that are secreted and do not accumulate inside the cells. The separation can be simplified by using plant cell cultures instead of whole tissues (Allwood et al., 2012), but with the same limitations. Nevertheless, for determination of early metabolic changes in plant cells elicited by microbes this approach can be useful.

A much more promising approach to distinguish origin of specific metabolites is isotope labeling. Stable isotope probing (SIP) approaches have been used to determine microbial activity in soil or water by using labeled substrates and measuring isotopic enrichment of microbial cells, products or biomarkers (Pett-Ridge & Firestone, 2017). SIP has been used to identify microorganisms in a complex community capable of metabolizing methane, after feeding $^{13}CH_4$, isolation of ^{13}C-enriched RNA, and metatranscriptomics (Dumont, Pommerenke, & Casper, 2013). Similarly, biotraps loaded with ^{13}C-labeled naphthalene or fluorene were used to collect bacterial biofilms in contaminated groundwater wells and ^{13}C-enriched proteins were identified by comparison with similar biotraps loaded with non-labeled hydrocarbons (Herbst et al., 2013). The analysis detected Burkholderiales, Actinomycetales, and Rhizobiales as the most active microorganisms for degradation of polycyclic aromatic hydrocarbon contaminants in the groundwater communities. In animal gut microbiome research SIP was used, e.g., to identify *Ruminococcus bromii* as the primary starch degrading bacterial strain (Kovatcheva-Datchary et al., 2009) or *Allobaculum* spp. as particularly active glucose utilizer (Herrmann et al., 2017). Similar approach can be used to study plant microbe interactions, where the source of labeled metabolites is the plant, through growing in $^{13}CO_2$ enriched atmosphere (Pett-Ridge & Firestone, 2017). Using this approach, several bacterial strains, mainly of the order Burkholderiales, have been found through strong ^{13}C labeling of their RNA, indicating rapid uptake of root exudates (Vandenkoornhuyse et al., 2007). Conceptually, heavy isotope labeling of one of the interacting partners should enable identification of origins of specific metabolites at the site of interaction, however, this has been only very rarely used in practice. In a pioneering study, Pang et al. (2018) used ^{13}C and ^{15}N labeled *Pseudomonas* cultures to study early metabolic events during plant immune response. Due to the isotope labeling the bacterial metabolites show a mass shift at the same retention times and can be distinguished from the plant derived counterparts. As a proof of principle, the phytotoxin coronatine that is exported from the bacteria, was measured and confirmed to be composed of the heavy isotopes, i.e., of bacterial origin (Pang et al., 2018). The analysis allowed to determine plant metabolites affected by interaction with the bacteria and showed their particular effect on amino acid synthesis. The promising technology thus awaits further testing, particularly in the rhizosphere environment.

7. Concluding remarks

In the last decade, it has become increasingly recognized that root-dwelling microorganisms play a major role in shaping plant performance. Metabolites are the primary means of communication between plants and their microbiota, but there is an incomplete mechanistic knowledge about how root exudates shape the microbiome. In our opinion, the next step change in understanding plant-microbe interactions will be dependent on applying metabolomics techniques to identify which specific rhizosphere metabolites are involved in key biological processes such as mutualist recruitment, pathogen suppression, and immune signaling. Several such signals have already been identified using metabolomics techniques, and the corresponding analytic pipelines are becoming more widespread and easy to use. Although there are inherent technical challenges to studying plant-microbe interactions using metabolomics, new approaches such as exometabolomics and stable isotope probing seem well-suited to deliver mechanistic insights. Assessing the current state of the field, there seems to be an opportune alignment of research questions with methodological tools, because many important processes in plant-microbe metabolic interactions are underpinned by poorly characterized metabolic exchanges, and newly emerging metabolomic approaches provide a high-throughput platform to identify the relevant molecules. Therefore, we are confident that the next decade will witness a dramatic surge of new discoveries found using metabolomics in plant-microbe interactions.

References

Abdel-Lateif, K., Vaissayre, V., Gherbi, H., Verries, C., Meudec, E., Perrine-Walker, F., et al. (2013). Silencing of the chalcone synthase gene in Casuarina glauca highlights the important role of flavonoids during nodulation. *New Phytologist, 199*(4), 1012–1021. https://doi.org/10.1111/nph.12326.

Adeniji, A. A., Babalola, O. O., & Loots, D. T. (2020). Metabolomic applications for understanding complex tripartite plant-microbes interactions: Strategies and perspectives. *Biotechnology Reports (Amsterdam, Netherlands), 25*, e00425. https://doi.org/10.1016/j.btre.2020.e00425.

Agtuca, B. J., Stopka, S. A., Tuleski, T. R., do Amaral, F. P., Evans, S., Liu, Y., et al. (2020). In-situ Metabolomic analysis of Setaria viridis roots colonized by beneficial Endophytic Bacteria. *Molecular Plant-Microbe Interactions, 33*(2), 272–283. https://doi.org/10.1094/MPMI-06-19-0174-R.

Akiyama, K., Matsuzaki, K., & Hayashi, H. (2005). Plant sesquiterpenes induce hyphal branching in arbuscular mycorrhizal fungi. *Nature, 435*(7043), 824–827. https://doi.org/10.1038/nature03608.

Allwood, J. W., Heald, J., Lloyd, A. J., Goodacre, R., & Mur, L. A. (2012). Separating the inseparable: The metabolomic analysis of plant-pathogen interactions. *Methods in Molecular Biology, 860*, 31–49. https://doi.org/10.1007/978-1-61779-594-7_3.

Arias, M. M., Leandro, L. F., & Munkvold, G. P. (2013). Aggressiveness of Fusarium species and impact of root infection on growth and yield of soybeans. *Phytopathology, 103*(8), 822–832. https://doi.org/10.1094/PHYTO-08-12-0207-R.

Auguy, F., Abdel-Lateif, K., Doumas, P., Badin, P., Guerin, V., Bogusz, D., et al. (2011). Activation of the isoflavonoid pathway in actinorhizal symbioses. *Functional Plant Biology, 38*(9), 690–696. https://doi.org/10.1071/FP11014.

Back, M. A., Haydock, P. P. J., & Jenkinson, P. (2002). Disease complexes involving plant parasitic nematodes and soilborne pathogens. *Plant Pathology, 51*(6), 683–697. https://doi.org/10.1046/j.1365-3059.2002.00785.x.

Bai, Y., Muller, D. B., Srinivas, G., Garrido-Oter, R., Potthoff, E., Rott, M., et al. (2015). Functional overlap of the Arabidopsis leaf and root microbiota. *Nature, 528*(7582), 364–369. https://doi.org/10.1038/nature16192.

Bednarek, P., Pislewska-Bednarek, M., Svatos, A., Schneider, B., Doubsky, J., Mansurova, M., et al. (2009). A glucosinolate metabolism pathway in living plant cells mediates broad-spectrum antifungal defense. *Science, 323*(5910), 101–106. https://doi.org/10.1126/science.1163732.

Bergelson, J., Mittelstrass, J., & Horton, M. W. (2019). Characterizing both bacteria and fungi improves understanding of the Arabidopsis root microbiome. *Scientific Reports, 9*(1), 24. https://doi.org/10.1038/s41598-018-37208-z.

Bladergroen, M. R., & Spaink, H. P. (1998). Genes and signal molecules involved in the rhizobia-leguminoseae symbiosis. *Current Opinion in Plant Biology, 1*(4), 353–359. https://doi.org/10.1016/1369-5266(88)80059-1.

Bones, A. M., & Rossiter, J. T. (1996). The myrosinase-glucosinolate system, its organisation and biochemistry. *Physiologia Plantarum, 97*, 194–208.

Bulgarelli, D., Rott, M., Schlaeppi, K., Ver Loren van Themaat, E., Ahmadinejad, N., Assenza, F., et al. (2012). Revealing structure and assembly cues for Arabidopsis root-inhabiting bacterial microbiota. *Nature, 488*(7409), 91–95. https://doi.org/10.1038/nature11336.

Bulgarelli, D., Schlaeppi, K., Spaepen, S., Ver Loren van Themaat, E., & Schulze-Lefert, P. (2013). Structure and functions of the bacterial microbiota of plants. *Annual Review of Plant Biology, 62(64)*, 807–838. https://doi.org/10.1146/annurev-arplant-050312-120106.

Carvalhais, L. C., Dennis, P. G., Fan, B., Fedoseyenko, D., Kierul, K., Becker, A., et al. (2013). Linking plant nutritional status to plant-microbe interactions. *PLoS One*, *8*(7), e68555. https://doi.org/10.1371/journal.pone.0068555.

Chagas, F. O., Pessotti, R. C., Caraballo-Rodriguez, A. M., & Pupo, M. T. (2018). Chemical signaling involved in plant-microbe interactions. *Chemical Society Reviews*, *47*(5), 1652–1704. https://doi.org/10.1039/c7cs00343a.

Chen, S., Waghmode, T. R., Sun, R., Kuramae, E. E., Hu, C., & Liu, B. (2019). Root-associated microbiomes of wheat under the combined effect of plant development and nitrogen fertilization. *Microbiome*, 7(1), 136. https://doi.org/10.1186/s40168-019-0750-2.

Clay, N. K., Adio, A. M., Denoux, C., Jander, G., & Ausubel, F. M. (2009). Glucosinolate metabolites required for an Arabidopsis innate immune response. *Science*, *323*(5910), 95–101. https://doi.org/10.1126/science.1164627.

Cotton, T. E. A., Petriacq, P., Cameron, D. D., Meselmani, M. A., Schwarzenbacher, R., Rolfe, S. A., et al. (2019). Metabolic regulation of the maize rhizobiome by benzoxazinoids. *The ISME Journal*, *13*, 1647–1658. https://doi.org/10.1038/s41396-019-0375-2.

Cullimore, J. V., Ranjeva, R., & Bono, J. J. (2001). Perception of lipo-chitooligosaccharidic nod factors in legumes. *Trends in Plant Science*, *6*(1), 24–30. https://doi.org/10.1016/s1360-1385(00)01810-0.

Dennis, P. G., Miller, A. J., & Hirsch, P. R. (2010). Are root exudates more important than other sources of rhizodeposits in structuring rhizosphere bacterial communities? *FEMS Microbiology Ecology*, *72*(3), 313–327. https://doi.org/10.1111/j.1574-6941.2010.00860.x.

Dietz, K. J., Sauter, A., Wichert, K., Messdaghi, D., & Hartung, W. (2000). Extracellular beta-glucosidase activity in barley involved in the hydrolysis of ABA glucose conjugate in leaves. *Journal of Experimental Botany*, *51*(346), 937–944.

Dumont, M. G., Pommerenke, B., & Casper, P. (2013). Using stable isotope probing to obtain a targeted metatranscriptome of aerobic methanotrophs in lake sediment. *Environmental Microbiology Reports*, *5*(5), 757–764. https://doi.org/10.1111/1758-2229.12078.

Eshelli, M., Qader, M. M., Jambi, E. J., Hursthouse, A. S., & Rateb, M. E. (2018). Current status and future opportunities of Omics tools in Mycotoxin research. *Toxins (Basel)*, *10*(11). https://doi.org/10.3390/toxins10110433.

Fahey, J. W., Zalcmann, A. T., & Talalay, P. (2001). The chemical diversity and distribution of glucosinolates and isothiocyanates among plants. *Phytochemistry*, *56*(1), 5–51.

Fei, F., diCenzo, G. C., Bowdish, D. M. E., McCarry, B. E., & Finan, T. M. (2016). Effects of synthetic large-scale genome reduction on metabolism and metabolic preferences in a nutritionally complex environment. *Metabolomics*, *23*, 23. https://doi.org/10.1007/s11306-015-0928-y.

Feussner, K., & Feussner, I. (2019). Comprehensive LC-MS-based metabolite fingerprinting approach for plant and fungal-derived samples. *Methods in Molecular Biology*, *1978*, 167–185. https://doi.org/10.1007/978-1-4939-9236-2_11.

Gamir, J., Pastor, V., Cerezo, M., & Flors, V. (2012). Identification of indole-3-carboxylic acid as mediator of priming against Plectosphaerella cucumerina. *Plant Physiology and Biochemistry*, *61*, 169–179. https://doi.org/10.1016/j.plaphy.2012.10.004.

Gemperline, E., Jayaraman, D., Maeda, J., Ane, J. M., & Li, L. (2015). Multifaceted investigation of metabolites during nitrogen fixation in Medicago via high resolution MALDI-MS imaging and ESI-MS. *Journal of the American Society for Mass Spectrometry*, *26*(1), 149–158. https://doi.org/10.1007/s13361-014-1010-0.

Genre, A., Chabaud, M., Balzergue, C., Puech-Pages, V., Novero, M., Rey, T., et al. (2013). Short-chain chitin oligomers from arbuscular mycorrhizal fungi trigger nuclear

Ca2+ spiking in Medicago truncatula roots and their production is enhanced by strigolactone. *New Phytologist*, *198*(1), 190–202. https://doi.org/10.1111/nph.12146.

Glawischnig, E. (2007). Camalexin. *Phytochemistry*, *68*(4), 401–406. https://doi.org/10.1016/j.phytochem.2006.12.005.

Gordon, T. R. (2017). Fusarium oxysporum and the Fusarium wilt syndrome. *Annual Review of Phytopathology*, *55*, 23–39. https://doi.org/10.1146/annurev-phyto-080615-095919.

Halkier, B. A., & Gershenzon, J. (2006). Biology and biochemistry of glucosinolates. *Annual Review of Plant Biology*, *57*, 303–333. https://doi.org/10.1146/annurev.arplant.57.032905.105228.

Haney, C. H., Samuel, B. S., Bush, J., & Ausubel, F. M. (2015). Associations with rhizosphere bacteria can confer an adaptive advantage to plants. *Nat Plants*, *1*(6), 15051. https://doi.org/10.1038/nplants.2015.51.

Hansen, E. M., & Goheen, E. M. (2000). Phellinus Weirii and other native root pathogens as determinants of Forest structure and process in Western North America. *Annual Review of Phytopathology*, *38*, 515–539. https://doi.org/10.1146/annurev.phyto.38.1.515.

Harrison, M. J., Dewbre, G. R., & Liu, J. (2002). A phosphate transporter from Medicago truncatula involved in the acquisition of phosphate released by arbuscular mycorrhizal fungi. *Plant Cell*, *14*(10), 2413–2429.

Hartman, K., van der Heijden, M. G. A., Wittwer, R. A., Banerjee, S., Walser, J. C., & Schlaeppi, K. (2018). Cropping practices manipulate abundance patterns of root and soil microbiome members paving the way to smart farming. *Microbiome*, *6*(1), 14. https://doi.org/10.1186/s40168-017-0389-9.

Herbst, F. A., Bahr, A., Duarte, M., Pieper, D. H., Richnow, H. H., von Bergen, M., et al. (2013). Elucidation of in situ polycyclic aromatic hydrocarbon degradation by functional metaproteomics (protein-SIP). *Proteomics*, *13*(18–19), 2910–2920. https://doi.org/10.1002/pmic.201200569.

Herrmann, E., Young, W., Rosendale, D., Reichert-Grimm, V., Riedel, C. U., Conrad, R., et al. (2017). RNA-based stable isotope probing suggests Allobaculum spp. as particularly active glucose assimilators in a complex murine microbiota cultured in vitro. *BioMed Research International*, *2017*. https://doi.org/10.1155/2017/1829685.

Huang, A. C., Jiang, T., Liu, Y. X., Bai, Y. C., Reed, J., Qu, B., et al. (2019). A specialized metabolic network selectively modulates Arabidopsis root microbiota. *Science*, *364*(6440), eaau6389. https://doi.org/10.1126/science.aau6389.

Imai, A., Lankin, D. C., Godecke, T., Chen, S. N., & Pauli, G. F. (2020). NMR based quantitation of cycloartane triterpenes in black cohosh extracts. *Fitoterapia*, *141*, 104467. https://doi.org/10.1016/j.fitote.2019.104467.

Iven, T., Konig, S., Singh, S., Braus-Stromeyer, S. A., Bischoff, M., Tietze, L. F., et al. (2012). Transcriptional activation and production of tryptophan-derived secondary metabolites in arabidopsis roots contributes to the defense against the fungal vascular pathogen Verticillium longisporum. *Molecular Plant*, *5*(6), 1389–1402. https://doi.org/10.1093/mp/sss044.

Jacoby, R. P., Chen, L., Schwier, M., Koprivova, A., & Kopriva, S. (2020). Recent advances in the role of plant metabolites in shaping the root microbiome. *F1000Res*, *9*. https://doi.org/10.12688/f1000research.21796.1.

Jacoby, R. P., & Kopriva, S. (2019). Metabolic niches in the rhizosphere microbiome: New tools and approaches to analyse metabolic mechanisms of plant-microbe nutrient exchange. *Journal of Experimental Botany*, *70*(4), 1087–1094. https://doi.org/10.1093/jxb/ery438.

Jacoby, R. P., Martyn, A., & Kopriva, S. (2018). Exometabolomic profiling of bacterial strains cultivated using Arabidopsis root extract as the sole carbon source. *Molecular Plant-Microbe Interactions*, *31*(8), 803–813. https://doi.org/10.1094/MPMI-10-17-0253-R.

Jacoby, R. P., Peukert, M., Succurro, A., Koprivova, A., & Kopriva, S. (2017). The role of soil microorganisms in plant mineral nutrition-current knowledge and future directions. *Frontiers in Plant Science*, *8*, 1617. https://doi.org/10.3389/fpls.2017.01617.

Jones, D. L., Nguyen, C., & Finlay, R. D. (2009). Carbon flow in the rhizosphere: Carbon trading at the soil–root interface. *Plant and Soil*, *321*, 5–33. https://doi.org/10.1007/s11104-009-9925-0.

Jung, S. C., Martinez-Medina, A., Lopez-Raez, J. A., & Pozo, M. J. (2012). Mycorrhiza-induced resistance and priming of plant defenses. *Journal of Chemical Ecology*, *38*(6), 651–664. https://doi.org/10.1007/s10886-012-0134-6.

Kamoun, S., Furzer, O., Jones, J. D., Judelson, H. S., Ali, G. S., Dalio, R. J., et al. (2015). The top 10 oomycete pathogens in molecular plant pathology. *Molecular Plant Pathology*, *16*(4), 413–434. https://doi.org/10.1111/mpp.12190.

Kong, C. H., Zhang, S. Z., Li, Y. H., Xia, Z. C., Yang, X. F., Meiners, S. J., et al. (2018). Plant neighbor detection and allelochemical response are driven by root-secreted signaling chemicals. *Nature Communications*, *9*(1), 3867. https://doi.org/10.1038/s41467-018-06429-1.

Koprivova, A., Schuck, S., Jacoby, R. P., Klinkhammer, I., Welter, B., Leson, L., et al. (2019). Root-specific camalexin biosynthesis controls the plant growth-promoting effects of multiple bacterial strains. *Proceedings of the National Academy of Sciences of the United States of America*, *116*(31), 15735–15744. https://doi.org/10.1073/pnas.1818604116.

Korenblum, E., Dong, Y., Szymanski, J., Panda, S., Jozwiak, A., Massalha, H., et al. (2020). Rhizosphere microbiome mediates systemic root metabolite exudation by root-to-root signaling. *Proceedings of the National Academy of Sciences of the United States of America*, *117*(7), 3874–3883. https://doi.org/10.1073/pnas.1912130117.

Kosina, S. M., Greiner, A. M., Lau, R. K., Jenkins, S., Baran, R., Bowen, B. P., et al. (2018). Web of microbes (WoM): A curated microbial exometabolomics database for linking chemistry and microbes. *BMC Microbiology*, *18*(1), 115. https://doi.org/10.1186/s12866-018-1256-y.

Kovatcheva-Datchary, P., Egert, M., Maathuis, A., Rajilic-Stojanovic, M., de Graaf, A. A., Smidt, H., et al. (2009). Linking phylogenetic identities of bacteria to starch fermentation in an in vitro model of the large intestine by RNA-based stable isotope probing. *Environmental Microbiology*, *11*(4), 914–926. https://doi.org/10.1111/j.1462-2920.2008.01815.x.

Kunej, U., Mikulic-Petkovsek, M., Radisek, S., & Stajner, N. (2020). Changes in the phenolic compounds of hop (Humulus lupulus L.) induced by infection with Verticillium nonalfalfae, the causal agent of hop Verticillium wilt. *Plants (Basel)*, *9*(7), 841. https://doi.org/10.3390/plants9070841.

Lahrmann, U., Strehmel, N., Langen, G., Frerigmann, H., Leson, L., Ding, Y., et al. (2015). Mutualistic root endophytism is not associated with the reduction of saprotrophic traits and requires a noncompromised plant innate immunity. *New Phytologist*, *207*(3), 841–857. https://doi.org/10.1111/nph.13411.

Lambers, H., Mougel, C., Jaillard, B., & Hinsinger, P. (2009). Plant-microbe-soil interactions in the rhizosphere: An evolutionary perspective. *Plant and Soil*, *321*, 83–115. https://doi.org/10.1007/s11104-009-0042-x.

Larsson, L., & Saraf, A. (1997). Use of gas chromatography-ion trap tandem mass spectrometry for the detection and characterization of microorganisms in complex samples. *Molecular Biotechnology*, 7(3), 279–287. https://doi.org/10.1007/BF02740818.

Lee, H. J., & Ryu, D. (2017). Worldwide occurrence of Mycotoxins in cereals and cereal-derived food products: Public health perspectives of their co-occurrence. *Journal of Agricultural and Food Chemistry*, *65*(33), 7034–7051. https://doi.org/10.1021/acs.jafc.6b04847.

Long, S. R. (1996). Rhizobium symbiosis: Nod factors in perspective. *Plant Cell*, *8*(10), 1885–1898. https://doi.org/10.1105/tpc.8.10.1885.

Lundberg, D. S., Lebeis, S. L., Paredes, S. H., Yourstone, S., Gehring, J., Malfatti, S., et al. (2012). Defining the core Arabidopsis thaliana root microbiome. *Nature*, *488*(7409), 86–90. https://doi.org/10.1038/nature11237.

Maillet, F., Poinsot, V., Andre, O., Puech-Pages, V., Haouy, A., Gueunier, M., et al. (2011). Fungal lipochitooligosaccharide symbiotic signals in arbuscular mycorrhiza. *Nature*, *469*(7328), 58–63. https://doi.org/10.1038/nature09622.

Mansfield, J., Genin, S., Magori, S., Citovsky, V., Sriariyanum, M., Ronald, P., et al. (2012). Top 10 plant pathogenic bacteria in molecular plant pathology. *Molecular Plant Pathology*, *13*(6), 614–629. https://doi.org/10.1111/j.1364-3703.2012.00804.x.

Mendes, R., Garbeva, P., & Raaijmakers, J. M. (2013). The rhizosphere microbiome: Significance of plant beneficial, plant pathogenic, and human pathogenic microorganisms. *FEMS Microbiology Reviews*, *37*(5), 634–663. https://doi.org/10.1111/1574-6976.12028.

Mierziak, J., Kostyn, K., & Kulma, A. (2014). Flavonoids as important molecules of plant interactions with the environment. *Molecules*, *19*(10), 16240–16265. https://doi.org/10.3390/molecules191016240.

Monchgesang, S., Strehmel, N., Schmidt, S., Westphal, L., Taruttis, F., Muller, E., et al. (2016). Natural variation of root exudates in Arabidopsis thaliana-linking metabolomic and genomic data. *Scientific Reports*, *6*, 29033. https://doi.org/10.1038/srep29033.

Montes-Grajales, D., Esturau-Escofet, N., Esquivel, B., & Martinez-Romero, E. (2019). Exo-metabolites of Phaseolus vulgaris-Nodulating Rhizobial strains. *Metabolites*, *9*(6), 105. https://doi.org/10.3390/metabo9060105.

Morris, P. F., Bone, E., & Tyler, B. M. (1998). Chemotropic and contact responses of phytophthora sojae hyphae to soybean isoflavonoids and artificial substrates. *Plant Physiology*, *117*(4), 1171–1178. https://doi.org/10.1104/pp.117.4.1171.

Muller, D. B., Vogel, C., Bai, Y., & Vorholt, J. A. (2016). The plant microbiota: Systems-level insights and perspectives. *Annual Review of Genetics*, *50*, 211–234. https://doi.org/10.1146/annurev-genet-120215-034952.

Nobori, T., Velasquez, A. C., Wu, J., Kvitko, B. H., Kremer, J. M., Wang, Y., et al. (2018). Transcriptome landscape of a bacterial pathogen under plant immunity. *Proceedings of the National Academy of Sciences of the United States of America*, *115*(13), E3055–E3064. https://doi.org/10.1073/pnas.1800529115.

Oburger, E., Gruber, B., Schindlegger, Y., Schenkeveld, W. D., Hann, S., Kraemer, S. M., et al. (2014). Root exudation of phytosiderophores from soil-grown wheat. *New Phytologist*, *203*(4), 1161–1174. https://doi.org/10.1111/nph.12868.

Oburger, E., & Jones, D. L. (2018). Sampling root exudates—Mission impossible? *Rhizosphere*, *6*, 116–133. https://doi.org/10.1016/j.rhisph.2018.06.004.

Oldroyd, G. E., Murray, J. D., Poole, P. S., & Downie, J. A. (2011). The rules of engagement in the legume-rhizobial symbiosis. *Annual Review of Genetics*, *45*, 119–144. https://doi.org/10.1146/annurev-genet-110410-132549.

Pang, Q., Zhang, T., Wang, Y., Kong, W., Guan, Q., Yan, X., et al. (2018). Metabolomics of early stage plant cell-microbe interaction using stable isotope Labeling. *Frontiers in Plant Science*, *9*, 760. https://doi.org/10.3389/fpls.2018.00760.

Papadopoulou, K., Melton, R. E., Leggett, M., Daniels, M. J., & Osbourn, A. E. (1999). Compromised disease resistance in saponin-deficient plants. *Proceedings of the National Academy of Sciences of the United States of America*, *96*(22), 12923–12928. https://doi.org/10.1073/pnas.96.22.12923.

Petriacq, P., Williams, A., Cotton, A., McFarlane, A. E., Rolfe, S. A., & Ton, J. (2017). Metabolite profiling of non-sterile rhizosphere soil. *Plant Journal*, *92*(1), 147–162. https://doi.org/10.1111/tpj.13639.

Pett-Ridge, J., & Firestone, M. K. (2017). Using stable isotopes to explore root-microbe-mineral interactions in soil. *Rhizosphere*, *3*(2), 244–253. https://doi.org/10.1016/j.rhisph.2017.04.016.

Pini, F., East, A. K., Appia-Ayme, C., Tomek, J., Karunakaran, R., Mendoza-Suarez, M., et al. (2017). Bacterial biosensors for in vivo spatiotemporal mapping of root secretion. *Plant Physiology*, *174*(3), 1289–1306. https://doi.org/10.1104/pp.16.01302.

Pinu, F. R., & Villas-Boas, S. G. (2017). Extracellular microbial metabolomics: The state of the art. *Metabolites*, 7(3), 43. https://doi.org/10.3390/metabo7030043.

Pusztahelyi, T., Holb, I. J., & Pocsi, I. (2015). Secondary metabolites in fungus-plant interactions. *Frontiers in Plant Science*, *6*, 573. https://doi.org/10.3389/fpls.2015.00573.

Rowe, H. C., & Kliebenstein, D. J. (2008). Complex genetics control natural variation in Arabidopsis thaliana resistance to Botrytis cinerea. *Genetics*, *180*(4), 2237–2250. https://doi.org/10.1534/genetics.108.091439.

Sasse, J., Martinoia, E., & Northen, T. (2018). Feed your friends: Do Plant exudates shape the root microbiome? *Trends in Plant Science*, *23*(1), 25–41. https://doi.org/10.1016/j.tplants.2017.09.003.

Schlaman, H. R., Spaink, H. P., Okker, R. J., & Lugtenberg, B. J. (1989). Subcellular localization of the nodD gene product in Rhizobium leguminosarum. *Journal of Bacteriology*, *171*(9), 4686–4693. https://doi.org/10.1128/jb.171.9.4686-4693.1989.

Schmitz, A. M., & Harrison, M. J. (2014). Signaling events during initiation of arbuscular mycorrhizal symbiosis. *Journal of Integrative Plant Biology*, *56*(3), 250–261. https://doi.org/10.1111/jipb.12155.

Schwachtje, J., Fischer, A., Erban, A., & Kopka, J. (2018). Primed primary metabolism in systemic leaves: A functional systems analysis. *Scientific Reports*, *8*(1), 216. https://doi.org/10.1038/s41598-017-18397-5.

Silva, L. P., & Northen, T. R. (2015). Exometabolomics and MSI: Deconstructing how cells interact to transform their small molecule environment. *Current Opinion in Biotechnology*, *34*, 209–216. https://doi.org/10.1016/j.copbio.2015.03.015.

Smith, S. E., & Smith, F. A. (2012). Fresh perspectives on the roles of arbuscular mycorrhizal fungi in plant nutrition and growth. *Mycologia*, *104*(1), 1–13. https://doi.org/10.3852/11-229.

Smith, S. E., Smith, F. A., & Jakobsen, I. (2003). Mycorrhizal fungi can dominate phosphate supply to plants irrespective of growth responses. *Plant Physiology*, *133*(1), 16–20.

Stahl, E., Bellwon, P., Huber, S., Schlaeppi, K., Bernsdorff, F., Vallat-Michel, A., et al. (2016). Regulatory and functional aspects of Indolic metabolism in plant systemic acquired resistance. *Molecular Plant*, *9*(5), 662–681. https://doi.org/10.1016/j.molp.2016.01.005.

Stopka, S. A., Agtuca, B. J., Koppenaal, D. W., Pasa-Tolic, L., Stacey, G., Vertes, A., et al. (2017). Laser-ablation electrospray ionization mass spectrometry with ion mobility separation reveals metabolites in the symbiotic interactions of soybean roots and rhizobia. *Plant Journal*, *91*(2), 340–354. https://doi.org/10.1111/tpj.13569.

Strehmel, N., Bottcher, C., Schmidt, S., & Scheel, D. (2014). Profiling of secondary metabolites in root exudates of Arabidopsis thaliana. *Phytochemistry*, *108*, 35–46. https://doi.org/10.1016/j.phytochem.2014.10.003.

Stringlis, I. A., Yu, K., Feussner, K., de Jonge, R., Van Bentum, S., Van Verk, M. C., et al. (2018). MYB72-dependent coumarin exudation shapes root microbiome assembly to promote plant health. *Proceedings of the National Academy of Sciences of the United States of America*, *115*(22), E5213–E5222. https://doi.org/10.1073/pnas.1722335115.

Tahara, S., Ohkawa, K., Takayama, T., & Ogawa, Y. (2001). The third naturally occurring attractant toward zoospores of phytopathogenic Aphanomyces cochlioides from the Spinacia oleracea host plant. *Bioscience, Biotechnology, and Biochemistry*, *65*(8), 1755–1760. https://doi.org/10.1271/bbb.65.1755.

Thimmappa, R., Geisler, K., Louveau, T., O'Maille, P., & Osbourn, A. (2014). Triterpene biosynthesis in plants. *Annual Review of Plant Biology*, *62*(65), 225–257. https://doi.org/10.1146/annurev-arplant-050312-120229.

Ton, J., & Mauch-Mani, B. (2004). Beta-amino-butyric acid-induced resistance against necrotrophic pathogens is based on ABA-dependent priming for callose. *Plant Journal*, *38*(1), 119–130. https://doi.org/10.1111/j.1365-313X.2004.02028.x.

Tsai, S. M., & Phillips, D. A. (1991). Flavonoids released naturally from alfalfa promote development of symbiotic glomus spores in vitro. *Applied and Environmental Microbiology*, *57*(5), 1485–1488. https://doi.org/10.1128/AEM.57.5.1485-1488.1991.

Udvardi, M., & Poole, P. S. (2013). Transport and metabolism in legume-rhizobia symbioses. *Annual Review of Plant Biology*, *62*(64), 781–805. https://doi.org/10.1146/annurev-arplant-050312-120235.

van der Heijden, M. G., de Bruin, S., Luckerhoff, L., van Logtestijn, R. S., & Schlaeppi, K. (2016). A widespread plant-fungal-bacterial symbiosis promotes plant biodiversity, plant nutrition and seedling recruitment. *The ISME Journal*, *10*(2), 389–399. https://doi.org/10.1038/ismej.2015.120.

Vandenkoornhuyse, P., Mahe, S., Ineson, P., Staddon, P., Ostle, N., Cliquet, J. B., et al. (2007). Active root-inhabiting microbes identified by rapid incorporation of plant-derived carbon into RNA. *Proceedings of the National Academy of Sciences of the United States of America*, *104*(43), 16970–16975. https://doi.org/10.1073/pnas.0705902104.

Velickovic, D., Agtuca, B. J., Stopka, S. A., Vertes, A., Koppenaal, D. W., Pasa-Tolic, L., et al. (2018). Observed metabolic asymmetry within soybean root nodules reflects unexpected complexity in rhizobacteria-legume metabolite exchange. *The ISME Journal*, *12*(9), 2335–2338. https://doi.org/10.1038/s41396-018-0188-8.

Vilchez, J. I., Yang, Y., He, D., Zi, H., Peng, L., Lv, S., et al. (2020). DNA demethylases are required for myo-inositol-mediated mutualism between plants and beneficial rhizobacteria. *Nat Plants.*, *6*(8), 983–995. https://doi.org/10.1038/s41477-020-0707-2.

Voges, M., Bai, Y., Schulze-Lefert, P., & Sattely, E. S. (2019). Plant-derived coumarins shape the composition of an Arabidopsis synthetic root microbiome. *Proceedings of the National Academy of Sciences of the United States of America*, *116*(25), 12558–12565. https://doi.org/10.1073/pnas.1820691116.

Voothuluru, P., Braun, D. M., & Boyer, J. S. (2018). An in vivo imaging assay detects spatial variability in glucose release from plant roots. *Plant Physiology*, *178*(3), 1002–1010. https://doi.org/10.1104/pp.18.00614.

Wang, M., Carver, J. J., Phelan, V. V., Sanchez, L. M., Garg, N., Peng, Y., et al. (2016). Sharing and community curation of mass spectrometry data with global natural products social molecular networking. *Nature Biotechnology*, *34*(8), 828–837. https://doi.org/10.1038/nbt.3597.

Wang, M., Schafer, M., Li, D., Halitschke, R., Dong, C., McGale, E., et al. (2018). Blumenols as shoot markers of root symbiosis with arbuscular mycorrhizal fungi. *eLife*, 7. https://doi.org/10.7554/eLife.37093.

Wang, E., Schornack, S., Marsh, J. F., Gobbato, E., Schwessinger, B., Eastmond, P., et al. (2012). A common signaling process that promotes mycorrhizal and oomycete colonization of plants. *Current Biology*, *22*(23), 2242–2246. https://doi.org/10.1016/j.cub.2012.09.043.

Wasson, A. P., Pellerone, F. I., & Mathesius, U. (2006). Silencing the flavonoid pathway in Medicago truncatula inhibits root nodule formation and prevents auxin transport regulation by rhizobia. *Plant Cell*, *18*(7), 1617–1629. https://doi.org/10.1105/tpc.105.038232.

Waters, M. T., Gutjahr, C., Bennett, T., & Nelson, D. C. (2017). Strigolactone Signaling and evolution. *Annual Review of Plant Biology*, *62*(68), 291–322. https://doi.org/10.1146/annurev-arplant-042916-040925.

Weidenhamer, J. D., Mohney, B. K., Shihada, N., & Rupasinghe, M. (2014). Spatial and temporal dynamics of root exudation: How important is heterogeneity in allelopathic interactions? *Journal of Chemical Ecology*, *40*(8), 940–952. https://doi.org/10.1007/s10886-014-0483-4.

Yuan, J., Zhao, J., Wen, T., Zhao, M., Li, R., Goossens, P., et al. (2018). Root exudates drive the soil-borne legacy of aboveground pathogen infection. *Microbiome*, *6*(1), 156. https://doi.org/10.1186/s40168-018-0537-x.

Zhalnina, K., Louie, K. B., Hao, Z., Mansoori, N., da Rocha, U. N., Shi, S., et al. (2018). Dynamic root exudate chemistry and microbial substrate preferences drive patterns in rhizosphere microbial community assembly. *Nature Microbiology*, *3*(4), 470–480. https://doi.org/10.1038/s41564-018-0129-3.

Ziegler, J., Schmidt, S., Chutia, R., Muller, J., Bottcher, C., Strehmel, N., et al. (2016). Non-targeted profiling of semi-polar metabolites in Arabidopsis root exudates uncovers a role for coumarin secretion and lignification during the local response to phosphate limitation. *Journal of Experimental Botany*, *67*(5), 1421–1432. https://doi.org/10.1093/jxb/erv539.

CHAPTER SIX

A practical guide to implementing metabolomics in plant ecology and biodiversity research

Henriette Uthe[a,b,*], Nicole M. van Dam[a,b], Maxime R. Hervé[c], Maria Sorokina[d], Kristian Peters[e], and Alexander Weinhold[a,b]
[a]German Centre for Integrative Biodiversity Research (iDiv) Halle-Jena-Leipzig, Jena, Germany
[b]Institute of Biodiversity, Friedrich Schiller University, Jena, Germany
[c]IGEPP, INRAE, Institut Agro, Univ Rennes, Rennes, France
[d]Institute for Inorganic and Analytical Chemistry, Friedrich Schiller University, Jena, Germany
[e]Bioinformatics and Scientific Data, Leibniz Institute of Plant Biochemistry, Halle (Saale), Germany
*Corresponding author: e-mail address: henriette.uthe@idiv.de

Contents

Advances in Botanical Research, Volume 98
ISSN 0065-2296
https://doi.org/10.1016/bs.abr.2020.09.019

Abstract

Ecologists study the interactions between plants and their changing biotic and abiotic environments. These interactions are directly mediated by the chemical phenotypes of the plants i.e. their metabolomes. In consequence, the study of these metabolomes (a branch of analytical chemistry called metabolomics) is a powerful tool for describing the chemical aspects of plant ecology. The application of metabolomics to ecology is growing because of current technical developments in high throughput and high-resolution mass spectrometry. Advanced metabolomics allows the combination of traditional plant phenotypic traits (biomass, flowering probability) with plant chemical phenotypes. The combination of ecology and the analytical chemistry of advanced metabolomics has given rise to a new discipline, that of "ecometabolomics." However, realizing the true potential of ecometabolomics is hindered because its constituent disciplines differ greatly in their experimental approaches, workflows and norms. These differences present challenges that must be overcome if the two constituent disciplines are to be successfully integrated and the potential of ecometabolomics successfully realized.

In our EcoMetabolomics facility at the German Centre for integrative Biodiversity Research (iDiv) Halle-Jena-Leipzig, we have, since 2016, built up expertise on bringing metabolomics into ecology research. Our experience tells us that the main challenges of combining the two disciplines pertain to technical and experimental issues or are related to differences in scientific background knowledge between metabolomicists and ecologists. Technical and experimental challenges in ecometabolomics occur mainly when experimental conditions are such that it is hard to take samples of good quality for metabolomic analysis. Moreover, ecologists taking metabolomic samples in the field often face challenges in obtaining enough replicates, providing proper storage conditions and meeting export regulations (Nagoya protocols, phytosanitary regulations). We have also noticed that ecologists lack knowledge of the dynamics of plant metabolomes, e.g. how they change over the day. And that ecologists often have unrealistic expectations about how fast the data can be processed and how many unknown compounds of interest it is possible to identify in non-model species. This lack of knowledge may cause unproductive collaborations between ecologists and metabolomicists. In this chapter, we share our experience in ecometabolomics projects by describing the main challenges in combining different areas of expertise. We aim to provide strategies and practical solutions, particularly for project management, training, advising on experimental workflows before sampling and communicating realistic outcomes of the metabolomic analysis. We believe these strategies will help ecologists and metabolomics experts collaborate more effectively and fruitfully.

1. Introduction

"We live by our eyes and ears and tend generally to be oblivious to the chemical happenings in our surrounds. Such happenings are ubiquitous. All organisms engender chemical signals, and all, in their respective ways, respond to the chemical emissions of others"

Eisner et al. (1995)

Metabolites are key components in both biochemical and ecological processes in all organisms. From the activity within individual cells to ecological processes and ecosystem functioning, everything is driven by changing levels of metabolites. Moreover, metabolites exist in a huge and complex array. It is estimated that species in the kingdom Plantae produce 200,000 to 1,000,000 metabolites (Dixon & Strack, 2003). Of these, about 51,000 are produced by Angiosperms (Afendi et al., 2012; Saito & Matsuda, 2010). Metabolomic approaches such as LC-MS, GC–MS and NMR promise to reveal the biochemical basis for many ecological interactions. In theory, they have the attractive advantage that they are able, from a single sample, to reveal large numbers of different metabolites and their abundances in a single analytical process. The sample can come from a single tissue, an organ, or the organism as a whole. Furthermore, metabolomics approaches can be applied to any species without previous knowledge of its biochemical or genetic composition (Peters et al., 2018). Because ecological research often deals with non-model organisms, metabolomics seems to be the perfect tool to unravel the molecular mechanisms of interaction in all ecosystems. The increased application of metabolomics techniques in ecological studies, both in the field and the greenhouse has created a new transdisciplinary research discipline: ecometabolomics. In ecometabolomics, metabolomic techniques are used to unravel molecular mechanisms governing the interactions of species with their abiotic and biotic environment across different spatial and temporal scales.

Despite its promise, this new transdisciplinary research discipline has both advantages and disadvantages. On the one hand, working within this discipline motivates scientists from the two different constituent disciplines to explore new analytical possibilities and research areas to answer highly relevant questions. On the other hand, it is characterized by needing considerable communication between the constituent disciplines and careful management of their divergent expectations if the ecometabolomic collaboration is not to fail and consume time and money without producing trustworthy results. We are very familiar with this need for communication and expectation management. Metabolomics is still rarely carried out in ecological research by the ecologists themselves. Therefore, ecologists often adopt the metabolomics approach without being familiar with its techniques, data structures and analyses. Furthermore, they often lack hands-on experience in chemical analytics. Because few laboratories run metabolomics units specializing in ecological research, these types of projects arise from collaborations. The collaboration partners providing the metabolomic infrastructure are usually academic facilities and unfortunately often considered as pure service facilities, which are supposed to add "another" molecular level of information to an existing approach. In many cases, experiments are often not designed in such a way that it is possible to apply metabolomics techniques. We still experience the acquisition of terabytes of metabolomic data without an appropriate experimental design, sample processing workflow, or clear data analysis strategy. The implementation of metabolomics in ecology is a multiple-step workflow (Fig. 1) and full of challenges caused by the complexities introduced by the involvement of the different scientific disciplines involved: ecology, analytical chemistry and bioinformatics.

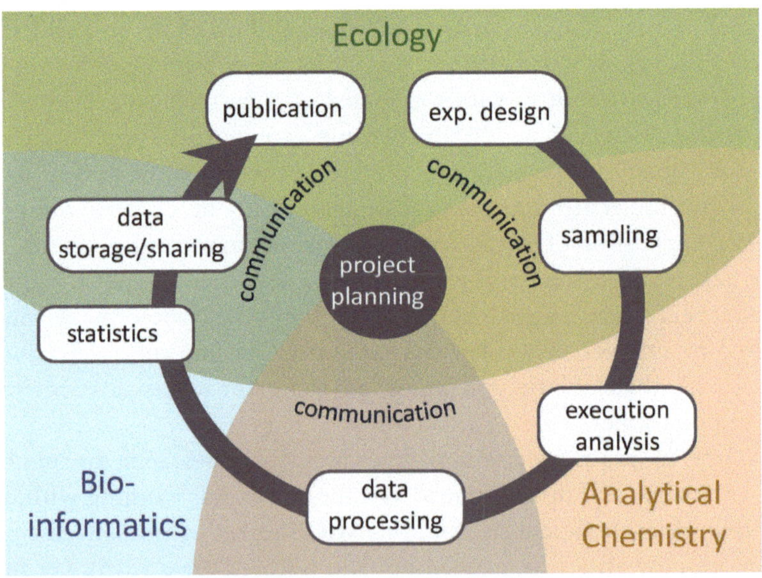

Fig. 1 Overview of the different steps of an ecometabolomic project. Each step requires the skills and experience of the respective scientific fields, Ecology (in green), Analytical Chemistry (orange) and Bioinformatics (blue). Communication and interactions among experts from these research fields, as indicated by the intersections, are necessary for a successful project planning.

We think a good project and expectation management can improve the cooperation of all three disciplines and is essential for a fruitful outcome. In this chapter, we name the main challenges in ecometabolomics and provide guidelines to help overcome them. We focus on plant metabolomics and the techniques HPLC- and GC–MS. These two methods are the ones we and the majority of scientists most frequently use and those in which we have the most expertise.

2. Project planning

2.1 Identify and manage the stakeholders

The integration of metabolomics with ecological and biodiversity needs the continuous involvement of different people who are interested, involved and affected by the scientific project. These are the stakeholders (Fig. 2). We recommend that all of the stakeholders are identified before the project starts. In many projects, PhD students are supposed to fill many of these positions during the 3–4 years of their doctoral study. Where this is so, they must be provided with good training and have an appropriate scientific and project-related background. Thus, it is also necessary to ensure the competence of the people responsible for their training.

As soon all stakeholders are identified, they should together discuss the experimental design and research question to ensure that the design, the technical approach and the statistical analysis fit together.

2.2 Time management

The duration of the steps in the workflow is often underestimated and should be discussed between all stakeholders. They can be estimated roughly, however (Fig. 2). Nevertheless, additional time might be necessary for student training and for the development and testing of specific methods for sample extraction and mass spectrometric analysis (see Section 5). Furthermore, ecological studies often accumulate very large numbers of samples. Indeed, hundreds to thousands of samples are not unusual. Because MS analysis of a sample takes 5–40 min, the running time for complete analysis rapidly adds up. This examples with 500 samples (extracts) and a 20 min gradient will give an impression:

$$(500 \text{ samples} + 50 \text{ QC} + 60 \text{ blanks}) * 2 \text{ modes(pos or neg)} * \text{gradient length } (20) = 406 \text{ h}$$

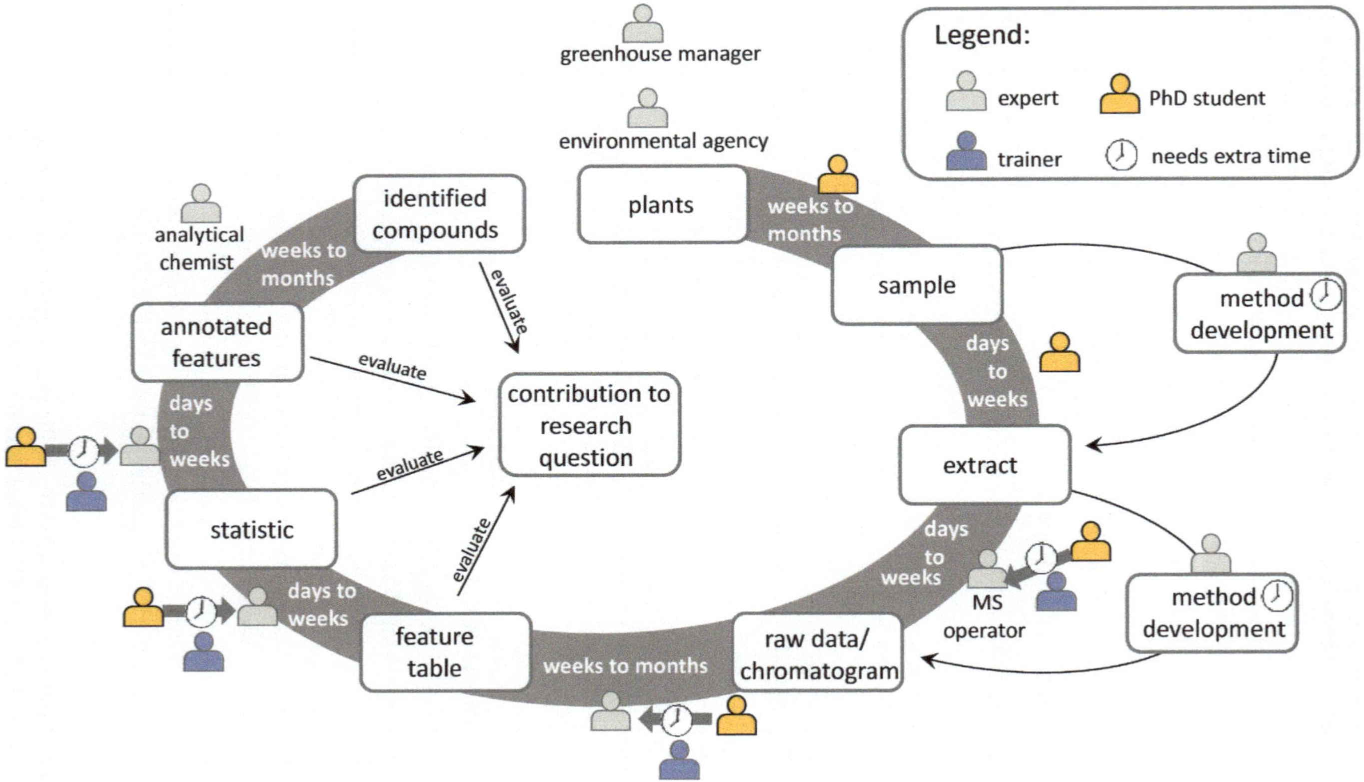

Fig. 2 Overview of the involved stakeholders and the approximate duration of every step. The gray figures symbolize the persons who are experts in their field or task at the beginning of the project. The yellow figures symbolize the untrained persons at the beginning of the project (usually PhD students), which need to gain knowledge and experience, which is transferred by a person with the function of a trainer (in blue), e.g. experts or facility operators. The clock symbolizes the extra time for training and method development in addition to the estimated duration in the gray circle.

This implies that the MS instrument of the cooperation partner would be busy and not available for anyone else for at least 17 days. The remaining part of the pipeline also requires both considerable effort and time neither of which can be calculated easily. There must be a reasonable understanding of the input for each project and a reasonable amount of time must also be allocated to the data analysis.

2.3 Expectation management

Due to technological and software development, metabolomic analysis in ecological and plant science is often considered to be standardized and routine analysis. The idea that untargeted mass spectrometry (MS)-based metabolomics analysis will result in a large list of "identified" small molecules that can be mapped to networks and pathways is often assumed. Furthermore, it is often expected that metabolomic analysis is straightforward, that it can be applied easily and rapidly, and that the output integrated easily with other results. This might be true for some approaches but not for most. The operators of the MS platform have their expectations as well. They expect a large number of biological and technical replicates, consistent harvesting and extracting procedures, and conception of the MS data structure. It is highly recommended to communicate these expectations at the very beginning of the project. The expectations of both sides must be aligned and finding answers to the following questions might help to clarify expectations:

- Are the hypotheses communicated clearly to all stakeholders?
- Which step in the workflow needs adaptation or method development?
- How long will every phase of the workflow last?
- What is the minimal and the best outcome?
- How many replicates are possible and reasonable?
- What are the realistic capacities of the platform operators? How many samples can realistically be measured given that other projects have to be run on the same instrument?
- How can samples be collected, transported, stored and prepared in a sufficiently consistent way to keep technical variations to an acceptable level?
- What is the best metabolomic approach to answering the biological question?
- What is the most appropriate statistical approach? What should the figures look like?
- Which level of identification is necessary to answer the biological question?

- What is necessary to reach that level of identification?
- What kind of journal should the results be published in?
- If several publications are planned, what is the thematic priority of each?

In our experience, expectations of the outcome can change during the project. Usually, once the data has been collected and the first data analysis has been performed. It is therefore recommended to repeat the expectation management process described at that time.

2.4 Legal considerations

Ecological and biodiversity research are global sciences, researchers often travel all over the world to explore different ecosystems and to sample in different places. This research rarely results in the identification of a pharmaceutically interesting compound and also rarely yields any financial profit. Nevertheless, researchers dealing with plant material from outside their home or host country must follow the legal procedures and government guidelines for the country from which the plant material originates. As scientists, we must respect the right of every state to regulate access to its genetic resources and to obtain a fair and equitable share in any benefits arising from them. To ensure this benefit sharing, in 2014 over 70 countries around the world agreed to the *Nagoya Protocol on Access to Genetic Resources and the Fair and Equitable Sharing of Benefits Arising from their Utilization (ABS)*.[a]

ABS acts as a tool to recognize more fully the value of biodiversity in research on and production of novel products while at the same time providing economic incentives for the conservation and sustainable use of nature through its valorization. In principle, contracting states are obliged to grant access to genetic resources by other contracting parties (not necessarily free of charge and unconditionally). At the same time, they have the power to make access to their genetic resources dependent on their consent and to demand a balanced and fair sharing of the benefits arising from the use of these resources. The implementation of the protocol is extremely different from country to country and can involve anything from complex long-lasting bureaucratic procedures to very short and efficient email contact. It is important to note that a specialized ABS system by the "Food and Agriculture Organization of the United Nations"[b] exists to regulate the

[a] www.cbd.int.
[b] www.fao.org.

utilization of plant genetic resources for food and agriculture. Many of these plants do not fall within the scope of the Nagoya Protocol. It is therefore recommended to:

- check if the used plant(s) fall within the scope of the Nagoya Protocol[a]
- contact the National Focal Point of the respective country
- contact your university legal department
- contact your national research association[c]
- (for EU scientists) use the DECLAIRE system to submit the due diligence declaration[d]

3. Experimental design and sampling

3.1 Experimental design and replicates

"Begin with the end in mind" is an important piece of advice you can find in influential management books such as those by Stephen R. Covey (Covey, 2004). This is even more so for projects that include metabolomic analyses. In the end, any metabolomic platform will produce "fat datasets." This term refers to datasets with relatively few rows, i.e. biological replicates, and many columns, i.e. variables, in this case, mass features or metabolites. Moreover, many variables in the dataset are interdependent, e.g., because they are in the same biosynthetic route. Most statistical models, such as Multiple Analysis of Variance (MANOVA), are not suitable for dealing with this kind of dataset (St»hle & Wold, 1990). For this reason, other multivariate analyses, such as Principal Component Analysis (PCA) or methods derived from Partial Least Squares (PLS) analyses are used. These models can visualize differences in metabolomic profiles and help to identify mass features explaining differences between treatment groups (see Section 7). Even so, these multivariate analyses also require a sufficient number of replicates. Most models require a minimum of 3 biological replicates to fit a model to the data. In order to obtain reliable results, 6 to 10 replicates are needed. In order to establish the optimal number of replicates, you should ideally run a pre-experiment with samples similar to those to be taken in the real experiments. Having this data allows the experimenter to estimate the variance in the full dataset and thereby the optimal number of replicates needed to reliably identify a defined difference between experimental groups. Trutschel et al. provide a

[c] http://www.dfg.de/foerderung/info_wissenschaft/2019/info_wissenschaft_19_90/index.html.

[d] http://ec.europa.eu/environment/nature/biodiversity/international/abs.

workflow and R-code to estimate the contribution of biological, technical and instrumental variation to the overall variation occurring in plant metabolomics (Trutschel, Schmidt, Grosse, & Neumann, 2015). They analyzed 108 samples of 27 genetically uniform *Arabidopsis thaliana* individuals, with two independent extractions per plant and two technical (injection) replicates per extract using liquid chromatography coupled mass spectrometry (LC-MS). Based on this hierarchical experiment, they were able to calculate how many biological replicates are needed to determine differences in chemical signals (features) with a given effect size and reliability (Trutschel et al., 2015). Instrumental and technical variation were generally moderate compared to the biological variation observed. Considering that obtaining biological samples is more costly than technical replicates (9:1), they calculated that in order to reliably (max. 5% false positives) observe a difference of 1 unit (delta [log2] = 0.5) for 80% of the metabolites, it is necessary to take 16 independent biological replicates with two technical replicates (Trutschel et al., 2015). In ecometabolomics, it is not very common to run a pre-experiment of such size. This may be due to lack of time and funding—the analysis of a single sample may cost up to 50 Euros if done externally. Moreover, the number of samples may be limited by circumstances, including the space available in growth rooms, maximum size of experimental field plots and, when sampling in the wild, how many individuals can be found during a sampling campaign. It is probably for these reasons that the number of replicates taken for metabolomic analyses is 5–15 per experimental treatment. Considering that it is usually possible to separate within-species variation from the effects of environmental variation or experimental treatments on metabolomes, these "gut feeling" numbers (Trutschel et al., 2015) usually suffice to detect ecological relevant differences.

3.2 What are "true replicates"?

Analytical chemists and ecologists may have different views on what is a true replicate. This stems in part from the fact that chemists are trained in minimizing (technical) variance in their measurements, whereas ecologists are primarily interested in assessing the (biological) variation. To pre-empt conflicts about the replicate number in collaborative ecometabolomic projects, it is of the utmost importance to specify the research question at the beginning. A well-defined research question guides the experimental design as well as the numbers and kinds of replicates. Experiments aiming to resolve well-defined questions, e.g., linking plant resistance to differences in metabolomic profiles, are commonly carried out with defined genotypes grown under controlled conditions in greenhouses or growth cabinets. Each plant commonly serves as a single replicate. This allows the estimation of individual variance within a species, population or treatment group. In addition, by removing other influences (biotic or abiotic), genetically determined metabolomic variation can be identified much better. In some cases, for example with small plants like *Arabidopsis thaliana* or pollen, it may be hard to obtain sufficient biomass for extraction. This is commonly solved by pooling several individuals into one "biological replicate." Whereas this makes it impossible to estimate variation among individuals, it still allows the identification of general shifts in chemical profiles due to treatments. However, when the genetic background of the seed stock used to grow the plants is not uniform, pooling may also dilute rare compounds that are present in only a few genotypes. The best advice is to avoid pooling and, if possible, adapt the extraction procedure so that sufficient material can be extracted from smaller amounts of biomass. Most important of all though is to talk to the stakeholder who is an expert on plant metabolomics at the very beginning of the experiment about how much biomass is needed for a high-quality extraction and analysis.

In ecological and biodiversity studies, it is not always possible to take "true replicates." When sampling in the wild, either the number of individuals of a (rare) species or the time taken to find enough individuals can be limiting factors. In addition, there may be a limit on how many samples can be taken to the laboratory due to export limitations. In that case, the expectations about which questions can be addressed must be clearly managed from the start. One consequence may be that the study remains rather descriptive if the number of replicates is insufficient for decent statistical analyses. "True replication" may also be difficult in biodiversity experiments such as the Jena Experiment (Weisser et al., 2017) or BEF China (Kröber, Heklau, & Bruelheide, 2015), where e.g. reduced designs or imperfect replicates become necessary for realistic planning.

3.3 Sampling and storage

In addition to optimizing replication, it is important to strive for uniformity. Even among leaves of the same plant, the metabolome varies strongly with leaf age and plant ontogeny. Thus, when comparing different treatments, samples should be taken from leaves of the same age. Better still, the leaves sampled should be of the same ontogenetic position (Papazian et al., 2019).

Fruits, flowers and roots have different metabolomes than leaves. Another important factor is the time of day at which sampling takes place. Plant metabolomes vary strongly over the day due to differences in light availability and temperature (Augustijn, de Groot, & Alia, 2019; Hildreth, Foley, Muday, Helm, & Winkel, 2020; Kim, Yon, Gaquerel, Gulati, & Baldwin, 2011).

When sampling large experiments, it is advisable to first label the plants or leaves to be sampled, prepare all materials (label bags, tubes) and then sample all plants in the shortest time possible. If sampling occurs over different days, make sure to randomize sampling of different treatment groups to evenly distribute the effects of sampling day conditions (rain, temperature fluctuations among days) on metabolomes.

After you sample, it is important to quench the plant metabolism. That is, in other words, to stop all enzyme activities by freezing or to destroy (denature) the enzymes by boiling or putting the material in an organic solvent (Villas-Bas, Roessner, Hansen, Smedsgaard, & Nielsen, 2007). Quenching by flash-freezing in liquid nitrogen right after harvest is the preferred method. This stops all biological processes immediately and prevents metabolite breakdown. It should be mentioned that the enzyme activity is just paused in this case. To avoid any reactivation the frozen materials can be processed as is—under liquid nitrogen—or after complete (!) freeze-drying (lyophilization). Freeze-drying removes all water for enzymatic reactions and thus eases further processing as the materials can then be processed at room temperature. Because freeze-drying takes place under high vacuum, highly volatile compounds may be lost in the process (Díaz-Maroto, Pérez-Coello, González Viñas, & Cabezudo, 2003). However, this is not necessarily avoided by processing frozen materials. There is always the risk of such materials thawing during processing and this too may lead to the loss of labile compounds. If liquid nitrogen or dry ice is not available at the sampling site, a second-best option is cooling the samples on ice packs until freezing or for short-term storage immersing them in solvents (methanol, ethanol). It should be mentioned, that long-term storage in methanol will lead to methylated artifacts (Sauerschnig, Doppler, Bueschl, & Schuhmacher, 2017). In last resort a practical solution preserving samples at remote field sites is to rapidly dry them in the dark, e.g. by air-drying or in silica powder or ground cat litter. Of course, the samples will experience strong drought stress and the metabolome will be modified accordingly. In addition, labile compounds may be lost during drying which is something to consider during the planning and when discussing the results.

At any rate, ensure you have sufficient material for the extraction and sufficient replicates for statistical analyses—talk to your local experts about this before going on your sampling expedition!

4. Preliminary experiments

A pre-experiment might not be possible in every collaboration. Nevertheless, even testing growing and experimental conditions of a small number of replicates brings the possibility to test the metabolomic approach on many different levels. First of all, it brings the stakeholders together and forces them to communicate about all the challenges named in this chapter. It also provides an opportunity to adapt to running protocols, develop workflows, and elaborate strategies for data analysis. This process is especially useful for familiarizing a team with the origin and structure of the data, particularly so when the approach is more or less new to the team. Preliminary data can also help adjust hypotheses and expectations. Such adjustment ultimately saves considerable time and energy during the main experiment.

For clear answers to our biological question, we aim for clear statistical effects on our plants caused by the observed factor of interest (e.g. stress, community level, etc.). However, there may be considerable variability in the data that has to be considered during the workflow (see Table 1). Variability arises from the plants or their location (biological variance) and from the workflow itself (technical variance). If this variance is not properly handled it might obscure any effect of the treatments. Pre-testing helps identify the factors causing biological and technical variability.

5. The metabolomic analysis

Metabolomics approaches are widely used across multiple research disciplines. They range from medicine (e.g. personalized medicine (Koen, Du Preez, & Loots, 2016)) over food science (e.g. food safety and food quality (Li et al., 2020)) and environmental research (Bundy, Davey, & Viant, 2009) to plant science (Weckwerth & Kahl, 2013).

In the same ways as these disciplines differ very much, so do the metabolomic analyses applied to each of them and the used equipment. This is also true for ecometabolomics. The analytical strategies can be divided roughly into four categories (Table 2). Researchers in ecology and biodiversity are highly recommended to align their research question with these categories so as to identify the appropriate strategy.

Table 1 List of the main causes of biological and technical variance and further obstacles of the metabolomic pipeline.

	Metabolomic issues	Recommendations
Exp. design	• Variance within species • Environmental variance • Number replicates too low • treatment variance • Variance in growth habit • Too little plant material for analysis	• At least five biological replicates • Pool biological replicates for sufficient material • Randomize the blocks • Record treatment • Reduce variance within species (cuttings, clones, etc.) • In the Greenhouse: o Preexperiment (~20 plants) o Assure consistent growing conditions • In the field: Presampling
Harvesting	• Metabolomic changes over time and due to stress • Variance in harvesting procedure • Incomplete quenching of the metabolome • Insufficient cooling	• greenhouse: o Preexperiment for handling improvement o Use liquid nitrogen or dry ice • Field: o Apply consistent air drying o Use silica pearls/sand or cat litter
Storage	• Decay of metabolites (enzymes, UV light, foul)	• If possible store at −20 °C or −80 °C • Store dried material in tight bags/exicator with silica • Store in the dark (apply vacuum)
Grinding	• Inhomogeneous grinding • Tough, oily or fibrous material • Decay of metabolites	• Choose grinding mill with low final fineness • Cryogenic grinding • Include defatting step (e.g. hexane) • Avoid thawing or heat development • Sequentially grinding + cooling

Extraction	• Variance by inconsistent procedure • Metabolism insufficiently quenched • Chemical identity not preserved • Cross contamination • Contamination with PEG	• Do technical replication (preexperiment) • If applicable: boiling for sufficient quenching • Use single use equipment (beads, tubes, etc.) • Use MS glassware only (NO dishwasher!) • Use MS-grade solvents only • Include "extraction blank"[a]
LC- or GCMS	• Low peak separation • Unstable retention time and intensity • Low peak number • High background • Accumulating peaks • Poor MS/MS fragmentation	• Adjust chromatographic gradient, solvents, temperature or separation column • Do technical replication (e.g. $20 \times$ QC injection) • Test positive and negative MS mode • Solvent blank and QC after every 10th injection • Analyze "extraction blank" • Adjust MS/MS method parameter
Data analysis	• Unusual software tools • Wrong statistical approach	• Use testing data set (preexperiment) to: o Familiarize with processing tools (training) o Test statistical approach o Evaluate biological and technical variance (computing distance matrix analyze with PCoA)
Identification	• Unusual software tools • Insufficient spectral libraries for non-model organisms • Unavailability of standards • Get "stuck" in feature annotation	• Use testing data set (preexperiment) to o Familiarize with identification workflow o Choose standards to order • Decide: compound ID or compound class? • Literature search to create a species-specific metabolite library (e.g. KNApSAcK[b])

[a]Applying the same solvents, chemicals and standard operating procedure, as for the test samples, but in the absence of any actual biological sample.
[b]http://www.knapsackfamily.com/MetaboliteEcology/top.php.
Given recommendations are highly case specific. This table does not claim to be complete.

Table 2 Overview about the general four metabolomic strategies.

		Strategy	Definition
Metabolomics	Validation	Targeted metabolomics	Absolute(relative) quantification and identification of one metabolite or a small number of metabolites
		Metabolomic profiling	Analysis to identify and quantify metabolites related through similar chemistries or metabolic pathways. Normally employing chromatographic separation before detection with minimal sample preparation. Analysis of a group of specific metabolites (e.g. lipids, sugar, amino acids, glucosinolates)
	Discovery	Metabolomic fingerprinting	A global, rapid and high-throughput analysis of the metabolites extracted from biological samples to identify *patterns*. It measures the metabolites that exist in a sample at a given time, this measurement is thought of as a "fingerprint," a unique pattern specific to that tissue at that time that may represent the outcome of a certain biological process
		Untargeted metabolomics	A global, rapid and high-throughput analysis of the metabolites extracted from biological samples to identify *patterns* and to *identify* metabolites of interest or metabolite classes

Untargeted metabolomics aims to detect as many molecules as possible and to create the most comprehensive possible data set. This can be achieved by two very distinct analytical techniques: mass spectrometry (MS) and Nuclear Magnetic Resonance (NMR). NMR has the advantage that it can provide the exact chemical structures of the molecules that need to be identified. NMR derives information of molecules from intrinsic resonances of molecules in a weakly oscillating electromagnetic field. MS derives structural information from the analysis of ionized molecules and their breakdown products (fragments). While NMR allows for very elaborate structural elucidation, this level of information is often too much for ecometabolomics. In addition, NMR usually requires higher sample amounts and concentrations. MS has the advantage of lower limits of detection and quantification but can only be used to some extent for structural elucidation. This chapter focuses only on MS and the application of the different metabolomics approaches (for more information about NMR—metabolomics see here (Emwas et al., 2019, Kim et al., 2011). Unfortunately, all approaches only cover a part of the real total metabolome. This leads to no "one size fits all" solution. Furthermore, ecologists might struggle to decide which instrument they should choose.

Mass spectrometry platforms cost a lot and require permanent maintenance. They also require expert operators. Ecologists therefore usually rely on cooperation partners. However, metabolomics might not be the main research focus of mass spectrometry platforms. Therefore, ecologists should use the following aspects to navigate, together with their collaborator, through the process of choosing a suitable method with the available equipment.

5.1 Targeted and untargeted approaches

In metabolomics two approaches can be distinguished. These are the targeted and untargeted approaches. A targeted approach is the analysis of known metabolites, also referred to as metabolic profiling (Fiehn, 2002). This means information is available on the exact chemical structures and chemical properties. In an optimal case, the metabolites are available (commercially, by synthesis, or by being isolated) as purified standards. This also allows for the development of specific analytical methods or the use of methods already described in the literature. An example would be the analysis of phytohormones as a response to environmental factors. This kind of targeted analysis has been described for combinations of many different plant species and environmental stressors (e.g. herbivory or salt, water, heat, and UV stress, respectively (Chu, Fang, Xin, Guo, & Chen, 2017; Davies, 2010). Standards for phytohormones, even isotopically labeled standards, are commercially available. In addition, a number of analytical methods on various platforms have been published (Schäfer, Brütting, Baldwin, & Kallenbach, 2016; Šimura et al., 2018). The targeted approach is therefore much more straightforward. It requires less expertise in data interpretation, metabolite identification, less time for data analysis and allows absolute quantification. On the other hand, it demands the availability of standard compounds and a-priori chemical knowledge about all target metabolites.

In an untargeted approach, little or no information about the metabolites is available. In consequence, as much information as possible about the metabolites must be collected within the metabolomic experiment. This results in more complex and larger data sets and requires much more elaborate data analysis (see Section 6). As for the targeted approach, a clear research question is crucial for successful and meaningful data analysis.

5.2 The platforms

Metabolomics analysis has two prerequisites: separation of metabolites and detection of metabolites. Separation is achieved by either gas (GC) or liquid chromatography (LC) (for more information see (Harris & Lucy, 2020). Gas chromatography requires samples that can be volatilized without decomposition. Samples that are suitable here are volatile organic compounds (e.g. monoterpenes, sesquiterpenes) or primary metabolites (e.g. sugars, amino acids, fatty acids) after derivatization. After chromatographic separation, the compounds are analyzed further by mass spectrometry. In metabolomics research, GC is rarely coupled with high-resolution mass spectrometry. Therefore, in this case, the resolution of the MS is of minor concern (Stettin, Poulin, & Pohnert, 2020). In some applications (rapid GC, GCxGC), however, a high acquisition rate is necessary and requires the use of fast ToF or Quadrupole instruments.

Liquid chromatography (LC) requires only that the sample is soluble in the solvents used in the chromatographic separation. This allows for a much wider range of metabolites. For this technique, a sample would normally be a liquid extract. When only specific compounds (e.g. phytohormones) need to be analyzed, a targeted approach is the method of choice. In this case, high-resolution MS is not needed.

In contrast, the untargeted approach requires high mass resolution and is not possible if only a medium or low-resolution instrument is available. High mass resolution is needed to calculate sum formulas. Next to the retention time and MS/MS fragmentation pattern the correct mass is necessary to distinguish between metabolites. Here is a short example to illustrate this. Coumaric acid and rhamnose are two common metabolites and as well as common moieties of secondary metabolites in various plants. Their monoisotopic masses (164.0473 Da vs. 164.0685 Da) differ by 21.2 mDa or 130 ppm. The metabolite themselves or as fragments of larger molecules can therefore only be distinguished by a mass spectrometer with an error less than 130 ppm. The choice of metabolomics method is therefore highly dependent on the research question or hypothesis, a priori knowledge of the plant system, and on the time, workforce and instrumentation available.

5.3 The need for tandem mass spectrometry (MS/MS)

One aim of metabolomic experiments is to visualize or to discover patterns between specific experimental factors (those factors must be identified in the design of the experiment). Another aim is to identify the metabolites that are responsible for this pattern. There are several methods that can provide structural information about molecules (see Section 6).

Tandem mass spectrometry (MS/MS) is able to help with structural identification. In brief, molecules are ionized and the resulting mass/charge ratio is measured (m/z). During this ionization step, the molecules are brought to higher energy states and become unstable. This leads to chemical bonds breaking and the fragmentation of the original molecule (precursor ion). The m/z values of the original molecule and the newly formed fragments (product ions) are visualized in the mass spectrum of the metabolite. A mass spectrum of fragments is characteristic for a metabolite and, to some degree, can be used to identify the original metabolite by comparing the spectrum to those stored in spectral libraries (See also Section 6). MS/MS makes it possible to look at specific precursors and the resulting product ions. This can be used to see if certain functionalities are conjugated to the precursor (e.g. sugar moieties). In addition, MS/MS can also aid in the classification of metabolites when complete identification is not possible (see Section 6).

Depending on the application, all of the samples or at minimum a pooled sample must be acquired in MS/MS mode. For an untargeted approach aiming to classify or identify, tandem mass spectrometry is absolutely necessary and requires high resolution (qToF, IMS-q-ToF, Orbitrap, FT-MS).

Another application of MS/MS is the quantification of metabolites in a targeted approach. By choosing a specific precursor and product ion combination, a specific and sensitive quantification can be achieved (Schäfer et al., 2016). There is, however, no need for high-resolution MS/MS (Fig. 3). Nevertheless, the quantification requires knowledge of the identity of the metabolites as well as of suitable precursor and product ion combinations. If no analytical method is available, this also requires time for method development. Because this application is for quantification, all samples must be analyzed in MS/MS mode.

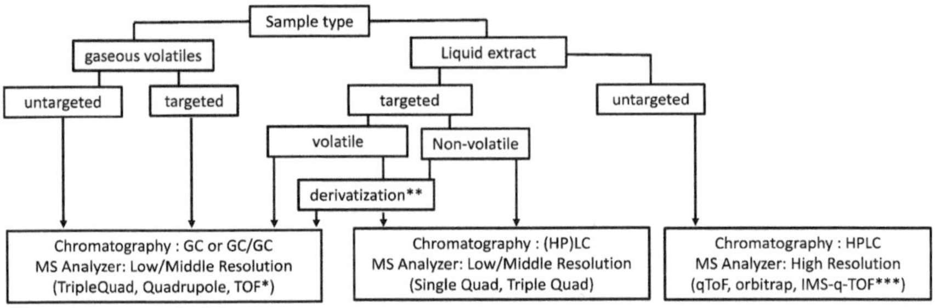

Fig. 3 Decision tree to choose the appropriate instruments for the planned metabolomic approach. (*) = high acquisition rate necessary for rapid GC, GCxGC (**) = e.g. silylation or esterification (Fatty acid methyl ester FAMES) (***) IMS-q-ToF: Ion mobility spectrometry—quadrupole–Time of Flight.

6. Data processing

Processing of (raw) data is often challenging to ecologists as it requires an integrative effort combining the knowledge of ecology, biochemistry and technical expertise of a data scientist. We, thus, recommend that ecologists seek support from skilled (bio)informaticians to efficiently process data. In general, (raw) data processing consists of several steps (see Fig. 4) that require different software to which the researchers need to be trained. There is a multitude of open-source processing tools (Peters et al., 2018; Stanstrup et al., 2019) available, with XCMS,[e] MS-DIAL (Tsugawa et al., 2015), MZmine 2 (Pluskal, Castillo, Villar-Briones, & Orešič, 2010) being some of the most popular tools.

Regardless of which tool is used, the output of processing raw data is a data matrix of the samples and intensities/abundances of peaks (rt_m/z) for each sample (see Fig. 4). This so-called feature table is being used to perform statistical analysis or to share the results with fellow researchers or the scientific community through the use of data repositories. However, there are also excellent platforms like Workflow4Metabolomics[f] or Metabo-Analyst[g] which can cover data processing entirely.

[e] https://xcmsonline.scripps.edu.
[f] https://workflow4metabolomics.org.
[g] https://www.metaboanalyst.ca.

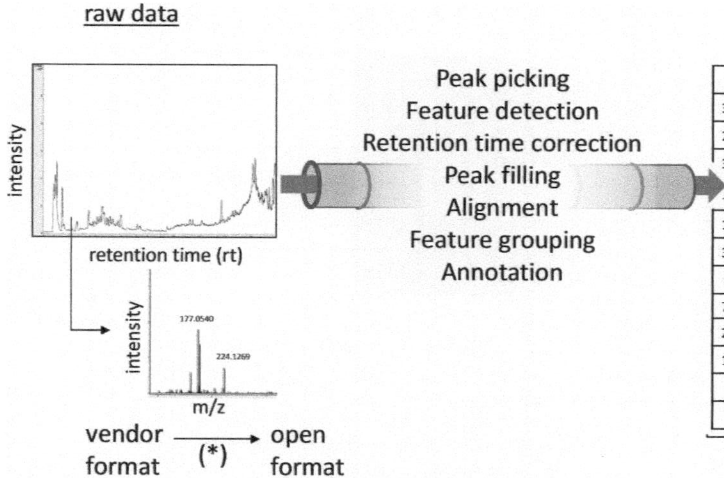

feature table

m/z	rt	CTRL_1	CTRL_2	CTRL_3	CTRL_4	...	...	ID
302.9099	6.95	291772.7	378984.4	191380.8	191380.8	191380.8	....	[M+H]+
271.1736	6.95	278.7003	4256.702	13672.27	13672.27	13672.27	....	[M+H]+
342.3931	6.95	320.6338	37710.5	44610.51	44610.51	44610.51	....	[M+Na]+
.63.1329	6.95	79.49969	41114.72	11549.39	11549.39	11549.39	....	[M-H_2O+H]+
115.3146	6.95	355.9445	43156.21	27335.64	27335.64	27335.64	....	[M+H]+
384.9483	6.98	394.1718	19173.02	24017.6	24017.6	24017.6	....	[M+H]+
88.2976	6.98	107.322	32584.6	30191.58	30191.58	30191.58	....	[M+H]+
79.49699	6.98	256.4272	45080.83	23496.16	23496.16	23496.16	....	[M+H]+
278.0636	6.98	375.6588	13908	19749.09	19749.09	19749.09	....	[M+Na]+
119.7327	6.98	259.6159	32671.13	27454.44	27454.44	27454.44	....	[M+H]+
340.48	6.98	405.5357	35534.26	23606.74	23606.74	23606.74	....	[M+Na]+
....	....	....	....	....	....	....	....	....

features — intensities — identity(**)

Fig. 4 Overview of a typical metabolomic data processing pipeline. Each step of the pipeline is further described in Karaman, Climaco Pinto, and Graça (2018); Kuhl, Tautenhahn, Böttcher, Larson, and Neumann (2012). (*) e.g. msconvert (https://bio.tools/msconvert), (**) e.g. adducts, isotope peaks, water loss.

6.1 Mass-spectrometry compound identification

The data matrix or feature table does not include any further information on the identity of the metabolite feature rather than a *m/z* value and a retention time. This level of information might be already enough for basic chemodiversity analysis and metabolite fingerprinting approaches. However, experimental design usually necessitates further information about the analyzed features (see Table 3). It is important to note that at this point (having only the MS feature table) it is only possible to annotate adducts, isotope peaks and in-source fragments of certain features (see Fig. 4) and the tool CAMERA (Kuhl et al., 2012). For a more comprehensive annotation, information at the MS/MS level is necessary. Due to different characteristics of the used instruments that influence the level of fragmentation, the parameter settings, annotation at the MS and the MS/MS level requires the knowledge of a skilled metabolomicist to answer the underlying research questions.

In addition to fragmentation spectra of the samples, chemical reference data is needed just as in compound identification. These are available either from in-house libraries, commercial databases or public databases such as MassBank. However, reference data (fully identified compounds) is really only available for model species. As ecometabolomics usually deals with non-model species reference data is, thus, rarely available for more

Table 3 Overview about the different levels of information achievable with different metabolomic facilities and a selection of MS software tools which can be used to achieve this level of information about a detected compound.

	Level of information						
Facility	***m/z***	**Structural class**	**Molecular formula**	**Functional groups**	**Molecular structure**	**Stereo-chemistry**	**Chirality**
Mass spectrometry (SingleQuad, TripleQuad)	✓	✓(**)	X	✓	X	X	X
High resolution Mass Spectrometry (Orbitrap, qToF, FT-Cyclotron)	✓	✓	✓	✓	✓(**)	X	X
NMR		✓	✓	✓	✓	✓(*)	✓(*)
MS software tools	XCMS MSDial	MetFamily, SIRIUS, ClassyFire	Smart Formular, SIRIUS	Compound Crawler, MetFrag	SIRIUS		

✓, achievable; X, not achievable; *, only with purified compounds and under certain circumstances derivatization is needed; **, case specific.

than a fraction of the fragmentation spectra. In ecometabolomics, the data matrix frequently contains many hundred or even to thousand metabolic features and it is nearly impossible to identify all of them. As a result, identification can be considered as a major bottleneck in ecometabolomic analysis as it is very elaborate and complex. To reduce the time, only a few relevant metabolic features are targeted that strongly relate to specific research question or ecological factor. The selection is carried out using variable selection methods such as Partial Least Square-Discriminant Analysis (PLS-DA) (see Section 7) or Random Forests (Scott et al., 2010). However, even here hundreds of features are often significantly related statistically.

It should be noted that the comparison of fragmentation spectra with spectral libraries can only be considered as a "first-level" annotation and needs further confirmation by other techniques or methods (e.g. injection of standard compounds or NMR) to confirm the compound identity. The same applies for the automated annotation based on the exact molecular mass. Both strategies, if not applied with reservations, may lead to false identifications. Table 3 shows the different levels of information which can be achieved with different facilities. The necessity of every level highly depends on the research question and should be discussed between the stakeholders.

6.2 Mass-spectrometry compound classification

In ecology, metabolomics is potentially able to reveal overwhelming amounts of information that are often too detailed to be used readily. An alternative that is considered viable is to not fully identify the molecular structure of the features but instead to relate them broadly to the chemical class. Ecological studies (and ecometabolomic studies are generally no exception) commonly aim to identify broader patterns instead of explaining molecular function in detail as in molecular biology. A more global description of the metabolomic differences is sufficient and, in this context, even more powerful because the information is abstracted and more generalized. In recent years, computational methods have been developed for classifying compounds in silico (Dührkop et al., 2019; Rogers et al., 2019; Treutler et al., 2016). These approaches are based on "ClassyFire" (Djoumbou Feunang et al., 2016) which uses the chemical ontology "ChemOnt" in which chemical classes are defined unambiguously. The evaluation of metabolic features on the level of classes is a promising alternative to detailed identification in ecometabolomics. It enables ecological roles or functions to be assigned to features and thus allows for functional interpretations at more coarse spatiotemporal scales (Peters et al., 2018).

For this approach, the traditional LC/MS-MS methods employing targeted precursor lists and measuring their spectra are often not efficient enough. Instead, Data Independent Acquisition (DIA) such as SWATH (Bonner & Hopfgartner, 2019) or Data Dependent Acquisition (DDA) can fragment a large fraction of the statistically relevant molecules and can cover a large magnitude of putative compounds of the feature table. In DIA all ions produced by one MS analyzer within a selected *m/z* range are fragmented and analyzed in a second analyzer (hence MS-MS or tandem mass spectrometry). In DDA, the basic method is to fragment in the second MS machine the largest peaks coming from the first analyzer within a certain retention time window. Both methods allow the acquisition of fragmentation spectra for the majority of ecologically relevant metabolite features and are perfectly suitable for typical classification approaches.

Classification of DIA and DDA data can be performed in the web-based tool MetFamily[h] (Treutler et al., 2016) which employs hierarchical clustering on spectra and clusters similar spectra more closely together. The nodes or leaves of the cluster are then run through a trained model that compares the spectrum with spectra in a database. These spectra are known to belong in certain compound classes. The model then assigns the unknown spectrum to the most closely related compound class. The software SIRIUS (Dührkop et al., 2019) takes a comparable computational approach although machine learning is done differently. For example, an ecological experiment might compare the metabolome of pathogen-infected plants with that of uninfected plants. The classification pipeline would quickly be able to demonstrate differences between the two, e.g. that compound classes phenolics and glucosinolates were more abundant in the infected plants than in uninfected ones. This difference might mean that plants build-up these compounds in order to defend themselves. These kinds of studies are perfectly suited for hypothesis generation and open the way for further studies that "zoom in" to an analysis of the molecular basis of the ecological relationship.

[h] https://msbi.ipb-halle.de/MetFamily/.

7. Statistical processing

The statistical analysis of metabolomic data is itself a challenge. Experimental designs in metabolomics are usually quite simple, but the complexity arises from a large number of compounds (i.e. dozens to thousands) that have to be analyzed relative to the number of samples, and from the high level of collinearity in the data (i.e. correlations between compounds). Many statistical tools that are used in the analysis of metabolomic data, especially multivariate methods, come from a field of statistics called chemometrics. Ecological data, however, often derive from complex experimental designs and consequently require analysis with elaborate statistical tools. Most of these multivariate tools come from another field of statistics called numerical ecology. Although there are many links in the mathematical foundations of multivariate methods developed in chemometrics and numerical ecology, there is almost no connection in practice between these two fields. Parallel methods have even been developed that aim to answer the same statistical questions. In the combined discipline of ecometabolomics, the data demands statistical analyses that embrace both complex experimental designs and medium to high-dimension data and uses statistical tools from the two different fields with their different cultures. For ecologists, this can easily look like too many challenges at once.

However, we show here that the statistical processing of ecometabolomic data is greatly facilitated by following a step-by-step workflow (Fig. 5). This workflow is not difficult to follow provided that the analyst always keeps clearly in mind the biological question at the origin of the study. Indeed, each of the steps of the workflow consists of a question or a series of questions that are mostly biological rather than technical.

7.1 Step 1—What is the relevant format for the data?

Metabolomic data is expressed in different formats, among which two are of interest in ecometabolomics. Each of these two formats addresses different biological questions.

If the question relies on the simple presence of compounds—In this case, a binary format is appropriate (0: absence of the compound in the sample, 1: presence of the compound). Such qualitative data are the poorest of the two formats in terms of information. However, it does not mean that this format is uninteresting. Indeed, quantitative data such as concentrations always show some random variation. When added to natural variation, which is

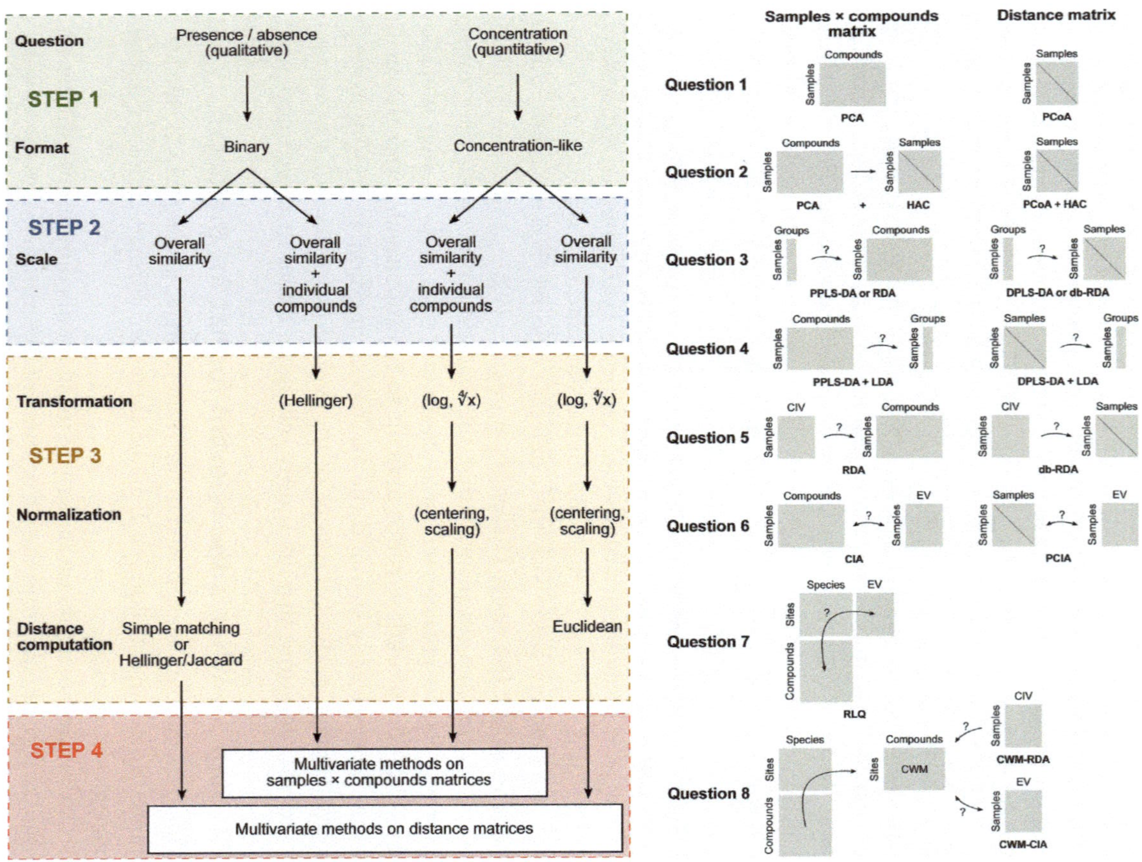

Fig. 5 Suggested statistical workflow for ecometabolomic studies. In brackets operations that are facultative. Chemically-informed distances are not included. CIV, controlled independent variables; EV, environmental variables. See text for more explanation.

likely to be important especially in semi-field or field experiments, this variation may blur the signal of interest. A binary format suppresses the noise introduced by quantification and can help reveal patterns of chemical diversity, especially in untargeted metabolomics where hundreds to thousands of compounds are considered at the same time.

If the question relies on the concentration of compounds—In this case, a "concentration-like" format is appropriate. We call "concentration-like" absolute concentrations of the compounds or any values that are proportional to these absolute concentrations (e.g. peak areas, ion intensities). From a statistical point of view, it makes no difference whether absolute concentrations are used, peak areas or ion intensities, as long as the compounds are expressed in the same way in all samples. For example, if the intensity of a given ion is twice in sample A than in sample B, the concentration of the metabolite is also twice in sample A than in sample B. In other words, for statistical analysis, there is no need for absolute quantification. In any way, absolute quantification, i.e. determination of the true value, is virtually impossible in untargeted metabolomics since it is not possible to have spiked labeled standard at known concentrations for all metabolites.

7.2 Step 2—Is the interest in individual compounds?

Many questions in ecometabolomics are answerable by analyzing only the overall metabolomic similarity between samples without any need to identify the individual compounds behind this similarity. Many other questions, however, are only answerable by analysis at a finer scale so as to be able to identify the compounds. These two different scales lead to different statistical strategies.

If the question relies on the overall similarity between samples only—The statistical analysis will only be multivariate, taking into account all compounds at the same time. It is recommended to perform this analysis not on a samples × compounds data matrix, but on a samples × samples matrix. Such a matrix is called a "distance matrix." It crosses all samples with all samples and each cell gives the distance value between the sample in row and the sample in column. Thus, a distance matrix is necessarily square and its diagonal (i.e. the distance between samples and themselves) is full of zeroes.

If the question relies on the overall similarity between samples as well as on individual compounds—The statistical analysis will be based on the usual samples × compounds matrix. Two strategies can then be conducted:

univariate analyses on one compound at a time, and multivariate analyses with all compounds considered at the same time. It is important to note that these two strategies are neither identical nor exclusive. As they do not look at the data from the same point of view, they are complementary (for a clear explanation see Saccenti, Hoefsloot, Smilde, Westerhuis, & Hendriks, 2014). Therefore, it is good practices (i) to perform both univariate and multivariate approaches and (ii) not to try to validate one approach by the other. The rest of this text will focus on multivariate analyses. However, we would like to emphasize one key aspect of the univariate approach: since multiple testing is performed (e.g. one correlation test, *t*-test or ANOVA per compound), it is essential to correct the *P*-values in order to avoid inflation of false-positive results (i.e. significant results that should not be). The most well-known correction method is the Bonferroni correction; however, this method is likely to inflate negative results (i.e. non-significant results that should be significant). It is recommended to use a correction method that is a better compromise between false positive and false negative inflations, such as the False Discovery Rate (FDR) of Benjamini & Hochberg (Benjamini & Hochberg, 1995; Saccenti et al., 2014).

7.3 Step 3 in analyses of samples × compounds matrices—Is it necessary to pre-treat data?

We call 'pre-treatment' two operations that can be performed, in this order, on the data matrix: transformation and normalization. Both are facultative and the way they are conducted depends on the format of the data.

The data are in a binary format (presence/absence)—Normalization cannot be applied and only transformation must be considered. The question relies entirely on a problem that is classical in community ecology and called the "double-zero problem" (Legendre & Legendre, 2012). In its traditional form, this problem lies in the following question: Are two sampling sites similar if a given species is absent in both? With metabolomic data, the question becomes: Are two samples similar if a given compound is absent in both? There is no absolute answer to this question because the answer depends on the research question. All multivariate methods that will be presented below (e.g. Principal Component Analysis) consider the double absence of a compound as a similarity between samples. If that is relevant to the study, no transformation of the data is needed. If it is more relevant not to consider the double absence of a compound as a similarity, transformation is needed. A good choice is the Hellinger transformation, which was originally proposed in community ecology (Legendre & Gallagher, 2001).

The data are in a concentration-like format—Both transformation and normalization can be envisaged. Transformation is not always needed, but it can improve the results of multivariate analyses by revealing clearer trends in the data. In practice, the decision to apply a transformation mostly relies on comparing an analysis with transformed data with the same analysis without transformation. If there is a notable improvement in the results, transforming the data is a good option. The logarithmic transformation is the most popular, however, it is not applicable when the data contain zero values. In that case, the fourth-root transformation is a possible alternative (Wold, Sjöström, & Eriksson, 2001). Normalization consists of centering and scaling, which, generally speaking, aim to give more or less the same weight to all compounds in the multivariate analysis. No normalization would lead to weights that increase with compound inter-sample variance. Since concentration variance usually increases with concentration mean (i.e. the higher the mean concentration of a compound, the higher its variance), no normalization gives generally more weight to compounds with higher concentrations (or peak areas, or ion intensities). This can be relevant for the research question. However, in most cases there is no prior hypothesis about the relationship between compound variance and biological importance. In such situations, normalization should be considered. Among the different scaling methods that can be used (autoscaling, Pareto scaling, range scaling, etc.), (Wold et al., 2001) and (van den Berg, Hoefsloot, Westerhuis, Smilde, & van der Werf, 2006) recommend autoscaling as a first choice. This method gives the exact same weight to all compounds.

7.4 Step 3 in analyses of distance matrices—How to compute inter-sample distances?

A distance matrix is obtained by applying a distance measure (or a dissimilarity measure) to a samples × compounds matrix. Alternatively, it can be obtained by applying a similarity index to the matrix, then computing 1—similarity (or a derived version of this formula) to get distances or dissimilarities. Note that the terms "distance" and "dissimilarity" are not strict synonyms, but the distinction is out of the scope of this chapter and the term "distance" will be used for convenience. Two questions are to be answered here: Which distance measure or similarity index to choose? Is it necessary to pre-treat data prior to distance computation? The answer to both questions depends on the format of the data.

The data are in a binary format (presence/absence)—Pre-treatment is not needed and the only question is about distance computation.

The double-zero problem is crucial here (see above). If the absence of a given compound in two samples should be considered as a similarity between these samples, a symmetric distance measure is to be preferred. Among many options, the most common one is the distance based on the simple matching coefficient (Sokal & Michener, 1958) which is a similarity index. If the double absence should not be considered as a similarity, an asymmetric distance measure is to be preferred. Among many options, the Hellinger distance (Rao, 1995) is a good choice, as well as the distance based on the popular Jaccard community coefficient (Jaccard, 1901) which is a similarity index too.

The data are in a concentration-like format—In this case, there is no real debate about which distance measure to prefer because the classical Euclidean distance is a natural choice. However, it is important to consider pre-treatment of the data before calculating the distance measure. The reasons for applying transformation and normalization do not differ from those explained above.

Finally, recent years have seen the development of a new type of distance matrices in metabolomics. These matrices are not based on purely statistical distance measures, which treat all compounds as independent variables, but on "chemically-informed distances." There are two kinds of such distances. Those based on mass spectra, such as the Chemical Structural and Compositional Similarity (Sedio, Rojas Echeverri, Boya, & Wright, 2017), and those based on putative or confirmed chemical structures, such as the BioSynthetically Informed Distance (Junker, 2018). Research is being conducted that promises to associate both spectral and structural components into integrative metabolomic distances (e.g. Ollivier et al., 2019).

7.5 Step 4—Which statistical method to choose?

The final step of the statistical workflow is the application of a multivariate method. Choosing the most appropriate method is often a challenge, for ecologists as well as for other life scientists. However, the choice of the relevant method is greatly facilitated if the research question is clear. Since there is no room here to expand on all possible methods, we will only list the most common questions asked in ecometabolomics and give methods of choice to address these questions. Further explanations, as well as real-life examples and R scripts, can be found for question 1 to 6 in (Hervé, Nicolè, & Lê Cao, 2018) and in (Kleyer et al., 2012) for questions 7 and 8. Ecologists are likely to be familiar with these methods, as many of them were initially developed for community ecology. The good news is that they are fully applicable to ecometabolomics.

Question 1—Do samples structure into continuous gradients? Answering this question requires an analysis that does not take as input any other information than the metabolomic data. Principal Component Analysis (PCA) is best suited for samples × compounds matrices, while Principal Coordinate Analysis (PCoA) is a good choice for distance matrices. We do not recommend Non-Metric Multidimensional Scaling (NMDS), which is very common in community ecology but causes a loss of information that is often unnecessary.

Question 2—Do samples structure into groups? For the same reason as above, PCA and PCoA are recommended. Both are well complemented with Hierarchical Agglomerative Clustering (HAC), which helps reveal objective groups. Note that whether or not the analysis is based on a samples × compounds matrix or a distance matrix, HAC is always based on a distance matrix.

Question 3—Does the metabolome differ between a priori groups of samples (e.g. functional groups)? When dealing with samples × compounds matrices, two analyses are interesting to answer this question. One comes from chemometrics and belongs to the Partial Least Squares (PLS, a.k.a. Projection to Latent Structures) family of methods: Powered Partial Least Squares—Discriminant Analysis (PPLS-DA). PPLS-DA is an extension of the widely used PLS-DA that has better discriminant abilities than PLS-DA and its derivative Orthogonal PLS-DA. The other method of choice comes from numerical ecology: Redundancy Analysis (RDA). Choosing between PPLS-DA and RDA is mostly a matter of convenience. Both analyses have extensions to deal with distance matrices, namely Dissimilarity PLS-DA (DPLS-DA) and distance-based RDA (db-RDA).

Question 4—Can the metabolome be used to predict a priori groups of samples (e.g. functional groups)? This question is very close to the previous one but requires a method that has a predictive aim. RDA is unsuitable for this and PPLS-DA is preferred. However, although PPLS-DA is designed for discrimination, it is not a classification tool. An interesting strategy is to combine PPLS-DA with a classification method such as Linear Discriminant Analysis (LDA). The first step is PPLS-DA used to distinguish groups and in the second step, LDA is used on the PPLS-DA scores for prediction and classification. The approach can be extended to distance matrices by combining DPLS-DA and LDA.

Question 5—Which variable or interactions influence the metabolome in a multifactorial experimental design? This question can be seen as an extension of Question 3 to multifactorial experimental designs.

We recommend RDA in this situation (or db-RDA in case of distance matrices) since it is very convenient to use for any ecologist familiar with linear models and ANOVAs.

Question 6—Does the metabolome of a given species respond to environmental gradients? In contrast to Questions 3 to 5, the context here is not an experimental design with controlled independent variables. The aim is rather explorative and is to find links between metabolomic data and a set of environmental data collected for the same samples. When working on samples × compounds matrices, Co-Inertia Analysis (CIA) is an interesting method since it allows metabolomic data to be linked to any kind of environmental data (quantitative, qualitative or a mix of both types). Procrustean CIA (PCIA) is an extension of CIA that is an equivalent for distance matrices.

Question 7—Does the metabolome of individual species in a community respond to environmental gradients? This question is an extension of Question 6 but here multiple species in a community are considered. Only samples × compounds matrices are concerned, which are reformatted as compounds × species matrices by computing e.g. mean compound concentrations per species. This type of question is common in community ecology, where the aim is to look for relationships between species functional traits and environmental gradients while considering the abundance of each species in sampled communities. RLQ analysis, which is a generalization of CIA aiming to answer this question, is fully suitable for ecometabolomics where functional traits are replaced with metabolomic data.

Question 8—Does the meta-metabolome of communities respond to environmental gradients? This question is asked at the highest possible scale, i.e. the community itself rather than individual species. It concerns samples × compounds matrices only (reformatted as compounds × species matrices as above) and necessitates an additional step prior to the application of a multivariate method: computing the meta-metabolome, i.e. the metabolome of the community. Community-Weighted Means (CWM), that are mainly used to compute mean functional traits of a community by weighting functional traits of individual species with their respective abundance in the community, seems an interesting option although it has probably never yet been applied to ecometabolomics. The new data matrices (i.e. communities × weighted metabolomic means) can then be used in analyses listed in the previous questions, namely RDA and CIA, leading to CWM-RDA and CWM-CIA.

8. Data storage, sharing and publication

After the very time-consuming work of acquiring data and processing it, the time comes to ensure its good management. This consists of safely storing the data, sharing it with peers and publishing it. Sharing research data, in addition to the results obtained from its analysis, is increasingly required by research institutions, grant providers, and journals. It is also an important part of good scientific practice, fostering transparency and trust within the scientific community. However, data storage, data sharing, and data publication, although natural in the life of data, present several challenges.

8.1 Challenges in data storage

Storing, or archiving, digital data after its acquisition and analysis consist of saving it on electromagnetic (hard drives), silicon-based (SSD-hard drives) storage media or on a remote (cloud) platform approved by the scientists' institution, in order to make them permanently available for future use or reuse. Inappropriate data storage or, worse, lack of it, can lead to data loss due to space unavailability over time or if the data producer becomes unavailable. Another major challenge is when stored data becomes unusable such as when the data producers themselves or their colleagues forget what the data is, how it was produced, processed or analyzed. Allowing data to become unusable is a waste of time, money and, importantly, of information. Furthermore, metabolomic data take a lot of space, one experiment might result in several terabytes. A long-time storage solution might be challenging for some institutions and need to be discussed in advance.

8.2 How to face the data storage challenges

First, it is necessary to be aware of the storage capacities and conditions of the institute. Sometimes storage solutions are already systematically implemented in research groups or the institute. Otherwise, it is necessary to contact the IT support of the institution that deals with it. The "3-2-1 rule" must be observed. Data needs to be stored in three copies, on two different supports (e.g. external hard drive and institutional servers) and one copy should be stored in a remote, perhaps virtual (e.g. cloud) location. In order to allow easy data reproduction and reuse, data should be organized, the organization should be documented, and there should be extensive metadata. Metadata, or "data about data," contains information that describes the data and gives context to it. It should be structured in order to be human and machine-readable, to comply with the Minimal Information Standard for metabolomic data,[i] be "FAIR (Wilkinson et al.)," and to be suitable for the use of appropriate metabolomic or ecological-specific controlled vocabularies (ontologies).

FAIR data principles[j] provide guidelines for improving the Findability, Accessibility, Interoperability, and Reusability of data. In this context, data is findable when metadata and data are easy to find for both humans and computers (e.g. when they have a permanent identifier, permanent access link, appropriate associated keywords). Next, data needs to be accessible, potential users need to know its permanent location and how it can be accessed and, if there are restrictions on its use, users need to know how to obtain authentication and authorization. In order to be interoperable, the data and metadata should be based on appropriate standard vocabularies, ontologies or thesauri so that they easily integrate with existing applications or workflows. To be reusable, metadata and data should be well-described so that they can be replicated, reproduced or combined in various research settings. More information and recommendations about FAIR data principles and on how to make data FAIR can be found in the resources of the international open science communities such as GO-FAIR,[k] FAIRsFAIR[l] or the Research Data Alliance.[m]

Minimal Information Standards (MIS) are a set of guidelines for reporting data derived from bioscience research. The Metabolomics Standards Initiative (MSI Board Members:, 2007) has developed MIS for different metabolomics types (e.g. LC-MS, GC–MS, NMR) and sources (e.g. plant, environmental…) and should be used as a reference for metabolomics metadata. In general, MIS ensures that data can be verified, analyzed and clearly interpreted by the broader scientific community. These standards, because agreed though a global metabolomics community process, also facilitate data retrieval from public databases, data integration, and the development of data analysis tools.

[i] https://www.ebi.ac.uk/ols/ontologies/msio.

[j] https://www.go-fair.org/fair-principles/.

[k] https://www.go-fair.org/.

[l] https://www.fairsfair.eu/.

[m] https://www.rd-alliance.org/.

8.3 Challenges in data sharing

Sharing data means making it available online in a digital format, free of charge, and without copyright restrictions, which is great for the scientific community as it enables data reuse, data repurpose, builds trust between scientists, facilitates methodological research, and saves money by avoiding reproducing the same experiments. Unfortunately, despite growing awareness of the value of data, papers are still the main way of evaluating the performance of scientists and therefore data is not often shared. Smith and Roberts (2016) investigated other reasons why data sharing is still challenging and identified the main reasons why it did not take place:

- fear of being scooped by other researchers working on the same subject
- fear of not generating enough publications from the data
- fear of exposing badly managed, flawed, inconsistent data (not replicable)
- stuck in the analytical or metabolite identification processes

Also, when data is shared, it is too often not reusable by the scientific community. The data reproducibility survey made for Nature in 2016 (Baker, 2016) found out that more than 70% of researchers have tried and failed to reproduce the experiments of other scientists. The principal reasons why data reuse is challenging reside in poorly annotated data, with inadequate or absent metadata, data in proprietary vendor formats, and in the lack of compatibility between datasets (Tenenbaum & Blach, 2018).

8.4 How to face the data-sharing challenges

Facing the challenges of data sharing starts with good data management and storage (as described above). Indeed, well-organized data with rich metadata using controlled vocabularies and following the MIS are very likely to be easily reusable and the corresponding studies to be reproducible. Early data sharing can even prevent being "scooped" because sharing creates a timestamp for those who produced the data. The original producers can even associate with their shared data their ideas and plans about the data in order to prove that "they were there first."

To enhance data reusability, they need to be shared only in open formats accepted by the community standards. Appropriate metabolomic data formats are discussed in the "Raw data conversion" section of Challenge. Metabolomic data is valuable and rich in content. Sharing it is particularly important because it can be used in a different context, especially when it comes to biodiversity and ecology, as it allows comparisons between different regions, different types of interactions, and between ecosystems at different times.

Data can be shared in the supplementary material of the corresponding publication and this for a long time has been the main way of sharing data. However, as it is good practice and an increasingly frequent request of journals, it is better to submit data to a specialized or general data platform that stores data permanently and provides a DOI for datasets. For metabolomic data, MetaboLights[n] and Metabolomics Workbench[o] are standards. However, it should be noted that MetaboLights is becoming increasingly popular because of its data and metadata validation processes, standardized metadata requirements, and more modern submission and browsing interfaces. Also, the Pangea[p] information system for georeferenced and ecological data is particularly recommended for complex data whether or not the data is metabolomic.

With the increasing value and quantity of data generated by researchers from all disciplines, and, at the same time, the continued pressure to publish in peer-reviewed journals because of the evaluation of researcher performance by a number of publications, some publishers now encourage the publishing of data as an article. For example, the publisher SpringerNature offers "Scientific Data." This journal publishes "data descriptor"[q] short papers describing why data was created and how it was generated. The data itself must be submitted to an appropriate data-sharing platform. For the journal GigaScience, published by OUP, the "data note"[r] can be a bit more extensive than in Scientific Data. The data must also be submitted to a sharing platform, either to that associated with the journal, GigaScienceDB, or an equally appropriate alternative. Other journals, such as "Ecology" also published by Springer, allow the publication of data articles alongside research papers. An article in any of these journals is a valid, peer-reviewed scientific publication centered on data. This new way of sharing and publishing data will hopefully grow in popularity and acknowledgement over the next few years.

[n] https://www.ebi.ac.uk/metabolights/.
[o] https://www.metabolomicsworkbench.org/.
[p] https://pangaea.de/.
[q] https://www.nature.com/sdata/publish.
[r] https://academic.oup.com/gigascience/pages/data_note.

9. Summary and concluding remarks

Today, ecometabolomics is the right tool to analyze biochemical relationships in the model as well as non-model species from greenhouse and field experiments. The reason is that ecometabolomic studies have become more comprehensive over time; they developed from targeted approaches for identifying single substances or substance classes to mainly untargeted approaches yielding metabolic fingerprints. This process was facilitated by the development of more powerful computer processors. This increased computing power is required among others for data processing using multivariate statistical analysis and machine learning. Lately, metabolomics has found application in large-scale ecological and biodiversity field experiments to find molecular mechanisms underlying fundamental ecological concepts. Effective application of metabolomics in ecological research requires good project and expectation management from the very start. This is particularly true when the different stakeholders come from different scientific disciplines and did not collaborate yet regularly. Communication with all stakeholders is key in the effective implementation of metabolomics in ecology and biodiversity research.

Ecometabolomics commonly yields metabolic fingerprints that can be related to ecological functions. Even though multivariate analyses and machine learning yield lists of explanatory mass features, their structures are commonly not fully identified. This may not always be necessary; depending on the ecological question at hand, compound classification (Dührkop et al., 2019; Treutler et al., 2016) or using metabolomic patterns as a measure of chemical diversity (Scherling, Roscher, Giavalisco, Schulze, & Weckwerth, 2010) might more appropriate. Such global analyses of plant metabolomes would also provide insights into the relation between metabolomic variation and ecological functions in a reasonable period.

In our experience, the untargeted metabolomics approach often serves as "hypothesis generator" and starting point for follow-up projects zooming in to specific sub-sections of the full metabolome. Often only a small fraction of the acquired metabolome is then fully exploited in a single project, although ecometabolomic datasets contain an overwhelming amount of information. This makes it all the more important to make the data available to the general public by sharing the complete metabolomic dataset with the scientific community. In conclusion, collegial and well-integrated collaborations between ecologists, chemists, and bioinformaticians in ecometabolomics will yield a treasure of novel information to explain interactions between plants and their environment.

Acknowledgments

This work was supported by the German Centre for Integrative Biodiversity Research (iDiv) Halle-Jena-Leipzig funded by the German Research Foundation (DFG–FZT 118, 202548816).

References

Afendi, F. M., Okada, T., Yamazaki, M., Hirai-Morita, A., Nakamura, Y., Nakamura, K., et al. (2012). KNApSAcK family databases: Integrated metabolite–plant species databases for multifaceted plant research. *Plant and Cell Physiology*, *53*(2), e1. https://doi.org/10.1093/pcp/pcr165.

Augustijn, D., de Groot, H. J. M., & Alia, A. (2019). A robust circadian rhythm of metabolites in Arabidopsis thaliana mutants with enhanced growth characteristics. *PLoS One*, *14*(6), e0218219. https://doi.org/10.1371/journal.pone.0218219.

Baker, M. (2016). 1,500 scientists lift the lid on reproducibility. *Nature*, *533*(7604), 452–454. https://doi.org/10.1038/533452a.

Benjamini, Y., & Hochberg, Y. (1995). Controlling the false discovery rate: A practical and powerful approach to multiple testing. *Journal of the Royal Statistical Society: Series B (Methodological)*, *57*(1), 289–300. https://doi.org/10.1111/j.2517-6161.1995.tb02031.x.

Bonner, R., & Hopfgartner, G. (2019). SWATH data independent acquisition mass spectrometry for metabolomics. *TrAC Trends in Analytical Chemistry*, *120*, 115278. https://doi.org/10.1016/j.trac.2018.10.014.

Bundy, J. G., Davey, M. P., & Viant, M. R. (2009). Environmental metabolomics: A critical review and future perspectives. *Metabolomics*, *5*(1), 3–21. https://doi.org/10.1007/s11306-008-0152-0.

Chu, J., Fang, S., Xin, P., Guo, Z., & Chen, Y. (2017). Quantitative analysis of plant hormones based on LC-MS/MS. In *Hormone metabolism and signaling in plants* (pp. 471–537). Elsevier. https://doi.org/10.1016/B978-0-12-811562-6.00014-1.

Covey, S. R. (2004). The 7 habits of highly effective people: [powerful lessons in personal change]; restoring the character ethic (rev. ed.). Free Press.

Davies, P. J. (Ed.). (2010). *Plant Hormones*. The Netherlands: Springer. https://doi.org/10.1007/978-1-4020-2686-7.

Díaz-Maroto, M. C., Pérez-Coello, M. S., González Viñas, M. A., & Cabezudo, M. D. (2003). Influence of drying on the flavor quality of spearmint (Mentha spicata L.). *Journal of Agricultural and Food Chemistry*, *51*(5), 1265–1269. https://doi.org/10.1021/jf020805l.

Dixon, R. A., & Strack, D. (2003). Phytochemistry meets genome analysis, and beyond. *Phytochemistry*, *62*(6), 815–816. https://doi.org/10.1016/S0031-9422(02)00712-4.

Djoumbou Feunang, Y., Eisner, R., Knox, C., Chepelev, L., Hastings, J., Owen, G., et al. (2016). ClassyFire: Automated chemical classification with a comprehensive, computable taxonomy. *Journal of Cheminformatics*, *8*(1), 61. https://doi.org/10.1186/s13321-016-0174-y.

Dührkop, K., Fleischauer, M., Ludwig, M., Aksenov, A. A., Melnik, A. V., Meusel, M., et al. (2019). SIRIUS 4: A rapid tool for turning tandem mass spectra into metabolite structure information. *Nature Methods*, *16*(4), 299–302. https://doi.org/10.1038/s41592-019-0344-8.

Eisner, T., Meinwald, J., & National Academy of Sciences (U.S.) (Eds.). (1995). *Chemical ecology: The chemistry of biotic interaction* National Academy Press.

Emwas, A.-H., Roy, R., McKay, R. T., Tenori, L., Saccenti, E., Gowda, G. A. N., et al. (2019). NMR spectroscopy for metabolomics research. *Metabolites*, *9*(7), 123. https://doi.org/10.3390/metabo9070123.

Fiehn, O. (2002). Metabolomics—The link between genotypes and phenotypes. *Plant Molecular Biology*, *48*(1–2), 155–171.

Harris, D. C., & Lucy, C. A. (2020). *Quantitative chemical analysis* (10th ed.). Macmillan learning.

Hervé, M. R., Nicolè, F., & Lê Cao, K.-A. (2018). Multivariate analysis of multiple datasets: A practical guide for chemical ecology. *Journal of Chemical Ecology*, *44*(3), 215–234. https://doi.org/10.1007/s10886-018-0932-6.

Hildreth, S. B., Foley, E. E., Muday, G. K., Helm, R. F., & Winkel, B. S. J. (2020). The dynamic response of the Arabidopsis root metabolome to auxin and ethylene is not predicted by changes in the transcriptome. *Scientific Reports*, *10*(1), 679. https://doi.org/10.1038/s41598-019-57161-9.

Jaccard, P. (1901). *Étude comparative de la distribution florale dans une portion des Alpes et du Jura [Text/html,application/pdf]*. https://doi.org/10.5169/SEALS-266450.

Junker, R. R. (2018). A biosynthetically informed distance measure to compare secondary metabolite profiles. *Chemoecology*, *28*(1), 29–37. https://doi.org/10.1007/s00049-017-0250-4.

Karaman, I., Climaco Pinto, R., & Graça, G. (2018). Metabolomics data preprocessing: From raw data to features for statistical analysis. In *Vol. 82. Comprehensive analytical chemistry* (pp. 197–225). Elsevier. https://doi.org/10.1016/bs.coac.2018.08.003.

Kim, H. K., Choi, Y. H., & Verpoorte, R. (2011). NMR-based plant metabolomics: Where do we stand, where do we go? *Trends in Biotechnology*, *29*(6), 267–275. https://doi.org/10.1016/j.tibtech.2011.02.001.

Kim, S.-G., Yon, F., Gaquerel, E., Gulati, J., & Baldwin, I. T. (2011). Tissue specific diurnal rhythms of metabolites and their regulation during herbivore attack in a native tobacco, Nicotiana attenuata. *PLoS One*, *6*(10), e26214. https://doi.org/10.1371/journal.pone.0026214.

Kleyer, M., Dray, S., Bello, F., Lepš, J., Pakeman, R. J., Strauss, B., et al. (2012). Assessing species and community functional responses to environmental gradients: Which multivariate methods? *Journal of Vegetation Science*, *23*(5), 805–821. https://doi.org/10.1111/j.1654-1103.2012.01402.x.

Koen, N., Du Preez, I., & Loots, D. T. (2016). Metabolomics and personalized medicine. In *Vol. 102. Advances in protein chemistry and structural biology* (pp. 53–78). Elsevier. https://doi.org/10.1016/bs.apcsb.2015.09.003.

Kröber, W., Heklau, H., & Bruelheide, H. (2015). Leaf morphology of 40 evergreen and deciduous broadleaved subtropical tree species and relationships to functional ecophysiological traits. *Plant Biology*, *17*(2), 373–383. https://doi.org/10.1111/plb.12250.

Kuhl, C., Tautenhahn, R., Böttcher, C., Larson, T. R., & Neumann, S. (2012). CAMERA: An integrated strategy for compound spectra extraction and annotation of liquid chromatography/mass spectrometry data sets. *Analytical Chemistry*, *84*(1), 283–289. https://doi.org/10.1021/ac202450g.

Legendre, P., & Gallagher, E. D. (2001). Ecologically meaningful transformations for ordination of species data. *Oecologia*, *129*(2), 271–280. https://doi.org/10.1007/s004420100716.

Legendre, P., & Legendre, L. (2012). *Numerical ecology (3. Engl. Ed)*. Elsevier.

Li, S., Tian, Y., Jiang, P., Lin, Y., Liu, X., & Yang, H. (2020). Recent advances in the application of metabolomics for food safety control and food quality analyses. *Critical Reviews in Food Science and Nutrition*, 1–22. https://doi.org/10.1080/10408398.2020.1761287.

MSI Board Members. (2007). The metabolomics standards initiative. *Nature Biotechnology*, *25*(8), 846–848. https://doi.org/10.1038/nbt0807-846b.
Ollivier, S., Olivier, D., Gerlach, A., Pellissier, L., Chollet-Krugler, M., Lohézic-Le Dévéhat, F., et al. (2019). Comprehensive chemotaxonomy: Mining data from tandem mass spectrometry of lichens. In *67th International Congress and Annual Meeting of the Society for Medicinal Plant and Natural Product Research (GA) in cooperation with the French Society of Pharmacognosy AFERP, s-0039-3399824*. https://doi.org/10.1055/s-0039-3399824.
Papazian, S., Girdwood, T., Wessels, B. A., Poelman, E. H., Dicke, M., Moritz, T., et al. (2019). Leaf metabolic signatures induced by real and simulated herbivory in black mustard (Brassica nigra). *Metabolomics*, *15*(10), 130. https://doi.org/10.1007/s11306-019-1592-4.
Peters, K., Worrich, A., Weinhold, A., Alka, O., Balcke, G., Birkemeyer, C., et al. (2018). Current challenges in plant eco-metabolomics. *International Journal of Molecular Sciences*, *19*(5). https://doi.org/10.3390/ijms19051385.
Pluskal, T., Castillo, S., Villar-Briones, A., & Orešič, M. (2010). MZmine 2: Modular framework for processing, visualizing, and analyzing mass spectrometry-based molecular profile data. *BMC Bioinformatics*, *11*(1), 395. https://doi.org/10.1186/1471-2105-11-395.
Rao, C. R. (1995). A review of canonical coordinates and an alternative to correspondence analysis using Hellinger distance. *Quaderns d'estadística i Investigació Operativa*, *19*(1–3), 23–63.
Rogers, S., Ong, C. W., Wandy, J., Ernst, M., Ridder, L., & van der Hooft, J. J. J. (2019). Deciphering complex metabolite mixtures by unsupervised and supervised substructure discovery and semi-automated annotation from MS/MS spectra. *Faraday Discussions*, *218*, 284–302. https://doi.org/10.1039/C8FD00235E.
Saccenti, E., Hoefsloot, H. C. J., Smilde, A. K., Westerhuis, J. A., & Hendriks, M. M. W. B. (2014). Reflections on univariate and multivariate analysis of metabolomics data. *Metabolomics*, *10*(3), 361–374. https://doi.org/10.1007/s11306-013-0598-6.
Saito, K., & Matsuda, F. (2010). Metabolomics for functional genomics, systems biology, and biotechnology. *Annual Review of Plant Biology*, *61*, 463–489. https://doi.org/10.1146/annurev.arplant.043008.092035.
Sauerschnig, C., Doppler, M., Bueschl, C., & Schuhmacher, R. (2017). Methanol generates numerous artifacts during sample extraction and storage of extracts in metabolomics research. *Metabolites*, *8*(1), 1. https://doi.org/10.3390/metabo8010001.
Schäfer, M., Brütting, C., Baldwin, I. T., & Kallenbach, M. (2016). High-throughput quantification of more than 100 primary- and secondary-metabolites, and phytohormones by a single solid-phase extraction based sample preparation with analysis by UHPLC–HESI–MS/MS. *Plant Methods*, *12*(1), 30. https://doi.org/10.1186/s13007-016-0130-x.
Scherling, C., Roscher, C., Giavalisco, P., Schulze, E.-D., & Weckwerth, W. (2010). Metabolomics unravel contrasting effects of biodiversity on the performance of individual plant species. *PLoS One*, *5*(9), e12569. https://doi.org/10.1371/journal.pone.0012569.
Scott, I. M., Vermeer, C. P., Liakata, M., Corol, D. I., Ward, J. L., Lin, W., et al. (2010). Enhancement of plant metabolite fingerprinting by machine learning. *Plant Physiology*, *153*(4), 1506–1520. https://doi.org/10.1104/pp.109.150524.
Sedio, B. E., Rojas Echeverri, J. C., Boya, P., & Wright, S. J. (2017). Sources of variation in foliar secondary chemistry in a tropical forest tree community. *Ecology*, *98*(3), 616–623. https://doi.org/10.1002/ecy.1689.
Šimura, J., Antoniadi, I., Široká, J., Tarkowská, D., Strnad, M., Ljung, K., et al. (2018). Plant hormonomics: Multiple phytohormone profiling by targeted metabolomics. *Plant Physiology*, *177*(2), 476–489. https://doi.org/10.1104/pp.18.00293.

Smith, R., & Roberts, I. (2016). Time for sharing data to become routine: The seven excuses for not doing so are all invalid. *F1000Research*, *5*, 781. https://doi.org/10.12688/f1000research.8422.1.

Sokal, R. R., & Michener, C. D. (1958). A statistical method for evaluating systematic relationships. *The University of Kansas Science Bulletin*, *38*(2), 1409–1438.

St»hle, L., & Wold, S. (1990). Multivariate analysis of variance (MANOVA). *Chemometrics and Intelligent Laboratory Systems*, *9*(2), 127–141. https://doi.org/10.1016/0169-7439(90)80094-M.

Stanstrup, J., Broeckling, C., Helmus, R., Hoffmann, N., Mathé, E., Naake, T., et al. (2019). The metaRbolomics toolbox in Bioconductor and beyond. *Metabolites*, *9*(10), 200. https://doi.org/10.3390/metabo9100200.

Stettin, D., Poulin, R. X., & Pohnert, G. (2020). Metabolomics benefits from Orbitrap GC–MS—Comparison of low- and high-resolution GC–MS. *Metabolites*, *10*(4), 143. https://doi.org/10.3390/metabo10040143.

Tenenbaum, J. D., & Blach, C. (2018). Best practices and lessons learned from reuse of 4 patient-derived metabolomics datasets in Alzheimer's disease. *Biocomputing*, *2018*, 280–291. https://doi.org/10.1142/9789813235533_0026.

Treutler, H., Tsugawa, H., Porzel, A., Gorzolka, K., Tissier, A., Neumann, S., et al. (2016). Discovering regulated metabolite families in untargeted metabolomics studies. *Analytical Chemistry*, *88*(16), 8082–8090. https://doi.org/10.1021/acs.analchem.6b01569.

Trutschel, D., Schmidt, S., Grosse, I., & Neumann, S. (2015). Experiment design beyond gut feeling: Statistical tests and power to detect differential metabolites in mass spectrometry data. *Metabolomics*, *11*(4), 851–860. https://doi.org/10.1007/s11306-014-0742-y.

Tsugawa, H., Cajka, T., Kind, T., Ma, Y., Higgins, B., Ikeda, K., et al. (2015). MS-DIAL: Data-independent MS/MS deconvolution for comprehensive metabolome analysis. *Nature Methods*, *12*(6), 523–526. https://doi.org/10.1038/nmeth.3393.

van den Berg, R. A., Hoefsloot, H. C., Westerhuis, J. A., Smilde, A. K., & van der Werf, M. J. (2006). Centering, scaling, and transformations: Improving the biological information content of metabolomics data. *BMC Genomics*, 7(1), 142. https://doi.org/10.1186/1471-2164-7-142.

Villas-Bas, S. G., Roessner, U., Hansen, M. A. E., Smedsgaard, J., & Nielsen, J. (2007). *Metabolome analysis*. John Wiley & Sons, Inc. https://doi.org/10.1002/0470105518.

Weckwerth, W., & Kahl, G. (2013). *The handbook of plant metabolomics: KAHL:PLANT METABOLOMICS O-BK*. KGaA: Wiley-VCH Verlag GmbH & Co. https://doi.org/10.1002/9783527669882.

Weisser, W. W., Roscher, C., Meyer, S. T., Ebeling, A., Luo, G., Allan, E., et al. (2017). Biodiversity effects on ecosystem functioning in a 15-year grassland experiment: Patterns, mechanisms, and open questions. *Basic and Applied Ecology*, *23*, 1–73. https://doi.org/10.1016/j.baae.2017.06.002.

Wold, S., Sjöström, M., & Eriksson, L. (2001). PLS-regression: A basic tool of chemometrics. *Chemometrics and Intelligent Laboratory Systems*, *58*(2), 109–130. https://doi.org/10.1016/S0169-7439(01)00155-1.

SECTION 3

Metabolomics applications in plant sciences

CHAPTER SEVEN

Plant metabolomics and breeding

Olivier Fernandez[a], Emilie J. Millet[b], Renaud Rincent[c], Sylvain Prigent[d], Pierre Pétriacq[d], and Yves Gibon[d,*]

[a]EA4707 Laboratoire Résistance Induite et Bioprotection des Plantes, Univ. Reims Champagne-Ardenne, Reims Cedex 2, France
[b]Department of Plant Sciences, Mathematical and Statistical Methods—Biometris, Wageningen, The Netherlands
[c]UMR 1095 Génétique, Diversité et Ecophysiologie des Céréales, INRAE, Clermont-Ferrand, France
[d]UMR1332, Université de Bordeaux, INRAE, Bordeaux Metabolome, MetaboHUB, PHENOME, Villenave d'Ornon, France
*Corresponding author: e-mail address: yves.gibon@inrae.fr

Contents

Abstract

Metabolomics has become essential in the field of phenotyping, providing information halfway between genomes and phenotypes that can be used to predict complex traits. Thus in plants, studies have shown that metabolomics can be used to predict complex traits such as leaf biomass or yield. From a fundamental point of view, these advances can be used to identify new mechanisms involved in plant performance, for example, through identifying and deciphering QTLs involved in the control of the underlying metabolic pathways. From a breeding perspective, "phenomic selection" protocols combining phenotyping under optimal and/or stressing conditions, and metabolomics can be used to construct models that are able to predict traits of interest. Such developments will only be relevant if high throughput and reasonable costs can be achieved, if possible not far from those of genomic selection. The combination of robotics and nontargeted mass spectrometry already allows to process thousands of samples whereas the current capabilities of bioinformatics make it possible to develop predictive models from thousands variables.

Advances in Botanical Research, Volume 98
ISSN 0065-2296
https://doi.org/10.1016/bs.abr.2020.09.020

1. Introduction

Centuries of breeding have reshaped a vast ensemble of species. From the first growers to modern breeding industry the selection process has been mostly driven by the improvement of performance (Fernandez et al., 2016). However, the definition of crop performance has long been restricted to yield, especially during the so-called "Green revolution" (Fernandez et al., 2016). Although spectacular improvements have certainly contributed to reduce hunger in the world, we now face a stagnation of yield in major crops (Long, Marshall-Colon, & Zhu, 2015), a situation that is further complicated by increasing uncertainties linked to climate changes. We are probably at a turning point for agriculture in general and breeding in particular. For example, we tended to rely excessively on single elite varieties that are regularly replaced by better performing ones. Additionally, by focusing on yield, we may have overlooked diversity and lost other important performance traits such as abiotic and biotic stress tolerance. Finally, while traditional agriculture sought to master the environment (pesticides, fertilizers, extreme mechanization) and used high-yielding varieties, it becomes clear that agriculture of the future will have to take much more account of local specificities (climate, soil, microflora, etc.), which will inevitably result in as many tailor-made needs in terms of varieties.

In this context, the ability to predict plant performance under any growth scenario will become increasingly important. Besides the great promises of genomic selection, plant phenomics has been proposed to deliver performance prediction based on low-cost and high-throughput phenotypic descriptors rather than DNA polymorphisms (Rincent et al., 2018). In this chapter we envisage metabolomics as a tool capable of contributing significantly to the phenotyping of plants and therefore to the selection of new varieties better suited to the world of the future. Indeed, we can consider that metabolomics is to biology what a supermarket inventory would be to sociology: by making an inventory of the metabolome we find metabolites involved in signaling (the multimedia department), those that serve to withstand biotic and abiotic stresses (the body care and pharmacy departments) and finally those that serve to build the biomass (the wholefood departments). Like the sociologist who can learn a lot by comparing the volumes of goods, the metabolome tells us a lot about the biology of the plant, so much so that we can now predict traits of interest. This still little-explored property of the metabolome opens up very interesting perspectives for breeding.

In this chapter, we will first give an overview of genomic selection and plant phenotyping. We will then consider metabolomics as an additional powerful tool for breeding. For this we will first review the results obtained with metabolomics in crops. We will describe the modeling principles used in predictive metabolomics, then discuss how highlighting the link between the metabolome and plant performance can lead to the discovery of mechanisms. Finally, we will propose to insert metabolomics into the breeding pipeline.

2. Genomic selection is an efficeint tool for breeding

For centuries and until recently plant breeding consisted in selecting the seeds from the plants with the best agronomical characteristics for the sowing of the next season. One limitation of this strategy mainly based on phenotypic data is that it is conducted without knowing the genes underlying the variation of the phenotypic trait and thus without knowing the favorable alleles that could be combined to produce an improved genotype. This is not necessarily a problem, but we can hope to increase genetic progress with a better knowledge of the genetic architecture of the traits of interest.

In the 1980s and later, new approaches appeared, enabling to detect polymorphism at the DNA level, in particular Single Nucleotide Polymorphism (SNP). These tools, combined with phenotypic data, allow detecting regions of the genome related to the trait of interest (QTL) using linkage-based detection (in segregating populations) or genome-wide association study (in panels with higher diversity). Once QTLs have been detected, markers can be used in breeding programs to follow the favorable alleles in a cross to select improved individuals. This marker-assisted selection (MAS) has typically been efficiently used to introgress resistance alleles in elite material through back-crosses (Randhawa et al., 2009; Riar, Kaur, Dhaliwal, Singh, & Chhuneja, 2012; Sanz-Alferez, Richter, Hulbert, & Bennetzen, 1995). This is more difficult when the trait is influenced by many genes, which is often the case for traits of interest. In that case, only a few QTLs are detected, and as a result only a fraction of the total genetic variability is explained. Moreover, cumulating several QTLs usually requires complex crossing schemes ("pyramiding") and large progenies (Servin, Martin, Mézard, & Hospital, 2004).

For these quantitative traits, it becomes interesting to estimate all the marker effects simultaneously to circumvent the limitations of QTL detection. Moreau, Charcosset, Hospital, and Gallais (1998) showed that using a lax threshold for detecting QTLs (i.e., allowing more false positive) led to higher predictive ability, likely through capturing more minor QTLs. A step further, first proposed by Whittaker, Thompson, and Denham (2000), and further formalized and extended by Meuwissen, Hayes, and Goddard (2001), consisted in estimating all the marker effects simultaneously without any significance threshold with the so-called genomic selection (GS) approach. GS can be applied as follows: in a first step the genotypes and phenotypes of reference individuals are combined to calibrate the chosen statistical model (mixed models, various priors in the Bayesian framework, or machine learning). In a second step, the calibrated model is used to predict the genotyped selection candidates, which can then be selected without being phenotyped. GS is expected to be more efficient than post-QTL MAS, because a more important part of the genetic variance is captured, reducing the amount of missing heritability (Yang et al., 2010). The main factors affecting GS predictive abilities are the marker density (Asoro, Newell, Beavis, Scott, & Jannink, 2011), although it has become nonlimiting with modern high throughput genotyping methods, either SNP-chips or genotyping-by-sequencing, the trait heritability (Combs & Bernardo, 2013) and genetic architecture (Spindel et al., 2015), the size of the calibration set (Lehermeier et al., 2014) and its composition with regard to the test set (Rincent et al., 2012), and the statistical model used for prediction (Heslot, Yang, Sorrells, & Jannink, 2012).

GS can be used in different ways to increase genetic progress. It can increase selection intensity, decrease the duration of the breeding cycles, increase the precision of the genetic values and predict the performance in situations of genotype by environment interactions (GxE). The optimal use strongly depends on the species biological characteristics, on the technological constraints and on the budget available for breeding. All these constraints specific to each species lead to numerous ways of applying GS in breeding schemes. Some authors propose to use GS for a prescreening of the early generations (Marulanda et al., 2016; Schrag, Schipprack, & Melchinger, 2019), for recurrent selection (Bassi, Bentley, Charmet, Ortiz, & Crossa, 2016; Gorjanc, Gaynor, & Hickey, 2018; Heffner, Lorenz, Jannink, & Sorrells, 2010), for optimal cross selection (Gorjanc et al., 2018), for the identification of segregating plants potentially generating doubled haploids of high value (Daetwyler, Hayden, Spangenberg, & Hayes, 2015), for sparse testing, while others propose to use GS to complement phenotyping of expensive traits in the late generations (Ben-Sadoun et al., 2020; Michel et al., 2016). GS is also an efficient tool for screening the huge collections of diversity material to identify promising genitors in prebreeding programs (Crossa et al., 2016; Yu et al., 2016). GS has to be used with care as it can result in the rapid fixation of alleles (Jannink, 2010), but it can at least be used to eliminate an important part of the varieties from the prebreeding process.

Most plant species are submitted to strong GxE, which makes the breeding process difficult because the best varieties are not the same in all environments. It is of course not possible to evaluate the performance of all candidates in all target environments, but it can be predicted by adapted GS models. This can be done by attributing environment-specific marker effects (Schulz-Streeck et al., 2013), or by introducing environmental covariates in the model (Heslot, Akdemir, Sorrells, & Jannink, 2014; Jarquín et al., 2014; Rincent et al., 2019), which allows predicting in new environments. Integrative approaches combining GS models and crop growth models involving genetic parameters and environmental covariates were also proposed and applied to predict GxE for various traits (Onogi et al., 2016; Technow, Messina, Totir, & Cooper, 2015).

GS has been successfully implemented in dairy cattle and has considerably increased genetic progress thanks to the replacement of progeny testing by genomic predictions (García-Ruiz et al., 2016), which is much cheaper and allowed to have genetic values on many more bulls. In plant breeding, simulations (Heslot et al., 2012; Jannink, 2010; Zhong, Dekkers, Fernando, & Jannink, 2009) and field experiments (Albrecht et al., 2011; Crossa et al., 2010; Hofheinz, Borchardt, Weissleder, & Frisch, 2012; Zhao et al., 2012) gave encouraging results in populations with variable levels of diversity. Bernardo and Yu (2007) showed that GS provides 18%–43% more genetic gain per cycle than traditional marker assisted recurrent selection in biparental populations. Crossa et al. (2010) confirmed the potential interest of GS in more diverse material. More recently, practical implementation of GS across breeding cycles gave encouraging results in various plant species (Albrecht et al., 2014; Asoro et al., 2011; Hofheinz et al., 2012; Michel et al., 2016), although GxE and genetic distance between the calibration set and the test set resulted in important variability in the predictive abilities and often lower than those obtained in within-year cross-validations.

Most GS models are "black-box" models, the objective being to predict, not to understand. However, it has been shown that the use of functional information such as annotation (MacLeod et al., 2016), results of QTL detection (Bernardo, 2014), or functional networks from different omics layers including metabolomics could increase prediction accuracy in some situations. It has indeed been shown that these approaches could be more efficient than reference models in challenging conditions such as the prediction of individuals genetically distant from the calibration set (Fang et al., 2017).

GS models predict the additive genetic values, and potentially the statistical epistasis (Varona, Legarra, Toro, & Vitezica, 2018) and GxE. But the combination of genomic and phenotypic information is not sufficient to reconstruct the biological processes which lead to a given phenotype. Predicting phenotypic values is not of interest for the species for which individuals are selected for their efficiency to produce performant offspring. But for most plant species, predicting a phenotype is interesting (Bernardo, 2020), as the underlying genotype can be infinitely replicated (e.g., clones, inbred). So for plant breeding the use of endophenotypes in addition to the genotypic information is of great interest. Models involving transcriptomics (Fu et al., 2012), small RNA (Seifert et al., 2018), metabolomics (Riedelsheimer et al., 2012; Schrag et al., 2018) or even NIRS (Rincent et al., 2018) were efficiently used for phenotype prediction, alone or in combination.

3. Plant phenomics, from phenotyping to prediction

The word "Phenotype" was firstly used by the Danish botanist Wilhelm Johannsen in 1909, more than 40 years after Mendel expressed its concept. From the Greek phainein, meaning "to show," and typos, meaning "type," it consists in all observable and quantifiable traits of an individual, that are the result of the combination of the genotype (the genetic background), the environment and the interaction between them (GxE), along the growth cycle of plants. Since the first studies, phenotyping has evolved "from bean weighting to image analysis" (Walter, Liebisch, & Hund, 2015) to become a part of plant science with a need for increase in throughput and precision. In the last decades, development of technologies with novel sensor, imaging techniques and automation made it possible to characterize plants at multiple scales, molecular level (proteomic, transcriptomic, metabolomic), organs, whole plant and canopy/community, and with a high dimensionality. Phenomics has emerged as a full discipline, mirroring and complementing genomic (Houle, 2009).

Phenotyping platforms have widely spread, with automated systems installed in greenhouses or growth chambers enabling a precise screening of thousands of plants at scales ranging from minutes to weeks. The first phenotyping platforms included, for example, automated growth chambers with a robotic arm to carry the pots, e.g., the Phenopsis platform, Montpellier, France (Granier et al., 2006) or the WIWAM xy platform, Ghent, Belgium (Skirycz et al., 2011), or greenhouses with conveyor belts that bring the plants to watering, weighing, and imaging stations, e.g., the PlantAccelerator, Adelaide, Australia or at the IPK Gatersleben, Germany (Junker et al., 2015). Automation and imaging also moved to the field enabling measurement at the canopy/ecosystem level, e.g., Pheno-Copter, Australia (Chapman et al., 2014) or Phenofield, France (Beauchêne et al., 2019). It is noteworthy that this was possible by borrowing tools from other disciplines, for example from physical sciences (hyperspectral imaging, e.g., Anderegg et al., 2020) or from biomedical imaging (root system characterization using MRI, e.g., Wasson, Nagel, Tracy, & Watt, 2020).

The diversity of platforms can be grouped into three main types (Tardieu, Cabrera-Bosquet, Pridmore, & Bennett, 2017): (i) the high-precision platforms with measurements at the organ level, (ii) the whole-plant, multi-environment platforms, highly instrumented greenhouses or fields, and (iii) the field multienvironment networks, classically used to study the genetic architecture of yield (for example multisite breeding trials). While classical genotype-by-environment (GxE) methods are widely used for the analysis of data coming from the last type (multisite experiments in field), the diversity of phenotyping techniques and the increasing amount of data points available in the other two types of platform raised new challenges to extract meaningful information for plant breeding purposes. Data management and storage, and data analysis appeared as the new bottlenecks.

Addressing these challenges is possible by generating synergies and shaping the phenotyping community. Individual platforms started as regional initiative, but they are now organized in networks or consortia, gathering different groups and facilities at the level of the country, e.g., DPPN (https://dppn.plant-phenotyping-network.de/), Phen-Italy (http://www.phen-italy.it/) or NPEC (https://www.npec.nl/). At the continent level, initiatives also emerged, e.g., NAPPN (http://nappn.plant-phenotyping.org/). In Europe two projects already helped the community progressing toward excellence across the whole phenotyping pipeline and facilitated their access: EPPN and EPPN^{2020} (https://eppn2020.plant-phenotyping.eu/). They are paving the way for the future infrastructure EMPHASIS, European Infrastructure for Multi-Scale Plant Phenotyping and Simulation for Food Security in a Changing Climate (https://emphasis.plant-phenotyping.eu/;

Roy, Tardieu, Tixier-Boichard, & Schurr, 2017). Finally, the International Plant Phenotyping Network (IPPN, https://www.plant-phenotyping.org/) facilitates the interaction between local consortia.

These networks and initiatives have already started to address the data management and storage issue by proposing MIAPPE standards, applying FAIR requirements (Findable, Accessible, Interoperable, Reusable) and developing integrated ontology-based information systems such as PHIS (Neveu et al., 2019). Methodologies for phenomic data analysis are also under development within these communities, from designing experiments to integrating multiple scales. Indeed, although not widely recognized, spatial and temporal heterogeneities are at least as important in controlled conditions as they are under field conditions (Cabrera-Bosquet et al., 2016; Konate et al., 2020). Therefore, it is essential that users of phenotyping platforms use appropriate statistical methods for designing and analyzing their experiments. New statistical models are required to make phenomics information useful either to increase genetic and biological understanding of yield dynamics and stability or to improve prediction accuracy as part of trait improvement in plant breeding. These models should be able to incorporate information at different measurement and organization scales and contain explicit formulations for GxE and GxM (genotype-by-management interactions). A general framework was developed for modelling GxE and GxM in the field and greenhouse (van Eeuwijk et al., 2019). The framework includes a hierarchy of types of phenotypic traits and a number of model classes to integrate phenotyping information from platforms and experimental fields into genotype-to-phenotype (G2P) models.

4. Toward metabolomics for breeding

4.1 Crop metabolomics

Whilst metabolomics has received inadequate attention in crop science, it has established itself as one of the essential breakthroughs in plant sciences by providing information halfway between genomes and phenotypes (Hall et al., 2002; Hong, Yang, Zhang, & Shi, 2016). Metabolomes not only describe the chemical composition of a biological sample but also accommodate integrative inferences from the genome, transcriptome, proteome and activome to the environmental influences. Empirically, metabolomics encompasses a plethora of analytical methods, usually based on MS or NMR (Deborde et al., 2017; Ernst, Silva, Silva, Vêncio, & Lopes, 2014) that aim at detecting and quantifying low-molecular-weight biochemicals corresponding to metabolites (Turner et al., 2016; Zhu et al., 2018). Hence, metabolomics offers a comprehensive view of the complexity of metabolic pathways, and provides insights into the biochemical mechanisms including those involved in developmental processes, tolerance to abiotic stress and microbial challenges, and further aspects of plant yield and quality (Razzaq, Sadia, Raza, Khalid Hameed, & Saleem, 2019; Roch et al., 2019; Tugizimana, Mhlongo, Piater, & Dubery, 2018; van Dam & Bouwmeester, 2016). Several chapters of this volume examine the contribution of plant metabolomics to such research fields.

Crop metabolomics exploits, e.g., mutants, transgenic lines, natural accessions and diversity panels, which are valuable material to investigate metabolic networks and unveil the underlying candidate genes or functions (Fernie & Schauer, 2009; Hong et al., 2016; Zhu et al., 2018). Over the past 30 years, the use of the model plant *Arabidopsis thaliana* and other reference species has played a pivotal role in providing a framework for investigating crops. Thus, metabolic genome-wide association studies (mGWAS) have been used to disentangle the genetic basis of metabolic networks through the discovering of new regulatory genes (or loci). This strategy has been applied to investigate the metabolic landscape of 309 Arabidopsis accessions grown under two independent environmental conditions (Wu et al., 2018). From several thousand semipolar metabolites, mGWAS resulted in the detection of 123 highly resolved quantitative traits, ca. 25% of which depended on the environment. The subsequent integration of transcript data allowed selecting 70 candidates for which 8 were validated experimentally. Furthermore, Arabidopsis mGWAS unraveled seed-specific interactions between the amino acid glutamine and glucosinolates supporting a role for metabolic homeostasis (Slaten et al., 2020). The genetic foundations of naturally occurring variation for flavonoids were also investigated in Arabidopsis through metabolite profiling and quantitative genetics (Routaboul et al., 2012). Besides, genetic dissection of metabolite changes evidenced mQTL hotspots and a master regulatory locus for seed metabolism in Arabidopsis (Knoch et al., 2017). In soybean, a legume model species, the genetic bases of oil biosynthesis were revealed by combining information from seed oil-related traits, metabolites and genes (Liu et al., 2020).

Fruit biology has considerably benefited from metabolomics strategies, which are examined in the chapter "Developmental metabolomics to decipher and improve fleshy fruit quality" by Allwood et al. of this volume. Besides, Kumar et al. reviewed the status and prospects of metabolomics for the improvement of fruit (Kumar, Bohra, Pandey, Pandey, & Kumar, 2017).

In this context, tomato fruit and grape are often the beneficiaries through a range of studies on growth, quality and stress responses (D'Angelo et al., 2019; Luna et al., 2020; Pinu, 2018; Zhu et al., 2018). For instance, tomato metabolomics was exploited to evaluate the impact of targeted genetic improvement (e.g., introgression of resistance genes from wild relatives in cultivars) on other unrelated traits including quality traits (Zhu et al., 2018).

Metabolite profiling in cereals often allows pinpointing the relations between yield prediction and specific responses to environmental constraints and growth. As for drought, metabolomics of durum wheat revealed that wheat spikes are essential upon grain filling, predominantly under contrasting water regimes (Vergara-Diaz et al., 2020). In Chinese wheat, higher growth indices in drought-tolerant genotypes were associated with metabolic rearrangements of primary (amino and organic acids, nucleotides) and specialized metabolites (phenolics, alkaloids; Guo et al., 2020). Postanthesis drought tolerance was also addressed by comparative GC-MS profiling in wheat leaves and roots, indicating greater changes for amino and organic acids in leaf than root tissues with resource allocation to leaves, principally (Kang et al., 2019). In wheat kernels, mGWAS enabled to dissect the flavonoid decoration pathway (Chen et al., 2020). In barley, phenolics profiles during grain development were combined to a weighted gene co-expression network analysis revealing two hub genes of the UDP-Glucuronosyltransferase family and three hub genes of the phenolics pathway (Han et al., 2020). In sunflower, metabolomics show potential for discriminating genotype groups and environmental conditions (Fernandez et al., 2019). In maize, hybrid metabotyping unraveled potential metabolic markers for breeding under early-sowing conditions (Lamari et al., 2018). Further, evolutionary metabolomics of maize and teosinte (a wild ancestor) identified metabolic divergence for 461 metabolites due to selection (Xu et al., 2019). High-quality maize improvement was also addressed through metabolite QTL analysis (Li et al., 2019). In rice grain, mQTL analysis also pinpointed a coordinated control for amino acids and flavonoids (Matsuda et al., 2012). Besides metabolomics, lipidomics allowed to pinpoint associations between rice quality traits (Concepcion, Calingacion, Garson, & Fitzgerald, 2020).

Overall, crop metabolomics show several advantages. It can help formulate smart strategies to (i) improve crop productivity including potential biomarkers for yield prediction, (ii) develop stress-tolerant genotypes and (iii) identify locus, gene or function associated with metabolic traits for crop improvement. Crop metabolomics is also essential to test evolutionary hypotheses on the evolution of plant metabolism, such as domestication or adaptation mechanisms (Batyrshina, Yaakov, Shavit, Singh, & Tzin, 2020; Fernie & Yan, 2019; Kroymann, 2011; Scossa & Fernie, 2020).

4.2 Principles of predictive metabolomics

Metabolomics combined or not to other omics, genetic information or environmental factors can be used to predict complex traits such as leaf biomass or yield. Here we give an overview of the existing methods that enable linking metabolomics and other omics data to complex traits. The aim of this part is not to describe into details the mathematical principles underlying the different methods, but instead to describe the requirements, limitations and outcomes of the different methods that can help predict phenotypes using metabolomics data.

A problem that often arises when dealing with metabolomic data is that samples size n is often very small compared to p, the number of variables that are measured in an experiment. Experiment are often dealing with tens (or at most hundreds) of samples, when thousands (or tens of thousands) different variables can be measured, often displaying high multicollinearity. From this observation, methods dealing with dimensionality reduction have been used for several years to describe and analyze the datasets. Principal Component Analysis (PCA) is often used to describe big datasets using its inherent capability to reduce dimensionality, by seeking for features that maximize the variance of the data (D'Angelo et al., 2019; Dan et al., 2019; Fernandez et al., 2019). Using this method it is possible to easily get an overview of the dataset and identify, for example, possible outliers or groups of individuals. The Partial Least Square (or Projection on Latent Structure, PLS) method is a supervised technique that enables a discrimination of the data by maximizing the covariance with a trait of interest, hence increasing the prediction capabilities. PLS has often been used to search for biomarkers in metabolomes, mainly by looking at the Variable Importance in Projection (VIP) score of variables. This score corresponds to the importance of the variable in the projection of the data in the multidimensional space. Since the early 2000, the development of Orthogonal PLS (OPLS) enabled a better interpretability of the models, without improvement of the prediction capabilities. If the predictive target is not numerical but a categorical variable, PLS Discriminant Analysis (PLS-DA) and OPLS-DA can also be used.

PLS-DA models have already been used in numerous plant studies to predict complex traits based on metabolomic data. One example is Dan et al. (2019) where the authors used LC-MS data of parental rice lines to predict thousand grain weight of an F1 hybrid population. PLS models have also been used in numerous studies to identify biomarkers (Fu et al., 2017; Sprenger et al., 2018; Steinfath et al., 2010).

PLS regression overcomes simple regression issues in the case of $n \ll p$ but it only acts on the reduction of dimensionality. When looking for biomarkers or when performing predictions, we are often only interested in a very small number of variables compared to the big size of p. It could then be interesting to use methods that can discard uninteresting variables while keeping only those linked with a given phenotype. Moreover, metabolomic variables often display a high level of collinearity, which can be a major drawback when looking for interesting variables. For example, a molecule possessing several adduct ions will appear as several highly correlated variables in the dataset. A good prediction method should enable the selection of only one variable among all of them without any pretreatment removing correlated variables. PLS based methods lack the ability to choose such a variable and discard others.

To achieve sparsity in the selection of variables, a penalty can be introduced to reduce the magnitude of coefficients associated with variables. Instead of choosing a-priori interesting variables highly linked with a given phenotype, methods using penalization decrease globally the weight of coefficients associated with the different variables. One of the most used penalization term is the L1-regularization that adds a L1 penalty equal to the absolute value of magnitude of coefficient (Park & Hastie, 2007). If this penalty is big, hence if the sum of the coefficient is forced to be less than a small value, it can make coefficients reach zero and remove completely noninteresting variables. The L2-regularization, instead, forces the square of the sum of coefficients to be smaller than a given value. With such a penalization, the coefficient associated with variables can only be shrunk close to zero, and all variables are always kept in the models. When applied to regression, L1- and L2-regularizations are implemented as Least Absolute Shrinkage and Selection Operator (LASSO) and ridge regression, respectively.

The sherry picking of variables inherent with the LASSO methods make it very interesting when looking for biomarkers and when performing prediction using omics data, which by essence contain a vast majority of non-interesting variables. It has been used successfully in several studies (Wei, Wang, Li, Qu, & Jia, 2018; Xu, Xu, Gong, & Zhang, 2016; Xu, Xu, & Xu, 2017). One big drawback of LASSO is that it can only accept at most n − 1 non-zero coefficients, with n the number of samples (de los Campos et al., 2009; Park & Casella, 2008). This can be a problem especially when the number of samples is very low, the constraint on the number of variables used in the model becoming very strong. Several methods have been designed to overcome this limitation, like the Bayesian LASSO (Park & Casella, 2008) or the Elastic Net method that is a mixture between ridge and LASSO regression that linearly combines L1 and L2-regularization. One example of prediction using Elastic Net is Gonzales and De Saeger (2018) where the authors used Elastic Net models to predict storage time of plasma based on its metabolome. Among 1229 and 1483 features measured in positive and negative mode, respectively, the methods enabled to pick 30 metabolites in both modes to perform accurate predictions ($R^2 > 0.8$ for both datasets). While predictions based on LASSO become more and more popular in plant metabolomics (Shi et al., 2020; Wang et al., 2019) and Elastic-net based predictions are already quite common in human metabolomics studies (Papandreou et al., 2019; Peña-Bautista et al., 2019; Trainor, Yampolskiy, & DeFilippis, 2018), Elastic-nets are only used as sparsely as their selection of variables in the plant metabolomics field where PLS models remain very popular. Given the high interpretability of Elastic-Net models (Loupit et al., 2020; Luna et al., 2020), and their prediction accuracies at least similar to the ones of PLS models, studies using such models will certainly appear soon.

4.3 From predictive metabolomics to mechanisms

Examples showing the possibility of predicting complex phenotypes from the metabolome are all hypotheses of the implication of metabolism in the establishment of plant performance. A number of publications have shown the existence of at least partial overlaps between metabolic traits and yield (e.g., Lisec et al., 2008; Mickelson et al., 2003), suggesting that researching the mechanisms controlling the metabolic traits associated with performance would consist of a sort of breakdown of performance. The discovery pipeline can then consist of (1) building and validating a predictive model, which associates one or more traits of interest with one or more metabolic variables; (2) extract the most predictive variables. These can be metabolic traits, analytes that correspond to unidentified metabolites such as *m/z* (untargeted mass spectrometry), buckets (untargeted nuclear magnetic resonance spectroscopy) or wavelengths (NIRS), or even composite variables when a reduction in the data complexity was operated, for example a principle component (Wilson et al., 2004) or even the variability of the metabolite concentrations (Alseekh et al., 2017); (3) demonstrate the mechanisms using independent evidence.

When the metabolic variables are unknown, it will first be of interest to elucidate their nature using an annotation strategy. This is however a major bottleneck, not least because many of the performance-associated metabolites are specialized and species-dependent (Yang et al., 2014).

When metabolic variables are known, an approach that has been used for decades is to directly test the molecule when it is available. This so-called pharmacological approach has produced numerous results. For example, the supply of allantoin, a metabolic intermediate that often accumulates in response to various stresses in plants, results in the activation of jasmonate-signaling, which induces defense responses (Takagi et al., 2016). Another example is the application of threitol promoting colonization of Eucalyptus roots by the pathogen that produces this metabolite (Wong et al., 2020). It is interesting to note that the accumulation of a metabolite in response to a stress does not necessarily mean that it is involved in resistance or tolerance. Furthermore, not all molecules are available or when they are, they may not be pure, which might lead to artefactual results due to contaminants (Figueroa & Lunn, 2016). Exogenous molecules may also not be absorbed or compartmentalized correctly once absorbed.

From the 1990s, reverse genetic strategies seeking to manipulate the synthetic (or degradative) pathways involved could therefore be considered in order to verify whether the metabolites in question are directly involved in performance. This approach has been used a lot in certain fields but with relatively few convincing results regarding plant performance. Thus, it has long been known that plants subjected to osmotic stress (drought, high salinity) accumulate so-called compatible metabolites such as proline, polyols, or glycine betaine, to which many protective properties have been attributed (Chen & Murata, 2002) but when plants were engineered to produce such metabolites the results were disappointing, in particular due to a trade-off with growth and thus biomass production (Bohnert & Shen, 1998). Although this finding is still not unanimous (Blum, 2017), it must be recognized that after more than 40 years of work, there is no GMO over-accumulating such metabolites on the market, and only one commercial cultivar achieving osmotic adjustment (Turner, 2018). Strikingly, it has been shown that plants with reduced proline synthesis (knockdown of pyrroline-5-carboxylate synthetase) are hypersensitive to stress, even when they are supplemented with proline (Nanjo et al., 1999). In addition to showing that the modalities of the synthesis of a compound can have more importance than the compound itself, these results have highlighted the importance of energy management, the accumulation of proline appearing like a redox valve (Székely et al., 2008). Moreover, the involvement of redox metabolism in tolerance to abiotic stress is supported by numerous studies. For example, in a panel of genetic diversity of rice subjected to drought, the ascorbate-glutathione cycle (concentrations of total ascorbate and glutathione but also capacities of the enzymes involved) was associated with resistance while the accumulation of malondialdehyde, which is a product of the peroxidation of lipids commonly used as a marker of oxidative stress, was associated with sensitivity (Melandri et al., 2020).

It is also possible to investigate the mechanisms underlying the association between metabolic traits and performance using a direct genetic approach. Thanks to significant advances in genetics (mutant populations, recombinant populations, GWAS, etc.) and molecular biology (NGS, CRISPR/CAS9, …) this approach now makes it possible to relatively quickly identify and validate candidate genes or even sequence modifications located in coding or non-coding regions that are capable of influencing the production of metabolites (e.g., Gemenet et al., 2020; Riedelsheimer, Lisec, et al., 2012). The advantage of this approach is that it is without preconceptions as to the role of the markers associated with performance. Thus, the identification in wheat of metabolites involved in wall synthesis, a determining element in resistance to *Fusarium graminearum*, has enabled the identification of a gene encoding an agmatine coumaroyl transferase that is involved in cell-wall synthesis (Kage, Karre, Kushalappa, & McCartney, 2017). The knockdown of this gene by virus-induced gene silencing (VIGS) then validated its involvement in tolerance to this pathogen. In tomato, a reverse genetics approach combined with analyses of gene expression, siRNA and DNA methylation has made it possible to elucidate a QTL associated with tocopherol content (Quadrana et al., 2014). The expression of the gene encoding 2-methyl-6-phytylquinol methyltransferase was found to be modulated by different levels of methylation of a retrotransposon located in its promoter region. This work, which has made it possible to understand the low heritability of tocopherol content, and therefore the difficulty in selecting lines enriched with vitamin E, shows all the interest in dissecting mechanisms. Although such examples of elucidation are still rare in the literature, we can expect that their number will dramatically increase in the years to come, in particular thanks to the increase in genetic resources (diversity panels, collections of mutants, etc.) and phenotyping, metabolomics and molecular biology tools available.

4.4 From predictive metabolomics to breeding

The predominant view of the contribution of metabolomics to breeding is that it is necessary to identify the genes involved in specific metabolic pathways that would in turn be involved in performance (Razzaq et al., 2019; Saito & Matsuda, 2010). Although this approach is commendable because leading to fundamental knowledge, it is probably very costly in terms of time and resources, and this might explain why metabolomics is still poorly used in breeding. Interestingly, pioneers of predictive metabolomics (Meyer et al., 2007; Riedelsheimer et al., 2012) largely agreed on the fact that metabolites also encompass various sources of variability emerging from the genome but also from the environment or the developmental stage of the crop. In particular, it was emphasized that tens or hundreds of metabolic variables can be almost as good as tens of thousands of SNPs in predicting traits of interest (Riedelsheimer, Czedik-Eysenberg, et al., 2012). Stated otherwise, predictive metabolomics could be used in plant phenomics to directly provide metabolic features (quantitative or semiquantitative data for metabolites or metabolic signatures respectively) as predictors and not necessarily through the elucidation of mQTLs or mGWAS.

Obviously, even if metabolomics is capable of delivering reliable performance predictions, its cost remains a deterrent. Putting metabolomics at the service of breeding will therefore imply finding the best compromise between costs of analyses and the benefit in terms of prediction. Two main aspects can be used to achieve such compromise. On the one hand, the cost of the analysis could drop if the throughput increases significantly. For example, the use of a low-toxic solvent in the extraction step (ethanol instead of the classically used methanol) made it possible to greatly increase the throughput (Bénard et al., 2018). More generally, the cost of metabolomics lies largely in the acquisition and maintenance of equipment. Increasing the throughput then means lowering the cost per sample. We estimate that it is possible to carry out nontargeted analyses for less than € 10, i.e., in a cost range equivalent to that of genotyping used in genomic selection. On the other hand, the predictive power of the metabolome could be optimized by applying relevant environmental scenarios (e.g., challenging water and nutrient supply, illumination and temperature regimes) the ideal being to obtain predictions at young stages and in controlled conditions, therefore repeatable and at low cost in terms of production of the plant material (Fig. 1). This point seems particularly important because the vast majority of analytes measured in nontargeted metabolomics come from secondary metabolism, whose role is to contribute to adaptation or

Fig. 1 Prediction and costs in omics-based breeding. The costs associated with cultivation are higher the longer it lasts. Obtaining predictive data therefore benefits from being as early as possible. In the case of genomic selection, the quality of the prediction remains unchanged throughout the life of the plant. Sampling for the genotype can therefore be done very early. Because genomic data can give less good predictions (for example when multiple factors interact in the field), it can be interesting to use proxies from phenotyping performed in controlled conditions. However, phenomics require quite long cultivation times while metabolomics can be used at younger stages or even from seed (e.g., Dan et al., 2019). In turn, metabolomics alone provide less good predictions. Predictivity which corresponds, for example, to the correlation coefficient between predicted data and measured data, is arbitrary.

acclimation to the environment. Overall, predictive metabolomics can be seen as complementary to genomic selection or even the use of non-invasive methods in order to predict performance.

Fig. 2 summarizes a tentative pipeline for the use of predictive metabolomics in breeding. Large panels of genotypes (up to 10,000) could be implemented in such system as long as there is (1) diversity for the chosen criteria of performance, (2) enough capacity to grow the plants under controlled conditions and (3) enough analytical capacity. Reference genotypes (50–100) are used to build a predictive performance model; they are the only ones grown in the field to assess performance and will be included in all panels evaluated under controlled conditions (Fig. 2). A preliminary step consisting in maximizing the predictive value of the metabolome by adjusting the culture conditions under controlled conditions may prove useful. The metabolome is typically obtained using HPLC coupled with and Orbitrap mass detector. Together with a short-length separation chromatography, such setup would provide a good balance between (i) number of processed samples (up to hundreds per week), (ii) the number of generated m/z that can be further used as metabolic variables (up to thousands) and (iii) a high mass accuracy, which facilitates the reuse of the data or further

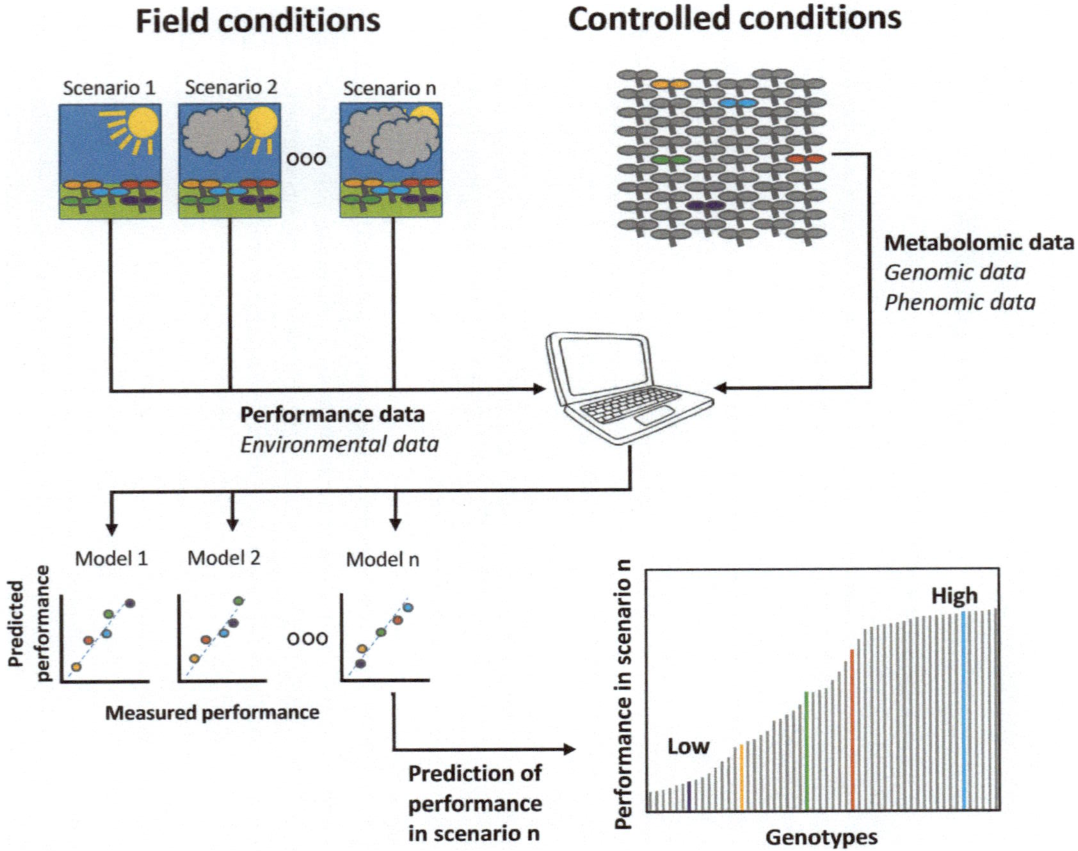

Fig. 2 A putative metabolomic selection pipeline for breeding. A large panel of genotypes bearing diversity for the chosen performance criteria (e.g., yield, yield resilience,…) is available. A subset of reference genotypes (colored plants) is grown in the field (in one to n growth scenarios) to assess their performance. Environmental data that characterize each growth scenario might also be collected. All genotypes including the reference genotypes are grown under controlled conditions, which maximize predictivity, to collect metabolomic data and, possibly, genomic and phenomic data. The data obtained with the reference genotypes is then used to build a post hoc model to predict performance for the scenario(s) of interest.

annotation. The most predictive model is then built with the performance and the metabolic data (Fig. 2). Noteworthy, the predictive model can eventually be improved using other inputs such as environmental, phenomic and/or genomic data.

Some specificity of metabolic features (especially when compared to SNPs) has to be taken into account. Thus, metabolic variables are prone to diurnal variations (Bénard et al., 2018) and can vary according to the developmental stage (Lecourieux et al., 2020; Li et al., 2019). These differences might affect the final prediction and should be accounted for. Moreover, while obtaining interoperable metabolomic data will certainly be a plus, it seems more wise to reconstruct a predictive model each time and therefore by systematically including the reference genotypes in cultures under controlled conditions.

5. Conclusion

Metabolomics could soon become a worthy part of the breeding process. Available analytical and modelling tools are now available for their use, alone or combined with genomics and phenomics, to predict plant performance. Metabolomics-based breeding could be valuable to address the stability of crop yield in the context of global change, with for example the ability to tolerate environmental stresses or diseases or maintain nutritional value under suboptimal growth conditions. Thus, it will be necessary to deploy significant efforts to be able to multiply the scenarios in controlled conditions that are capable of delivering tailor-made proxies. However, in order to make metabolomic selection a reality, moderate cost solution still need to be developed. For this, (1) predictive data should be collected as early as possible in the plant life cycle; (2) analytics should be robotized, from harvest to measurements, in order to enable processing thousands to tens of thousands of samples at low costs; (3) calculation capacity should be increased to enable faster development of predictive models from thousands of variables.

References

Albrecht, T., Auinger, H.-J., Wimmer, V., Ogutu, J. O., Knaak, C., Ouzunova, M., et al. (2014). Genome-based prediction of maize hybrid performance across genetic groups, testers, locations, and years. *Theoretical and Applied Genetics*, *127*(6), 1375–1386. https://doi.org/10.1007/s00122-014-2305-z.

Albrecht, T., Wimmer, V., Auinger, H.-J., Erbe, M., Knaak, C., Ouzunova, M., et al. (2011). Genome-based prediction of testcross values in maize. *Theoretical and Applied Genetics*, *123*(2), 339. https://doi.org/10.1007/s00122-011-1587-7.

Alseekh, S., Tong, H., Scossa, F., Brotman, Y., Vigroux, F., Tohge, T., et al. (2017). Canalization of tomato fruit metabolism. *The Plant Cell*, *29*(11), 2753–2765. https://doi.org/10.1105/tpc.17.00367.

Anderegg, J., Yu, K., Aasen, H., Walter, A., Liebisch, F., & Hund, A. (2020). Spectral vegetation indices to track senescence dynamics in diverse wheat germplasm. *Frontiers in Plant Science*, *10*. https://doi.org/10.3389/fpls.2019.01749.

Asoro, F. G., Newell, M. A., Beavis, W. D., Scott, M. P., & Jannink, J.-L. (2011). Accuracy and training population design for genomic selection on quantitative traits in elite North American oats. *The Plant Genome*, *4*(2), 132–144. https://doi.org/10.3835/plantgenome2011.02.0007.

Bassi, F. M., Bentley, A. R., Charmet, G., Ortiz, R., & Crossa, J. (2016). Breeding schemes for the implementation of genomic selection in wheat (Triticum spp.). *Plant Science*, *242*, 23–36. https://doi.org/10.1016/j.plantsci.2015.08.021.

Batyrshina, Z. S., Yaakov, B., Shavit, R., Singh, A., & Tzin, V. (2020). Comparative transcriptomic and metabolic analysis of wild and domesticated wheat genotypes reveals differences in chemical and physical defense responses against aphids. *BMC Plant Biology*, *20*(1), 19. https://doi.org/10.1186/s12870-019-2214-z.

Beauchêne, K., Leroy, F., Fournier, A., Huet, C., Bonnefoy, M., Lorgeou, J., et al. (2019). Management and characterization of abiotic stress via PhénoField®, a high-throughput field phenotyping platform. *Frontiers in Plant Science*, *10*. https://doi.org/10.3389/fpls.2019.00904.

Bénard, C., Acket, S., Rossez, Y., Fernandez, O., Berton, T., Gibon, Y., et al. (2018). Untargeted analysis of semipolar compounds by LC-MS and targeted analysis of fatty acids by GC-MS/GC-FID: From plant cultivation to extract preparation. *Methods in Molecular Biology*, *1778*, 101–124. https://doi.org/10.1007/978-1-4939-7819-9_8.

Ben-Sadoun, S., Rincent, R., Auzanneau, J., Oury, F. X., Rolland, B., Heumez, E., et al. (2020). Economical optimization of a breeding scheme by selective phenotyping of the calibration set in a multi-trait context: Application to bread making quality. *Theoretical and Applied Genetics*, *133*(7), 2197–2212. https://doi.org/10.1007/s00122-020-03590-4.

Bernardo, R. (2014). Genomewide selection when major genes are known. *Crop Science*, *54*(1), 68–75. https://doi.org/10.2135/cropsci2013.05.0315.

Bernardo, R. (2020). Reinventing quantitative genetics for plant breeding: Something old, something new, something borrowed, something BLUE. *Heredity*, 1–11. https://doi.org/10.1038/s41437-020-0312-1.

Bernardo, R., & Yu, J. (2007). Prospects for genomewide selection for quantitative traits in maize. *Crop Science*, *47*(3), 1082–1090. https://doi.org/10.2135/cropsci2006.11.0690.

Blum, A. (2017). Osmotic adjustment is a prime drought stress adaptive engine in support of plant production. *Plant, Cell & Environment*, *40*(1), 4–10. https://doi.org/10.1111/pce.12800.

Bohnert, H. J., & Shen, B. (1998). Transformation and compatible solutes. *Scientia Horticulturae*, *78*(1), 237–260. https://doi.org/10.1016/S0304-4238(98)00195-2.

Cabrera-Bosquet, L., Fournier, C., Brichet, N., Welcker, C., Suard, B., & Tardieu, F. (2016). High-throughput estimation of incident light, light interception and radiation-use efficiency of thousands of plants in a phenotyping platform. *New Phytologist*, *212*(1), 269–281. https://doi.org/10.1111/nph.14027.

Chapman, S. C., Merz, T., Chan, A., Jackway, P., Hrabar, S., Dreccer, M. F., et al. (2014). Pheno-copter: A low-altitude, autonomous remote-sensing robotic helicopter for high-throughput field-based phenotyping. *Agronomy*, *4*(2), 279–301. https://doi.org/10.3390/agronomy4020279.

Chen, J., Hu, X., Shi, T., Yin, H., Sun, D., Hao, Y., et al. (2020). Metabolite-based genome-wide association study enables dissection of the flavonoid decoration pathway of wheat kernels. *Plant Biotechnology Journal*, *18*(8), 1722–1735. https://doi.org/10.1111/pbi.13335.

Chen, T. H. H., & Murata, N. (2002). Enhancement of tolerance of abiotic stress by metabolic engineering of betaines and other compatible solutes. *Current Opinion in Plant Biology*, *5*(3), 250–257. https://doi.org/10.1016/S1369-5266(02)00255-8.

Combs, E., & Bernardo, R. (2013). Accuracy of genomewide selection for different traits with constant population size, heritability, and number of markers. *The Plant Genome*, *6*(1). https://doi.org/10.3835/plantgenome2012.11.0030.

Concepcion, J. C. T., Calingacion, M., Garson, M. J., & Fitzgerald, M. A. (2020). Lipidomics reveals associations between rice quality traits. *Metabolomics*, *16*(5), 54. https://doi.org/10.1007/s11306-020-01670-6.

Crossa, J., de los Campos, G., Pérez, P., Gianola, D., Burgueño, J., Araus, J. L., et al. (2010). Prediction of genetic values of quantitative traits in plant breeding using pedigree and molecular markers. *Genetics*, *186*(2), 713–724. https://doi.org/10.1534/genetics.110.118521.

Crossa, J., Jarquín, D., Franco, J., Pérez-Rodríguez, P., Burgueño, J., Saint-Pierre, C., et al. (2016). Genomic prediction of Gene Bank wheat landraces. *G3: Genes, Genomes, Genetics*, *6*(7), 1819–1834. https://doi.org/10.1534/g3.116.029637.

D'Angelo, M., Zanor, M. I., Burgos, E., Asprelli, P. D., Boggio, S. B., Carrari, F., et al. (2019). Fruit metabolic and transcriptional programs differentiate among Andean tomato (*Solanum lycopersicum* L.) accessions. *Planta*, *250*(6), 1927–1940. https://doi.org/10.1007/s00425-019-03274-4.

Daetwyler, H. D., Hayden, M. J., Spangenberg, G. C., & Hayes, B. J. (2015). Selection on optimal haploid value increases genetic gain and preserves more genetic diversity relative to genomic selection. *Genetics*, *200*(4), 1341–1348. https://doi.org/10.1534/genetics.115.178038.

Dan, Z., Chen, Y., Xu, Y., Huang, J., Huang, J., Hu, J., et al. (2019). A metabolome-based core hybridisation strategy for the prediction of rice grain weight across environments. *Plant Biotechnology Journal*, *17*(5), 906–913. https://doi.org/10.1111/pbi.13024.

de los Campos, G., Naya, H., Gianola, D., Crossa, J., Legarra, A., Manfredi, E., et al. (2009). Predicting quantitative traits with regression models for dense molecular markers and pedigree. *Genetics*, *182*(1), 375–385. https://doi.org/10.1534/genetics.109.101501.

Deborde, C., Moing, A., Roch, L., Jacob, D., Rolin, D., & Giraudeau, P. (2017). Plant metabolism as studied by NMR spectroscopy. *Progress in Nuclear Magnetic Resonance Spectroscopy*, *102–103*, 61–97. https://doi.org/10.1016/j.pnmrs.2017.05.001.

Ernst, M., Silva, D. B., Silva, R. R., Vêncio, R. Z. N., & Lopes, N. P. (2014). Mass spectrometry in plant metabolomics strategies: From analytical platforms to data acquisition and processing. *Natural Product Reports*, *31*(6), 784–806. https://doi.org/10.1039/C3NP70086K.

Fang, L., Sahana, G., Ma, P., Su, G., Yu, Y., Zhang, S., et al. (2017). Use of biological priors enhances understanding of genetic architecture and genomic prediction of complex traits within and between dairy cattle breeds. *BMC Genomics*, *18*(1), 604. https://doi.org/10.1186/s12864-017-4004-z.

Fernandez, O., Urrutia, M., Bernillon, S., Giauffret, C., Tardieu, F., Le Gouis, J., et al. (2016). Fortune telling: Metabolic markers of plant performance. *Metabolomics*, *12*(10), 158. https://doi.org/10.1007/s11306-016-1099-1.

Fernandez, O., Urrutia, M., Berton, T., Bernillon, S., Deborde, C., Jacob, D., et al. (2019). Metabolomic characterization of sunflower leaf allows discriminating genotype groups or stress levels with a minimal set of metabolic markers. *Metabolomics*, *15*(4), 56. https://doi.org/10.1007/s11306-019-1515-4.

Fernie, A. R., & Schauer, N. (2009). Metabolomics-assisted breeding: A viable option for crop improvement? *Trends in Genetics*, *25*(1), 39–48. https://doi.org/10.1016/j.tig.2008.10.010.

Fernie, A. R., & Yan, J. (2019). De novo domestication: An alternative route toward new crops for the future. *Molecular Plant*, *12*(5), 615–631. https://doi.org/10.1016/j.molp.2019.03.016.

Figueroa, C. M., & Lunn, J. E. (2016). A tale of two sugars: Trehalose 6-phosphate and sucrose. *Plant Physiology*, *172*(1), 7–27. https://doi.org/10.1104/pp.16.00417.

Fu, J., Falke, K. C., Thiemann, A., Schrag, T. A., Melchinger, A. E., Scholten, S., et al. (2012). Partial least squares regression, support vector machine regression, and transcriptome-based distances for prediction of maize hybrid performance with gene expression data. *Theoretical and Applied Genetics*, *124*(5), 825–833. https://doi.org/10.1007/s00122-011-1747-9.

Fu, H.-Y., Hu, O., Zhang, Y.-M., Zhang, L., Song, J.-J., Lu, P., et al. (2017). Mass-spectra-based peak alignment for automatic nontargeted metabolic profiling analysis for biomarker screening in plant samples. *Journal of Chromatography A*, *1513*, 201–209. https://doi.org/10.1016/j.chroma.2017.07.044.

García-Ruiz, A., Cole, J. B., VanRaden, P. M., Wiggans, G. R., Ruiz-López, F. J., & Van Tassell, C. P. (2016). Changes in genetic selection differentials and generation intervals in US Holstein dairy cattle as a result of genomic selection. *Proceedings of the National Academy of Sciences of the United States of America*, *113*(28), E3995–E4004. https://doi.org/10.1073/pnas.1519061113.

Gemenet, D. C., da Silva Pereira, G., De Boeck, B., Wood, J. C., Mollinari, M., Olukolu, B. A., et al. (2020). Quantitative trait loci and differential gene expression analyses reveal the genetic basis for negatively associated β-carotene and starch content in hexaploid sweetpotato [Ipomoea batatas (L.) Lam.]. *Theoretical and Applied Genetics*, *133*(1), 23–36. https://doi.org/10.1007/s00122-019-03437-7.

Gonzales, G. B., & De Saeger, S. (2018). Elastic net regularized regression for time-series analysis of plasma metabolome stability under sub-optimal freezing condition. *Scientific Reports*, *8*(1), 3659. https://doi.org/10.1038/s41598-018-21851-7.

Gorjanc, G., Gaynor, R. C., & Hickey, J. M. (2018). Optimal cross selection for long-term genetic gain in two-part programs with rapid recurrent genomic selection. *Theoretical and Applied Genetics*, *131*(9), 1953–1966. https://doi.org/10.1007/s00122-018-3125-3.

Granier, C., Aguirrezabal, L., Chenu, K., Cookson, S. J., Dauzat, M., Hamard, P., et al. (2006). PHENOPSIS, an automated platform for reproducible phenotyping of plant responses to soil water deficit in Arabidopsis thaliana permitted the identification of an accession with low sensitivity to soil water deficit. *New Phytologist*, *169*(3), 623–635. https://doi.org/10.1111/j.1469-8137.2005.01609.x.

Guo, X., Xin, Z., Yang, T., Ma, X., Zhang, Y., Wang, Z., et al. (2020). Metabolomics response for drought stress tolerance in Chinese wheat genotypes (Triticum aestivum). *Plants*, *9*(4), 520. https://doi.org/10.3390/plants9040520.

Hall, R., Beale, M., Fiehn, O., Hardy, N., Sumner, L., & Bino, R. (2002). Plant metabolomics: The missing link in functional genomics strategies. *The Plant Cell*, *14*(7), 1437–1440. https://doi.org/10.1105/tpc.140720.

Han, Z., Ahsan, M., Adil, M. F., Chen, X., Nazir, M. M., Shamsi, I. H., et al. (2020). Identification of the gene network modules highly associated with the synthesis of phenolics compounds in barley by transcriptome and metabolome analysis. *Food Chemistry*, *323*. https://doi.org/10.1016/j.foodchem.2020.126862.

Heffner, E. L., Lorenz, A. J., Jannink, J.-L., & Sorrells, M. E. (2010). Plant breeding with genomic selection: Gain per unit time and cost. *Crop Science*, *50*(5), 1681–1690. https://doi.org/10.2135/cropsci2009.11.0662.

Heslot, N., Akdemir, D., Sorrells, M. E., & Jannink, J.-L. (2014). Integrating environmental covariates and crop modeling into the genomic selection framework to predict genotype by environment interactions. *Theoretical and Applied Genetics*, *127*(2), 463–480. https://doi.org/10.1007/s00122-013-2231-5.

Heslot, N., Yang, H.-P., Sorrells, M. E., & Jannink, J.-L. (2012). Genomic selection in plant breeding: A comparison of models. *Crop Science*, *52*(1), 146–160. https://doi.org/10.2135/cropsci2011.06.0297.

Hofheinz, N., Borchardt, D., Weissleder, K., & Frisch, M. (2012). Genome-based prediction of test cross performance in two subsequent breeding cycles. *Theoretical and Applied Genetics*, *125*(8), 1639–1645. https://doi.org/10.1007/s00122-012-1940-5.

Hong, J., Yang, L., Zhang, D., & Shi, J. (2016). Plant metabolomics: An indispensable system biology tool for plant science. *International Journal of Molecular Sciences*, *17*(6), 767. https://doi.org/10.3390/ijms17060767.

Houle, D. (2009). Numbering the hairs on our heads: The shared challenge and promise of phenomics. *Proceedings of the National Academy of Sciences*. https://doi.org/10.1073/pnas.0906195106.

Jannink, J.-L. (2010). Dynamics of long-term genomic selection. *Genetics Selection Evolution*, *42*(1), 35. https://doi.org/10.1186/1297-9686-42-35.

Jarquín, D., Crossa, J., Lacaze, X., Du Cheyron, P., Daucourt, J., Lorgeou, J., et al. (2014). A reaction norm model for genomic selection using high-dimensional genomic and environmental data. *Theoretical and Applied Genetics*, *127*(3), 595–607. https://doi.org/10.1007/s00122-013-2243-1.

Junker, A., Muraya, M. M., Weigelt-Fischer, K., Arana-Ceballos, F., Klukas, C., Melchinger, A. E., et al. (2015). Optimizing experimental procedures for quantitative evaluation of crop plant performance in high throughput phenotyping systems. *Frontiers in Plant Science*, *5*. https://doi.org/10.3389/fpls.2014.00770.

Kage, U., Karre, S., Kushalappa, A. C., & McCartney, C. (2017). Identification and characterization of a fusarium head blight resistance gene TaACT in wheat QTL-2DL. *Plant Biotechnology Journal*, *15*(4), 447–457. https://doi.org/10.1111/pbi.12641.

Kang, Z., Babar, M. A., Khan, N., Guo, J., Khan, J., Islam, S., et al. (2019). Comparative metabolomic profiling in the roots and leaves in contrasting genotypes reveals complex mechanisms involved in post-anthesis drought tolerance in wheat. *Plos One*, *14*(3). https://doi.org/10.1371/journal.pone.0213502.

Knoch, D., Riewe, D., Meyer, R. C., Boudichevskaia, A., Schmidt, R., & Altmann, T. (2017). Genetic dissection of metabolite variation in Arabidopsis seeds: Evidence for mQTL hotspots and a master regulatory locus of seed metabolism. *Journal of Experimental Botany*, *68*(7), 1655–1667. https://doi.org/10.1093/jxb/erx049.

Konate, M., Wilkinson, M. J., Taylor, J., Scott, E. S., Berger, B., & Lopez, C. M. R. (2020). Greenhouse spatial effects detected in the barley (Hordeum vulgare L.) epigenome underlie stochasticity of DNA methylation. *BioRxiv*. https://doi.org/10.1101/2020.04.20.051326. 2020.04.20.051326.

Kroymann, J. (2011). Natural diversity and adaptation in plant secondary metabolism. *Current Opinion in Plant Biology*, *14*(3), 246–251. https://doi.org/10.1016/j.pbi.2011.03.021.

Kumar, R., Bohra, A., Pandey, A. K., Pandey, M. K., & Kumar, A. (2017). Metabolomics for plant improvement: Status and prospects. *Frontiers in Plant Science*, *8*. https://doi.org/10.3389/fpls.2017.01302.

Lamari, N., Zhendre, V., Urrutia, M., Bernillon, S., Maucourt, M., Deborde, C., et al. (2018). Metabotyping of 30 maize hybrids under early-sowing conditions reveals potential marker-metabolites for breeding. *Metabolomics*, *14*(10), 132. https://doi.org/10.1007/s11306-018-1427-8.

Lecourieux, D., Kappel, C., Claverol, S., Pieri, P., Feil, R., Lunn, J. E., et al. (2020). Proteomic and metabolomic profiling underlines the stage- and time-dependent effects of high temperature on grape berry metabolism. *Journal of Integrative Plant Biology*. https://doi.org/10.1111/jipb.12894.

Lehermeier, C., Krämer, N., Bauer, E., Bauland, C., Camisan, C., Campo, L., et al. (2014). Usefulness of multiparental populations of Maize (Zea mays L.) for genome-based prediction. *Genetics*, *198*(1), 3–16. https://doi.org/10.1534/genetics.114.161943.

Li, K., Wang, D., Gong, L., Lyu, Y., Guo, H., Chen, W., et al. (2019). Comparative analysis of metabolome of rice seeds at three developmental stages using a recombinant inbred line population. *The Plant Journal*, *100*(5), 908–922. https://doi.org/10.1111/tpj.14482.

Li, K., Wen, W., Alseekh, S., Yang, X., Guo, H., Li, W., et al. (2019). Large-scale metabolite quantitative trait locus analysis provides new insights for high-quality maize improvement. *The Plant Journal*, *99*(2), 216–230. https://doi.org/10.1111/tpj.14317.

Lisec, J., Meyer, R. C., Steinfath, M., Redestig, H., Becher, M., Witucka-Wall, H., et al. (2008). Identification of metabolic and biomass QTL in Arabidopsis thaliana in a parallel analysis of RIL and IL populations. *The Plant Journal*, *53*(6), 960–972. https://doi.org/10.1111/j.1365-313X.2007.03383.x.

Liu, J.-Y., Li, P., Zhang, Y.-W., Zuo, J.-F., Li, G., Han, X., et al. (2020). Three-dimension genetic networks among seed oil-related traits, metabolites and genes reveal the genetic foundations of oil synthesis in soybean. *The Plant Journal*. https://doi.org/10.1111/tpj.14788.

Long, S. P., Marshall-Colon, A., & Zhu, X.-G. (2015). Meeting the global food demand of the future by engineering crop photosynthesis and yield potential. *Cell*, *161*(1), 56–66. https://doi.org/10.1016/j.cell.2015.03.019.

Loupit, G., Prigent, S., Franc, C., De Revel, G., Richard, T., Cookson, S. J., et al. (2020). Polyphenol profiles of just pruned grapevine canes from wild vitis accessions and vitis vinifera cultivars. *Journal of Agricultural and Food Chemistry*. https://doi.org/10.1021/acs.jafc.9b08099. acs.jafc.9b08099.

Luna, E., Flandin, A., Cassan, C., Prigent, S., Chevanne, C., Kadiri, C. F., et al. (2020). Metabolomics to exploit the primed immune system of tomato fruit. *Metabolites*, *10*(3), 96. https://doi.org/10.3390/metabo10030096.

MacLeod, I. M., Bowman, P. J., Vander Jagt, C. J., Haile-Mariam, M., Kemper, K. E., Chamberlain, A. J., et al. (2016). Exploiting biological priors and sequence variants enhances QTL discovery and genomic prediction of complex traits. *BMC Genomics*, *17*(1), 144. https://doi.org/10.1186/s12864-016-2443-6.

Marulanda, J. J., Mi, X., Melchinger, A. E., Xu, J.-L., Würschum, T., & Longin, C. F. H. (2016). Optimum breeding strategies using genomic selection for hybrid breeding in wheat, maize, rye, barley, rice and triticale. *Theoretical and Applied Genetics*, *129*(10), 1901–1913. https://doi.org/10.1007/s00122-016-2748-5.

Matsuda, F., Okazaki, Y., Oikawa, A., Kusano, M., Nakabayashi, R., Kikuchi, J., et al. (2012). Dissection of genotype–phenotype associations in rice grains using metabolome quantitative trait loci analysis. *The Plant Journal*, *70*(4), 624–636. https://doi.org/10.1111/j.1365-313X.2012.04903.x.

Melandri, G., AbdElgawad, H., Riewe, D., Hageman, J. A., Asard, H., Beemster, G. T. S., et al. (2020). Biomarkers for grain yield stability in rice under drought stress. *Journal of Experimental Botany*, *71*(2), 669–683. https://doi.org/10.1093/jxb/erz221.

Meuwissen, T. H. E., Hayes, B. J., & Goddard, M. E. (2001). Prediction of total genetic value using genome-wide dense marker maps. *Genetics*, *157*(4), 1819–1829.

Meyer, R. C., Steinfath, M., Lisec, J., Becher, M., Witucka-Wall, H., Törjék, O., et al. (2007). The metabolic signature related to high plant growth rate in Arabidopsis thaliana. *Proceedings of the National Academy of Sciences of the United States of America*, *104*(11), 4759–4764. https://doi.org/10.1073/pnas.0609709104.

Michel, S., Ametz, C., Gungor, H., Epure, D., Grausgruber, H., Löschenberger, F., et al. (2016). Genomic selection across multiple breeding cycles in applied bread wheat breeding. *Theoretical and Applied Genetics*, *129*(6), 1179–1189. https://doi.org/10.1007/s00122-016-2694-2.

Mickelson, S., See, D., Meyer, F. D., Garner, J. P., Foster, C. R., Blake, T. K., et al. (2003). Mapping of QTL associated with nitrogen storage and remobilization in barley (Hordeum vulgare L.) leaves. *Journal of Experimental Botany*, *54*(383), 801–812. https://doi.org/10.1093/jxb/erg084.

Moreau, L., Charcosset, A., Hospital, F., & Gallais, A. (1998). Marker-assisted selection efficiency in populations of finite size. *Genetics*, *148*(3), 1353–1365.

Nanjo, T., Kobayashi, M., Yoshiba, Y., Sanada, Y., Wada, K., Tsukaya, H., et al. (1999). Biological functions of proline in morphogenesis and osmotolerance revealed in antisense transgenic Arabidopsis thaliana. *The Plant Journal*, *18*(2), 185–193. https://doi.org/10.1046/j.1365-313X.1999.00438.x.

Neveu, P., Tireau, A., Hilgert, N., Nègre, V., Mineau-Cesari, J., Brichet, N., et al. (2019). Dealing with multi-source and multi-scale information in plant phenomics: The ontology-driven Phenotyping Hybrid Information System. *New Phytologist*, *221*(1), 588–601. https://doi.org/10.1111/nph.15385.

Onogi, A., Watanabe, M., Mochizuki, T., Hayashi, T., Nakagawa, H., Hasegawa, T., et al. (2016). Toward integration of genomic selection with crop modelling: The development of an integrated approach to predicting rice heading dates. *Theoretical and Applied Genetics*, *129*(4), 805–817. https://doi.org/10.1007/s00122-016-2667-5.

Papandreou, C., Bulló, M., Ruiz-Canela, M., Dennis, C., Deik, A., Wang, D., et al. (2019). Plasma metabolites predict both insulin resistance and incident type 2 diabetes: A metabolomics approach within the Prevención con Dieta Mediterránea (PREDIMED) study. *The American Journal of Clinical Nutrition*, *109*(3), 626–634. https://doi.org/10.1093/ajcn/nqy262.

Park, T., & Casella, G. (2008). The Bayesian Lasso. *Journal of the American Statistical Association*, *103*(482), 681–686. https://doi.org/10.1198/016214508000000337.

Park, M. Y., & Hastie, T. (2007). L 1-regularization path algorithm for generalized linear models. *Journal of the Royal Statistical Society: Series B (Statistical Methodology)*, *69*(4), 659–677. https://doi.org/10.1111/j.1467-9868.2007.00607.x.

Peña-Bautista, C., Roca, M., Hervás, D., Cuevas, A., López-Cuevas, R., Vento, M., et al. (2019). Plasma metabolomics in early Alzheimer's disease patients diagnosed with amyloid biomarker. *Journal of Proteomics*, *200*, 144–152. https://doi.org/10.1016/j.jprot.2019.04.008.

Pinu, F. R. (2018). Grape and wine metabolomics to develop new insights using untargeted and targeted approaches. *Fermentation*, *4*(4), 92. https://doi.org/10.3390/fermentation4040092.

Quadrana, L., Almeida, J., Asís, R., Duffy, T., Dominguez, P. G., Bermúdez, L., et al. (2014). Natural occurring epialleles determine vitamin E accumulation in tomato fruits. *Nature Communications*, *5*(1), 4027. https://doi.org/10.1038/ncomms5027.

Randhawa, H. S., Mutti, J. S., Kidwell, K., Morris, C. F., Chen, X., & Gill, K. S. (2009). Rapid and targeted introgression of genes into popular wheat cultivars using marker-assisted background selection. *PLos One*, *4*(6), e5752. https://doi.org/10.1371/journal.pone.0005752.

Razzaq, A., Sadia, B., Raza, A., Khalid Hameed, M., & Saleem, F. (2019). Metabolomics: A way forward for crop improvement. *Metabolites*, *9*(12), 303. https://doi.org/10.3390/metabo9120303.

Riar, A. K., Kaur, S., Dhaliwal, H. S., Singh, K., & Chhuneja, P. (2012). Introgression of a leaf rust resistance gene from Aegilops caudata to bread wheat. *Journal of Genetics*, *91*(2), 155–161. https://doi.org/10.1007/s12041-012-0161-7.

Riedelsheimer, C., Czedik-Eysenberg, A., Grieder, C., Lisec, J., Technow, F., Sulpice, R., et al. (2012). Genomic and metabolic prediction of complex heterotic traits in hybrid maize. *Nature Genetics*, *44*(2), 217–220. https://doi.org/10.1038/ng.1033.

Riedelsheimer, C., Lisec, J., Czedik-Eysenberg, A., Sulpice, R., Flis, A., Grieder, C., et al. (2012). Genome-wide association mapping of leaf metabolic profiles for dissecting complex traits in maize. *Proceedings of the National Academy of Sciences of the United States of America, 109*(23), 8872–8877. https://doi.org/10.1073/pnas.1120813109.

Rincent, R., Charpentier, J.-P., Faivre-Rampant, P., Paux, E., Gouis, J. L., Bastien, C., et al. (2018). Phenomic selection is a low-cost and high-throughput method based on indirect predictions: Proof of concept on wheat and poplar. *G3: Genes, Genomes, Genetics, 8*(12), 3961–3972. https://doi.org/10.1534/g3.118.200760.

Rincent, R., Laloë, D., Nicolas, S., Altmann, T., Brunel, D., Revilla, P., et al. (2012). Maximizing the reliability of genomic selection by optimizing the calibration set of reference individuals: Comparison of methods in two diverse groups of Maize Inbreds (Zea mays L.). *Genetics, 192*(2), 715–728. https://doi.org/10.1534/genetics.112.141473.

Rincent, R., Malosetti, M., Ababaei, B., Touzy, G., Mini, A., Bogard, M., et al. (2019). Using crop growth model stress covariates and AMMI decomposition to better predict genotype-by-environment interactions. *Theoretical and Applied Genetics, 132*(12), 3399–3411. https://doi.org/10.1007/s00122-019-03432-y.

Roch, L., Dai, Z., Gomès, E., Bernillon, S., Wang, J., Gibon, Y., et al. (2019). Fruit salad in the lab: Comparing botanical species to help deciphering fruit primary metabolism. *Frontiers in Plant Science, 10*. https://doi.org/10.3389/fpls.2019.00836.

Routaboul, J.-M., Dubos, C., Beck, G., Marquis, C., Bidzinski, P., Loudet, O., et al. (2012). Metabolite profiling and quantitative genetics of natural variation for flavonoids in Arabidopsis. *Journal of Experimental Botany, 63*(10), 3749–3764. https://doi.org/10.1093/jxb/ers067.

Roy, J., Tardieu, F., Tixier-Boichard, M., & Schurr, U. (2017). European infrastructures for sustainable agriculture. *Nature Plants, 3*(10), 756–758. https://doi.org/10.1038/s41477-017-0027-3.

Saito, K., & Matsuda, F. (2010). Metabolomics for functional genomics, systems biology, and biotechnology. *Annual Review of Plant Biology, 61*, 463–489. https://doi.org/10.1146/annurev.arplant.043008.092035.

Sanz-Alferez, S., Richter, T. E., Hulbert, S. H., & Bennetzen, J. L. (1995). The Rp3 disease resistance gene of maize: Mapping and characterization of introgressed alleles. *Theoretical and Applied Genetics, 91*(1), 25–32. https://doi.org/10.1007/BF00220854.

Schrag, T. A., Schipprack, W., & Melchinger, A. E. (2019). Across-years prediction of hybrid performance in maize using genomics. *Theoretical and Applied Genetics, 132*(4), 933–946. https://doi.org/10.1007/s00122-018-3249-5.

Schrag, T. A., Westhues, M., Schipprack, W., Seifert, F., Thiemann, A., Scholten, S., et al. (2018). Beyond genomic prediction: Combining different types of omics data can improve prediction of hybrid performance in Maize. *Genetics, 208*(4), 1373–1385. https://doi.org/10.1534/genetics.117.300374.

Schulz-Streeck, T., Ogutu, J. O., Gordillo, A., Karaman, Z., Knaak, C., & Piepho, H.-P. (2013). Genomic selection allowing for marker-by-environment interaction. *Plant Breeding, 132*(6), 532–538. https://doi.org/10.1111/pbr.12105.

Scossa, F., & Fernie, A. R. (2020). The evolution of metabolism: How to test evolutionary hypotheses at the genomic level. *Computational and Structural Biotechnology Journal, 18*, 482–500. https://doi.org/10.1016/j.csbj.2020.02.009.

Seifert, F., Thiemann, A., Schrag, T. A., Rybka, D., Melchinger, A. E., Frisch, M., et al. (2018). Small RNA-based prediction of hybrid performance in maize. *BMC Genomics, 19*(1), 371. https://doi.org/10.1186/s12864-018-4708-8.

Servin, B., Martin, O. C., Mézard, M., & Hospital, F. (2004). Toward a theory of marker-assisted gene pyramiding. *Genetics, 168*(1), 513–523. https://doi.org/10.1534/genetics.103.023358.

Shi, T., Zhu, A., Jia, J., Hu, X., Chen, J., Liu, W., et al. (2020). Metabolomic analysis and metabolite-agronomic trait associations using kernels of wheat (Triticum aestivum) recombinant inbred lines. *The Plant Journal*. https://doi.org/10.1111/tpj.14727. tpj.14727.

Skirycz, A., Vandenbroucke, K., Clauw, P., Maleux, K., De Meyer, B., Dhondt, S., et al. (2011). Survival and growth of Arabidopsis plants given limited water are not equal. *Nature Biotechnology*, *29*(3), 212–214. https://doi.org/10.1038/nbt.1800.

Slaten, M. L., Yobi, A., Bagaza, C., Chan, Y. O., Shrestha, V., Holden, S., et al. (2020). MGWAS uncovers Gln-glucosinolate seed-specific interaction and its role in metabolic homeostasis. *Plant Physiology*. https://doi.org/10.1104/pp.20.00039.

Spindel, J., Begum, H., Akdemir, D., Virk, P., Collard, B., Redoña, E., et al. (2015). Genomic selection and association mapping in rice (Oryza sativa): Effect of trait genetic architecture, training population composition, marker number and statistical model on accuracy of rice genomic selection in elite, tropical rice breeding lines. *PLos Genetics*, *11*(2). https://doi.org/10.1371/journal.pgen.1004982.

Sprenger, H., Erban, A., Seddig, S., Rudack, K., Thalhammer, A., Le, M. Q., et al. (2018). Metabolite and transcript markers for the prediction of potato drought tolerance. *Plant Biotechnology Journal*, *16*(4), 939–950. https://doi.org/10.1111/pbi.12840.

Steinfath, M., Strehmel, N., Peters, R., Schauer, N., Groth, D., Hummel, J., et al. (2010). Discovering plant metabolic biomarkers for phenotype prediction using an untargeted approach: Discovering plant metabolic biomarkers. *Plant Biotechnology Journal*, *8*(8), 900–911. https://doi.org/10.1111/j.1467-7652.2010.00516.x.

Székely, G., Ábrahám, E., Cséplő, Á., Rigó, G., Zsigmond, L., Csiszár, J., et al. (2008). Duplicated P5CS genes of Arabidopsis play distinct roles in stress regulation and developmental control of proline biosynthesis. *The Plant Journal*, *53*(1), 11–28. https://doi.org/10.1111/j.1365-313X.2007.03318.x.

Takagi, H., Ishiga, Y., Watanabe, S., Konishi, T., Egusa, M., Akiyoshi, N., et al. (2016). Allantoin, a stress-related purine metabolite, can activate jasmonate signaling in a MYC2-regulated and abscisic acid-dependent manner. *Journal of Experimental Botany*, *67*(8), 2519–2532. https://doi.org/10.1093/jxb/erw071.

Tardieu, F., Cabrera-Bosquet, L., Pridmore, T., & Bennett, M. (2017). Plant phenomics, from sensors to knowledge. *Current Biology*, *27*(15), R770–R783. https://doi.org/10.1016/j.cub.2017.05.055.

Technow, F., Messina, C. D., Totir, L. R., & Cooper, M. (2015). Integrating crop growth models with whole genome prediction through approximate bayesian computation. *PLos One*, *10*(6). https://doi.org/10.1371/journal.pone.0130855.

Trainor, P. J., Yampolskiy, R. V., & DeFilippis, A. P. (2018). Wisdom of artificial crowds feature selection in untargeted metabolomics: An application to the development of a blood-based diagnostic test for thrombotic myocardial infarction. *Journal of Biomedical Informatics*, *81*, 53–60. https://doi.org/10.1016/j.jbi.2018.03.007.

Tugizimana, F., Mhlongo, M. I., Piater, L. A., & Dubery, I. A. (2018). Metabolomics in plant priming research: The way forward? *International Journal of Molecular Sciences*, *19*(6), 1759. https://doi.org/10.3390/ijms19061759.

Turner, N. C. (2018). Turgor maintenance by osmotic adjustment: 40 Years of progress. *Journal of Experimental Botany*, *69*(13), 3223–3233. https://doi.org/10.1093/jxb/ery181.

Turner, M. F., Heuberger, A. L., Kirkwood, J. S., Collins, C. C., Wolfrum, E. J., Broeckling, C. D., et al. (2016). Non-targeted metabolomics in diverse Sorghum breeding lines indicates primary and secondary metabolite profiles are associated with plant biomass accumulation and photosynthesis. *Frontiers in Plant Science*, 7. https://doi.org/10.3389/fpls.2016.00953.

van Dam, N. M., & Bouwmeester, H. J. (2016). Metabolomics in the Rhizosphere: Tapping into belowground chemical communication. *Trends in Plant Science*, *21*(3), 256–265. https://doi.org/10.1016/j.tplants.2016.01.008.

van Eeuwijk, F. A., Bustos-Korts, D., Millet, E. J., Boer, M. P., Kruijer, W., Thompson, A., et al. (2019). Modelling strategies for assessing and increasing the effectiveness of new phenotyping techniques in plant breeding. *Plant Science*, *282*, 23–39. https://doi.org/10.1016/j.plantsci.2018.06.018.

Varona, L., Legarra, A., Toro, M. A., & Vitezica, Z. G. (2018). Non-additive effects in genomic selection. *Frontiers in Genetics*, *9*. https://doi.org/10.3389/fgene.2018.00078.

Vergara-Diaz, O., Vatter, T., Vicente, R., Obata, T., Nieto-Taladriz, M. T., Aparicio, N., et al. (2020). Metabolome profiling supports the key role of the spike in wheat yield performance. *Cells*, *9*(4), 1025. https://doi.org/10.3390/cells9041025.

Walter, A., Liebisch, F., & Hund, A. (2015). Plant phenotyping: From bean weighing to image analysis. *Plant Methods*, *11*(1), 14. https://doi.org/10.1186/s13007-015-0056-8.

Wang, S., Wei, J., Li, R., Qu, H., Chater, J. M., Ma, R., et al. (2019). Identification of optimal prediction models using multi-omic data for selecting hybrid rice. *Heredity*, *123*(3), 395–406. https://doi.org/10.1038/s41437-019-0210-6.

Wasson, A. P., Nagel, K. A., Tracy, S., & Watt, M. (2020). Beyond digging: Noninvasive root and rhizosphere phenotyping. *Trends in Plant Science*, *25*(1), 119–120. https://doi.org/10.1016/j.tplants.2019.10.011.

Wei, J., Wang, A., Li, R., Qu, H., & Jia, Z. (2018). Metabolome-wide association studies for agronomic traits of rice. *Heredity*, *120*(4), 342–355. https://doi.org/10.1038/s41437-017-0032-3.

Whittaker, J. C., Thompson, R., & Denham, M. C. (2000). Marker-assisted selection using ridge regression. *Genetics Research*, *75*(2), 249–252. https://doi.org/10.1017/S0016672399004462.

Wilson, L. M., Whitt, S. R., Ibáñez, A. M., Rocheford, T. R., Goodman, M. M., & Buckler, E. S. (2004). Dissection of Maize Kernel composition and starch production by candidate gene association. *The Plant Cell*, *16*(10), 2719–2733. https://doi.org/10.1105/tpc.104.025700.

Wong, J. W.-H., Plett, K. L., Natera, S. H. A., Roessner, U., Anderson, I. C., & Plett, J. M. (2020). Comparative metabolomics implicates threitol as a fungal signal supporting colonization of Armillaria luteobubalina on eucalypt roots. *Plant, Cell & Environment*, *43*(2), 374–386. https://doi.org/10.1111/pce.13672.

Wu, S., Tohge, T., Cuadros-Inostroza, Á., Tong, H., Tenenboim, H., Kooke, R., et al. (2018). Mapping the Arabidopsis metabolic landscape by untargeted metabolomics at different environmental conditions. *Molecular Plant*, *11*(1), 118–134. https://doi.org/10.1016/j.molp.2017.08.012.

Xu, G., Cao, J., Wang, X., Chen, Q., Jin, W., Li, Z., et al. (2019). Evolutionary metabolomics identifies substantial metabolic divergence between maize and its wild ancestor, teosinte. *The Plant Cell*, *31*(9), 1990–2009. https://doi.org/10.1105/tpc.19.00111.

Xu, S., Xu, Y., Gong, L., & Zhang, Q. (2016). Metabolomic prediction of yield in hybrid rice. *The Plant Journal*, *88*(2), 219–227. https://doi.org/10.1111/tpj.13242.

Xu, Y., Xu, C., & Xu, S. (2017). Prediction and association mapping of agronomic traits in maize using multiple omic data. *Heredity*, *119*(3), 174–184. https://doi.org/10.1038/hdy.2017.27.

Yang, J., Benyamin, B., McEvoy, B. P., Gordon, S., Henders, A. K., Nyholt, D. R., et al. (2010). Common SNPs explain a large proportion of the heritability for human height. *Nature Genetics*, *42*(7), 565–569. https://doi.org/10.1038/ng.608.

Yang, Z., Nakabayashi, R., Okazaki, Y., Mori, T., Takamatsu, S., Kitanaka, S., et al. (2014). Toward better annotation in plant metabolomics: Isolation and structure elucidation of 36 specialized metabolites from Oryza sativa (rice) by using MS/MS and NMR analyses. *Metabolomics*, *10*(4), 543–555. https://doi.org/10.1007/s11306-013-0619-5.

Yu, X., Li, X., Guo, T., Zhu, C., Wu, Y., Mitchell, S. E., et al. (2016). Genomic prediction contributing to a promising global strategy to turbocharge gene banks. *Nature Plants*, *2*(10), 1–7. https://doi.org/10.1038/nplants.2016.150.

Zhao, Y., Gowda, M., Liu, W., Würschum, T., Maurer, H. P., Longin, F. H., et al. (2012). Accuracy of genomic selection in European maize elite breeding populations. *Theoretical and Applied Genetics*, *124*(4), 769–776. https://doi.org/10.1007/s00122-011-1745-y.

Zhong, S., Dekkers, J. C. M., Fernando, R. L., & Jannink, J.-L. (2009). Factors affecting accuracy from genomic selection in populations derived from multiple inbred lines: A Barley Case Study. *Genetics*, *182*(1), 355–364. https://doi.org/10.1534/genetics.108.098277.

Zhu, G., Wang, S., Huang, Z., Zhang, S., Liao, Q., Zhang, C., et al. (2018). Rewiring of the fruit metabolome in tomato breeding. *Cell*, *172*(1), 249–261.e12. https://doi.org/10.1016/j.cell.2017.12.019.

CHAPTER EIGHT

Plant genome-scale metabolic networks

Léo Gerlin[a], Clément Frainay[b], Fabien Jourdan[b], Caroline Baroukh[a], and Sylvain Prigent[c,*]
[a]LIPM, Université de Toulouse, INRAE, CNRS, Castanet-Tolosan, France
[b]Toxalim (Research Center in Food Toxicology), Université de Toulouse, INRAE, ENVT, INP-Purpan, UPS, Toulouse, France
[c]UMR1332, Université de Bordeaux, INRAE, Bordeaux Metabolome, MetaboHUB, PHENOME, Villenave d'Ornon, France
*Corresponding author: e-mail address: sylvain.prigent@inrae.fr

Contents

Abstract

With the development of plant metabolomics, projects like the 10,000 plants genome sequencing project and the growth of other omics, the amount of data describing plant metabolism has never been so big. Genome-scale metabolic models (GEMs) are widely used to integrate and study the available information, and to better understand global responses to metabolic changes.

In the first part of this chapter, we will focus on the existing plant GEMs, their history, and the biological questions associated with their reconstruction. Those GEMs were initially reconstructed based on cell suspension systems, but current reconstructions are widely focused on multiorgans and multitissues models, requiring some model integration.

Model integration is the focus of the second part of this chapter, with the study of biotic interactions with the integration of GEMs coming from plant-interacting microorganisms and their host plants, modeling either pathogenic and symbiotic interactions. Emphasis will be placed on modeling quantitative interaction between plants and microorganisms.

Advances in Botanical Research, Volume 98
ISSN 0065-2296
https://doi.org/10.1016/bs.abr.2020.09.021

Finally, the third part of this chapter describes how the visualization of networks could be used to improve the understanding of the models and the integration of different omics data, and how good representation of metabolic networks can help the user to perform diverse tasks and put knowledge in context.

1. An introduction on existing genome-scale metabolic models

Genome-scale metabolic networks provide a representation of all metabolic actors present in the cell, by describing gene-protein-reaction associations, based on genome annotations and experimental data (Thiele & Palsson, 2010). Such networks contain sets of metabolic reactions, ideally mass-balanced, that can be transformed into mathematical models, the genome-scale metabolic models (GEMs) that allow for modeling of complex behavior. Hence, GEMs are of great interest to study the metabolism of species possessing complex metabolic pathways, and plants are such species. It is therefore not surprising that plant GEMs are being developed quite extensively in recent years.

This development is driven by the rise of the omics era in plants, making it easier to obtain genomic, transcriptomic, proteomic and metabolomic data during the past few years. The genome sequence of *Arabidopsis thaliana* was released in 2000 (The Arabidopsis Genome Initiative, 2000), and 3 years later was published the first version of AraCyc, a database containing more than 170 biochemical pathways linked with *A. thaliana*. It was not before 2009 that was published the first curated plant GEM (Poolman, Miguet, Sweetlove, & Fell, 2009), enabling the simulation of growth and biomass formation of a heterotrophic cell suspension system. Since this first publication, a number of different GEMs has been published for a variety of species. We identified 30 different models reconstructed for 12 different plant species (Table 1). Given the fact that *A. thaliana* remains a model organism for plant metabolism, it is not surprising that GEMs reconstructed to study different metabolic features of this species represent more than one-third of identified plant GEMs. Quite interestingly, reconstructed plant GEMs cover a wide array of phylogenetic clades, with species from monocots to eudicots clades. At least one representative of Poaceae, Brassicaceae, Solanaceae, Fabaceae, Lamiaceae and Salicaceae families are present. This diversity of species could be very interesting, in particular to compare metabolic features for the identification of conserved or diverse patterns across species. It is interesting to note that most of the reconstructed models initially come from two data

Table 1 Genome-scale metabolic models of plant species.

Species	Classification	Publication	Reactions	Metabolites	Genes	Applications
Arabidopsis thaliana	Eudicot, Rosids, Brassicaceae	Poolman et al. (2009)	1406	1253	–	First plant GEM, modeling heterotrophic cell growth
		de Oliveira Dal'Molin, Quek, Palfreyman, Brumbley, and Nielsen (2010a)	1567	1748	1419	AraGEM, modeling photosynthesis, photorespiration and respiration
		Saha, Suthers, and Maranas (2011)	1798	1820	1597	GEM build alongside a rice model
		Mintz-Oron et al. (2012)	1363	1079	1791	Emphasis on compartmentalization. Pipeline to reconstruct tissue specific models
		Bekaert, Edger, Hudson, Pires, and Conant (2012)	1785	1914	–	Focus on glucosinolate biosynthesis, in particular on the metabolic cost of that biosynthesis
		Cheung et al. (2013)	2769	2739	2857	Study maintenance cost, in particular in stress conditions, using Pareto optimality analysis

Continued

Table 1 Genome-scale metabolic models of plant species.—cont'd

Species	Classification	Publication	Reactions	Metabolites	Genes	Applications
		Cheung, Poolman, Fell, Ratcliffe, and Sweetlove (2014)	5889	5515	–	Comparison of C3/CAM photosynthesis. Modeling of day/night metabolism for leaves
		de Oliveira Dal'Molin, Quek, Saa, and Nielsen (2015)	1607	1782	1409	Multi-tissue model (leaf, stem, root) across diurnal cycle. Study differences between source and sink tissues in particular on C and N allocation
		Seaver et al. (2015)	2801	2864	4262	Creation of a so-called "Evidence-Based Model" from KEGG, AraCyc and existing *A. Thaliana* GEMs. A particular focus was put on compartmentalization
		Shaw and Cheung (2018)	11,320	10,664	–	Multi-tissue model (leaf and root) across diurnal cycle. Study metabolism through different growth stages and resource availability with dFBA
		Scheunemann, Brady, and Nikoloski (2018)	2199	1813		Root-specific model constructed using transcriptomics and proteomics data, from the PlantSEED based Arabidopsis model

Oryza sativa	Monocot, Commelinids, Poaceae	Liu et al. (2013)	3316	2986	4426	Integration of GEM, protein-protein interaction, and gene regulatory networks to study phenotype-genotype interactions in rice
		Poolman, Kundu, Shaw, and Fell (2013)	1736	1484	–	Leaf model to study redox, photorespiration and mitochondrial metabolism in response to varying light levels
		Lakshmanan et al. (2015)	2283	1999	2164	Using transcriptomic data to study metabolism under different light conditions (blue, green, red, white and dark)
		Chatterjee and Kundu (2015)	1721	1544	–	Focus on nitrogen assimilation and chlorophyll biosynthesis under different photorespiration conditions
		Chatterjee, Huma, Shaw, and Kundu (2017)	1136	1380	3602	Simulating leaf metabolism across different light and enzymatic cost conditions, and for varying RuBisCO activity

Continued

Table 1 Genome-scale metabolic models of plant species.—cont'd

Species	Classification	Publication	Reactions	Metabolites	Genes	Applications
Zea mays	Monocot, Commelinids, Poaceae	de Oliveira Dal'Molin, Quek, Palfreyman, Brumbley, and Nielsen (2010b)	1588	1755	11,623	C4GEM: comparison between maize, sorghum and sugarcane metabolism. Study flux distributions in mesophyll cells and the bundle sheath cells
		Saha et al. (2011)	1985	1825	1563	FBA simulation under different conditions (photosynthesis, photorespiration, respiration)
		Simons et al. (2014)	8525	9153	5824	Study carbon fixation and nitrogen assimilation under different nitrate conditions, and comparison with mutant phenotypes
		Seaver et al. (2015)	2629	2634	5540	Three models for leaf, embryo and endosperm tissues, based on transcriptomic data
		Bogart and Myers (2016)	2720 (635)	2725 (603)	5204 (2140)	A comprehensive model and a high-confidence subset of this model (in parentheses). Introduce non-linear relationship between carbon dioxide and oxygen levels, and reaction rates

Sorghum bicolor	Monocot, Commelinids, Poaceae	de Oliveira Dal'Molin et al. (2010b)	1588	1755	3557	C4GEM: comparison between maize, sorghum and sugarcane metabolism. Study flux distributions in mesophyll cells and the bundle sheath cells
Saccharum officinarum	Monocot, Commelinids, Poaceae	de Oliveira Dal'Molin et al. (2010b)	1588	1755	3881	C4GEM: comparison between maize, sorghum and sugarcane metabolism. Study flux distributions in mesophyll cells and the bundle sheath cells
Setaria italica	Monocot, Commelinids, Poaceae	de Oliveira Dal'Molin et al. (2016)	1515	1690	–	Multi-omics data integration (transcriptome, proteome, metabolome) in mature and immature stem/leaf phytomers
Solanum lycopersicum	Eudicot, Asterids, Solanaceae	Yuan, Cheung, Poolman, Hilbers, and van Riel (2016)	2143	1998	3410	Simulate tomato leaf metabolism under heterotrophic and phototrophic conditions, and under drought stress
Solanum tuberosum	Eudicot, Asterids, Solanaceae	Botero, Restrepo, and Pinzón (2018)	2072	1938	2751	Simulate response to a pathogen by integrating transcriptomic data as constraints into fluxes of reactions

Continued

Table 1 Genome-scale metabolic models of plant species.—cont'd

Species	Classification	Publication	Reactions	Metabolites	Genes	Applications
Medicago truncatula	Eudicot, Rosids, Fabaceae	Pfau et al. (2018)	2909	2780	3403	Study the effect of symbiosis on metabolic fluxes, in particular regarding oxygen availability
Mentha x *piperita*	Eudicot, Asterids, Lamiaceae	Johnson, Lange, Srividya, and Lange (2017)	624	466	757	Transcriptomics used to reconstruct the model and to perform modeling. Focus on monoterpenes biosynthesis
Populus trichocarpa	Eudicot, Rosids, Salicaceae	Sarkar and Maranas (2020)	3282	2502	7188	First GEM for a perennial woody tree. Reconstructed to help interpreting results of SNP analysis
Glycine max	Eudicot, Rosids, Fabaceae	Moreira et al. (2019)	3001	2814	6127	Studying the mobilization of soybean seed reserves during growth over 4 days. Highlight metabolic switches

sources: KEGG (Kanehisa, 2000) and MetaCyc (Caspi et al., 2016), via the Plant Metabolic Network (Schläpfer et al., 2017). Having so little data-sources could become an issue, in particular regarding plant species, since compartmentalization is of great importance in plant metabolism. Those databases often lack precise information regarding the subcellular localization of reactions, and therefore the need for transport reactions across intracellular membranes. Such transport reactions are enabled by protein transporters that are often poorly characterized. The lack of information on subcellular localization of reactions and transporter protein can lead to modeling inconsistencies.

During the reconstruction of GEMs, metabolomics could play a determinant role. Indeed, the reactions present in a metabolic network describe how metabolites are produced and consumed. Theoretically, metabolomics could one day produce a comprehensive list of molecules present in a given tissue. If, in practice, the list of molecules is never complete, it can still give a large set of molecules either preponderant in the biomass, hence helping for the formulation of the biomass function, or present in very small quantities, as results of the secondary metabolism pathways, for instance. Such lists can be very useful to perform gap-filling, either from a topological (Prigent et al., 2017) or flux-based (Thiele, Vlassis, & Fleming, 2014) point of view.

2. Bases and specificities of plant metabolic modeling

Constraint-based modeling is a classical approach to study flux distributions in GEMs. Indeed, by assuming that the models are at quasi-steady states, flux-based methods such as Flux Balance Analysis (FBA) (Orth, Thiele, & Palsson, 2010) enable exploration of possible fluxes going throughout a model. Flux-based approaches have been widely used to study plant metabolism in conjunction with GEMs. Given the Quasi-Steady-State Assumption (QSSA), performing accurate modeling using GEMs is largely based on a good definition of influxes and outfluxes of the models. Those are usually defined as the entrance of metabolites from outside the metabolic system leading to the production of biomass. Plants possess a big specificity here, since they are gaining most of their energy and carbon from photosynthesis. This mechanism has thus been widely studied using GEMs. This aspect was tackled at first with the AraGEM model (de Oliveira Dal'Molin et al., 2010a) where photosynthesis and photorespiration were accurately modeled.

Plants are complex organisms possessing several organs and tissues, and some of them are not capable of photosynthesis. Hence, to model the biological reality, several multi-organ and multi-tissue models have been built and analyzed. We can, for example, cite (Seaver et al., 2015) where the authors used transcriptomic data to build specialized GEMs for maize leaves, endosperm and embryos, and specialized biomass functions for each model. More recently, Shaw and Cheung developed a multi-tissue model for *A. thaliana* by reconstructing a model including leaves and root metabolism during day and night phases (Shaw & Cheung, 2018). Those tissues could cross-talk via a common pool of metabolites representing the phloem. A similar approach was used previously to model interactions between leaves, stems and roots (de Oliveira Dal'Molin et al., 2015).

To perform accurate modeling, precise biomass and nutrient definitions are required but it is often not sufficient since the solution space remains very big. Hence, several types of data can be incorporated into GEMs to improve the quality of the models. Since the cost of generating omics data continues to decrease while the quality of such data increases, it is not surprising that transcriptomics, proteomics and metabolomics are more and more used to reduce the number of possible flux distributions in a GEM. While gene expression, protein concentration and fluxes do not always correlate, transcriptomics and proteomics are often used to build tissue/environment/time-specific GEMs that contain reactions that could be catalyzed only in some cells or conditions. On the other hand, metabolomics (including isotope labelling) gives information that can be directly used in simulations. For example, targeted metabolomics has been previously used to obtain semi-quantitative information on 100 metabolites (de Oliveira Dal'Molin et al., 2016), in conjunction with transcriptomics and proteomics data, to compare the abundance of metabolites in young and mature tissues of *Setaria italica*. Isotopic labeling such as ^{13}C can also be used to obtain very precise information on subsets of the metabolic models. Metabolic Flux Analysis (Wiechert, 2001) can be used to identify precise fluxes that would lead to a given distribution of labeled isotope metabolites. Such fluxes put big constraints on the other parts of the metabolism, as what was performed by Hay et al. (2014) on a quite small model where this technique improved the predictive power of the simulations.

We have seen previously how metabolomics can be used to improve the quality of GEMs, in particular during the gap-filling process and during simulations. GEMs could also be used to improve the quality of metabolomics data, in particular of Mass Spectrometry (MS) data. Indeed, metabolic networks give access to a large number of metabolic reactions involving reactants and products. Hence, GEMs give access to thousands of metabolites that should be present at least under some conditions in the metabolome of a given species. Depending on the metabolites and the technique used, the signatures associated with those molecules should be recovered, in MS data for instance. This information could then be used to annotate some of the peaks, thus participating in the identification effort that remains highly challenging. This could lead to a virtuous loop where GEMs reactions coverage would be improved by metabolomics, and those GEMs would, in turn, improve metabolomics annotation.

All the GEMs presented in this part have been successfully used to model the metabolism of plant cells or tissues, sometimes interacting together. Yet, plants never live alone. Bacteria and fungi permanently accompany them during their normal or pathological development. The modeling of the metabolism of bacteria and fungi is far ahead of that of plants. It is now becoming necessary to integrate plant models with the symbionts, parasites and pathogens that accompany them.

3. Metabolic modeling of plant-microorganism interactions

3.1 Microorganisms involved in biotic interactions

GEMs were initially developed, since 1999, to study microbial strains for applications in medicine and biotechnology (Kim, Sohn, Kim, Kim, & Lee, 2012). Nowadays, a wide diversity of GEMs for microorganisms exist for more diverse applications, and the available GEMs include several microorganisms interacting with plants such as symbionts and pathogens, from bacteria and eukarya domains (Table 2).

Extension of GEMs to bacterial plant pathogens started in 2015 for the causal agent of soft-rot disease: *Pectobacterium carotovorum* (Wang et al., 2015). The metabolic model of *P. carotovorum* was reconstructed based on genomic annotation and *Escherichia coli* metabolic model. It was confronted with Biolog Phenotype Microarrays (PMs) experiments and applied to the discovery of agricultural bactericide targets. The next GEM, published in 2016 (Peyraud et al., 2016) was the model of *Ralstonia solanacearum*, the agent of bacterial wilt. It was reconstructed using a broader range of reference organisms (*E. coli*, *Pseudomonas aeruginosa*, *Ralstonia eutropha*, *Bacillus subtilis*), and included a macromolecule module representing the biosynthesis of macromolecular components (539 reactions), such as a wide range of

Table 2 Genome-scale metabolic models of plant-interacting microorganisms.

Species	Classification	Biotic interaction	Publication	Metabolites	Reactions	Genes	Applications
Rhizobium etli	α-Proteobacteria	Symbiotic	Resendis-Antonio, Reed, Encarnación, Collado-Vides, and Palsson (2007)	371	387	363	Metabolic pathways involved in nitrogen fixation
Pectobacterium carotovorum	γ-Proteobacteria	Pathogenic	Wang et al. (2015)	1113	2235	1209	Discovery of agricultural bactericide targets
Ralstonia solanacearum	β-Proteobacteria	Pathogenic	Peyraud, Cottret, Marmiesse, Gouzy, and Genin (2016)	2644*	2086*	1476	Resource allocation trade-off between growth and virulence
Sinorhizobium meliloti	α-Proteobacteria	Symbiotic	diCenzo et al. (2016)	1579	1825	1575	Metabolic reprogramming in different environments
Phytophtora infestans	Eukaryota, oomycete	Pathogenic	Rodenburg, Seidl, de Ridder, and Govers (2018)	2394	2685	1301	Mapping of transcriptomics data at different life-stages
Pectobacterium parmentieri	γ-Proteobacteria	Pathogenic	Zoledowska et al. (2019)	2080	2182	1245	Metabolic behaviors at different steps
Sclerotinia sclerotiorum	Eukaryota, fungus	Pathogenic	Peyraud, Mbengue, Barbacci, and Raffaele (2019)	1525	1495	1039	Metabolic heterogeneity/division of labor
Xanthomonas oryzae	γ-Proteobacteria	Pathogenic	Koduru et al. (2020)	808	839	673	Effects of nitrogen fertilizers, relation between quorum sensing and metabolism
Sinorhizobium fredii	α-Proteobacteria	Symbiotic	Contador, Lo, Chan, and Lam (2020)	508	535	541	Symbiosis yields in different symbiotic organisms/upon different environmental conditions
Xylella fastidiosa	γ-Proteobacteria	Pathogenic	Gerlin et al. (2020)	1107	1158	473	Metabolic determinants of the slow-growing (fastidious) phenotype

[a]Including 539 reactions and 367 metabolites corresponding to the biosynthesis of macromolecules (e.g. type-three effectors, plant cell-wall degrading enzymes).

virulence factors. The model was also intensively calibrated: Biolog PMs, bacterial cultures, chemical analysis of supernatants, virulence factors quantification, both on the wild-type strain and mutants reduced in virulence. The authors demonstrated that a resource allocation trade-off controlled by the population level occurs between the production of virulence factors and fast proliferation. Metabolic modeling thus helps understanding the metabolic adaptation of pathogens to plant environments. Three other bacterial plant pathogen GEMs were published:

(i) *Pectobacterium parmentieri* (Zoledowska et al., 2019). The reconstruction used the online tool KBase (Arkin et al., 2018), and Biolog PMs data. It predicted the metabolic behaviors of the bacterium in two conditions: persistence in soil and plant colonization in the rhizosphere.

(ii) *Xanthomonas oryzae* (Koduru et al., 2020). The reconstruction was based on KEGG (Kanehisa, 2000) and MetaCyc (Caspi et al., 2016) databases and refined by adding specific pathways of the organism. The authors studied the impact of quorum sensing deletion on metabolic pathways. Experiments were performed to validate the model: bacterial cultures, excretion quantifications and amino acid consumption.

(iii) *Xylella fastidiosa* (Gerlin et al., 2020). The GEM was reconstructed using five reference organisms using the same method as *R. solanacearum*, and was refined with Biolog PMs. This work showed how metabolism was involved in the self-limiting behavior of the bacterium inside the host. In particular, *X. fastidiosa* lacks several key metabolic reactions, which contribute to the slow-growing phenotype of the organism and is suspected to be an advantage to remain under the radar of the plant immune system.

So far, fewer GEMs are available for bacterial symbionts. In 2007, a metabolic model of a legume symbiont (*Rhizobium etli*) was reconstructed using the KEGG database (Resendis-Antonio et al., 2007). This bacterium, located in legume roots, transforms atmospheric nitrogen into ammonium and thus provides nitrogen to the plant in exchange of organic matter. The qualification of the model as a "genome-scale" metabolic model is questionable as only 387 reactions are included in the model, versus between 1000 and 2500 reactions for most metabolic models. This model compared in silico metabolic pathways involved in nitrogen fixation to experiments. More recently, diCenzo et al. (diCenzo et al., 2016) applied metabolic modeling to another legume symbiont *Sinorhizobium meliloti*, a major model organism of nitrogen fixation by bacteria, using KBase (Arkin et al., 2018) for reconstruction. This model was used to simulate metabolic reprogramming in different niches, and the importance of the symbiotic plasmid for growth in the rhizosphere environment. *S. meliloti* metabolic network was also used as a reference organism to generate the metabolic model of another legume symbiont, *Sinorhizobium fredii* (Contador et al., 2020). On this study, the authors compared the symbiotic yields between existing legume symbionts models.

Metabolic modeling is more easily applicable and extended to prokaryotes than eukaryotes, because of the added complexity of eukaryotic cell compartmentalization, which increases the curation and validation duration. Thus, to our knowledge, only two eukaryotic microbe models related to plants are available. A model of the pathogenic oomycete *Phytophtora infestans*, the causal agent of the late blight on potato and tomato (Rodenburg et al., 2018) was generated using hidden Markov models and KEGG database and was used to map transcriptomics data. It was followed by the metabolic model of *Sclerotinia sclerotiorum* (Peyraud et al., 2019), based on *Saccharomyces cerevisiae* and *Aspergillus niger* as reference organisms. It investigated the division of labor and metabolic heterogeneity *in planta*.

3.2 Plant-microorganism interactions

Several metabolic models of plant-interacting microorganisms are now available to the scientific community (Table 2). On the other hand, plant GEMs were also largely developed for both model organisms and crops (Table 1). It is thus tempting to reuse these constructions to build plant-microorganism interaction metabolic models, in order to better understand trophic interactions, the effect of one organism on the other and the ways to favor or restrict an interaction. However, this raises the question of how to model multi-organism systems. Metabolic modeling has already been applied to other multi-organism systems. The first published model studied the cooperation of two microorganisms for degrading organic substrates in the context of anaerobic digestion (Stolyar et al., 2007). In a biotechnological context, it helped to understand and quantify how cooperation between strains could improve the yield of a desired compound (Salimi, Zhuang, & Mahadevan, 2010). An immune cell-pathogen system was also built to provide medical solutions against *Mycobacterium tuberculosis* (Bordbar, Lewis, Schellenberger, Palsson, & Jamshidi, 2010).

For plant-microorganism interactions, three models were published so far. The first studied the process of nitrogen fixation through legume-rhizobia symbiosis (Pfau et al., 2018), with the model organisms *Medicago truncatula* (legume) and *S. meliloti* (symbiont). The authors combined a two-compartment model of *M. truncatula* (root and shoot) and a model of *S. meliloti*. In the model, *S. meliloti* is able to acquire atmospheric nitrogen (N_2) and convert it to ammonium and amino acids. These nitrogen sources are subsequently provided to the plant root. In exchange, the plant root provides oxygen and organic carbon to the symbiont. Simulations showed that symbiosis was not beneficial for the global growth rate (plant and symbiont) if enough nitrogen was available in the soil as nitrate (NO_3^-) and ammonium (NH_4^+). On the contrary, the growth increased when low nitrogen was available in the soil. In the same study, the authors investigated the impact of varying oxygen supply on the symbiont central metabolism, as oxygen is suspected to be a limiting factor for this symbiosis. In this interaction model, a curated metabolic model of the plant *M. truncatula* was used. However, the symbiont was not modeled using a curated and published metabolic model (as the one presented in the precedent section), but an automatic extraction from an online metabolic database (MetaCyc). This approximation limits the precision and the reliability of the study to predict fluxes inside the bacterium, and to associate bacterial phenotypes to metabolic genes. A new model of *M. truncatula-S. meliloti* interaction was published by diCenzo and collaborators (diCenzo, Tesi, Pfau, Mengoni, & Fondi, 2020), named ViNE (*Virtual Nodule Environment*). The bacterium part of the model was based on a curated and refined GEM (diCenzo et al., 2018, 2016). The multi-tissue plant part of the model was the previously published GEM of *M. truncatula* (Pfau et al., 2018) and its multi-tissue modeling methodology. Some quantitative aspects of the interaction in accordance with experimental data were included in the simulations (shoot and root representing 98% of the biomass versus 2% for the nodule). This allowed to validate the model prediction (growth rates, nitrogen fixation rates, mutant phenotypes). It also provided novel insights on the symbiotic interaction such as costs of nitrogen fixation and nodule formation, impacts of nodule size/fixation efficiency on the system, or bacteroid carbon sources (diCenzo et al., 2020).

Rodenburg et al. (Rodenburg et al., 2019) developed a metabolic model for a plant-pathogen interaction. They combined a metabolic model of tomato (*S. lycopersicum*) leaf and *P. infestans*. The two-organism metabolic model was manually curated to represent the interaction in agreement with the biological knowledge, such as the dependence of *P. infestans* to thiamine produced by the plant. To study flux distribution, different potential objective functions were tested: minimization of import fluxes from the leaf to the pathogen, minimization/maximization of the number of metabolic reactions used by the pathogen, and minimization of both the pathogen's import fluxes and the pathogen's metabolic reactions. Flux coupling analysis methodology suggested plant or pathogen pathways that could be used to control the pathogen spread. Finally, transcriptomics and metabolomics data at different stages of the infection were mapped onto the metabolic network to highlight the dynamics of metabolism during the infection process. This mapping illustrated the shutting down of alternative pathways as the infection progressed, and identified potential pathogen's targets and plant pathways to engineer to block the infection. However, these study results are limited by the lack of physiological/metabolomic experiments on *P. infestans*-tomato leaf interaction. Quantitative data (e.g. growth rates, organisms' biomass ratio) and evidence on the exchanged molecules would provide additional insights on tomato late blight.

3.3 Toward a quantitative and dynamic representation of plant-microorganism interaction

Metabolic modeling of plant-microorganisms interaction is already emerging (diCenzo et al., 2020; Pfau et al., 2018; Rodenburg et al., 2019), and many will be in the future, since the number of plant and microorganism models increases each year. The model developed by diCenzo et al. (2020) included a quantitative representation of the interaction and addressed the question of the availability of the resources at the interaction site. The numerous insights provided by the study confirms that these considerations are central to investigate metabolic interactions.

Thus, we propose three main methodological points that should be addressed in the future metabolic models of interactions: the quantitative aspect of the interaction, the application of QSSA and the availability of the resources between organisms (Fig. 1).

3.3.1 Quantitative representation of the interaction

So far, most of the models available lack the integration of plant and microorganism quantifications; the analysis of the interaction is only qualitative. Most of the approaches unraveled which metabolic fluxes were activated in the presence or absence of the organism interacting with plants, but they did not estimate whether these flux variations were significant or not at the scale of whole plant metabolism. Being able to quantify the metabolic effect of one organism on the other would help to better understand the interaction. For example, Rodenburg et al. were able to highlight by FCA the plant

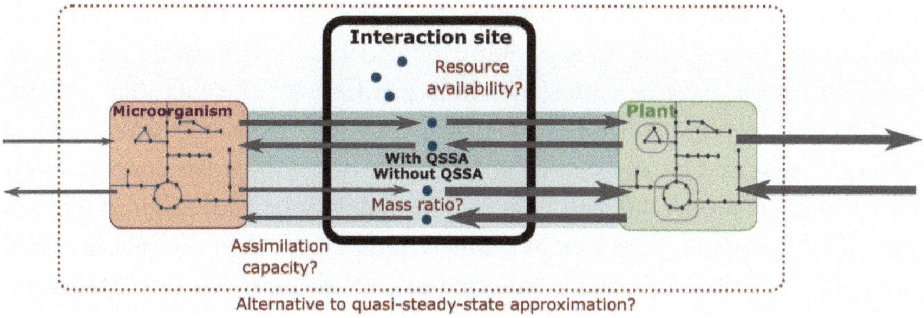

Fig. 1 Toward a quantitative and dynamic metabolic model of biotic interactions. The principal methodological and experimental issues that must be addressed are represented in red: quantification of the microorganism/plant ratio, use of a methodology that relaxes the QSSA growth hypothesis, and accurate representation of the resources available in the interaction site. Exchange fluxes are represented in two scenarios: assuming QSSA, metabolic flux from plant to interaction site and interaction site to microorganism must be equal (in mol·time unit^{-1}) while without QSSA, this constraint is relaxed. Relaxing equality in exchange fluxes could be relevant in several cases, as the microorganism and the plant can have contrasted assimilation fluxes.

metabolites potentially assimilated by the pathogen as nutritive sources (Rodenburg et al., 2019). Implementation of quantitative data (ratios, growth rates) would allow verifying if the proposed scenarios were realistic. Being able to quantify the metabolic effect of one organism on the other would help to understand better the interaction. It would be interesting to estimate at which population density the pathogen or symbiont impacts negatively or positively plant growth, as in the *Virtual Nodule Environment* metabolic model. It would also be interesting to study how much the microorganism's growth relies on virulence mechanisms, and if it can coerce the plant to provide nutrients in the interaction site. Ultimately, this study would allow unraveling how symbiont organisms could be optimized and selected to favor plant growth and agricultural yields or conversely, how pathogens could be controlled in their growth and virulence to avoid being deleterious to the plant. Yet, building a quantitative metabolic model would need experimental work to calibrate the model. Still, some of the experiments could be simple to perform, such as mass estimations of the ratio between the plant and the interacting organism, ideally at different stages of plant growth and development.

3.3.2 Quasi-steady-state approximation in a multiorganism system

To estimate fluxes, metabolic modeling relies on simplifications, as the numerous kinetic parameters required at the global cellular scale are majorly unknown (Orth et al., 2010). One of the hypotheses behind the simplification of the model, the QSSA, stipulates that concentrations of internal metabolites are assumed constant. It is justified by the fact that internal metabolites do not accumulate, they are usually consumed immediately as they are produced by the cell to synthesize macromolecules such as membrane lipids or proteins (Orth et al., 2010). Metabolic modeling of a multi-organism system at QSSA would imply that metabolites in the whole system (metabolites of the plant, the interacting organism and in the interaction site) are consumed immediately as they are produced, and consequently that both plant and interacting organism grow at the same velocity. While this could potentially be a reasonable hypothesis for symbiosis, balanced growth seems inadequate to simulate the dynamics of pathogenesis. Indeed, in several cases, the pathogen's proliferation will accelerate as it deploys its virulence mechanisms against the host, and oppositely, the plant growth will decelerate. To tackle this problem and simulate the dynamics of plant-pathogen interaction, different options could be considered. First, it is possible to perform the simulation at a precise instant and consider that on a relatively short time interval, the system could be considered at quasi-steady-state. This instantaneous simulation would then be repeated at different time steps to access the dynamics of the interaction, by updating the biomass of each organism at each time step. This process would, however, require the acquisition of at least growth of the two organisms throughout the experiment. Another solution is to consider that separately, the organisms are at QSSA, and allow non-QSSA behaviors for the exchanged metabolites. This would require implementing kinetics in the model, and thus estimate their parameters if it is possible.

Alternatively, metabolic modeling frameworks were developed to overpass the limitations of the QSSA hypothesis. To represent the dynamics of microbial communities, Zomorrodi et al. (Zomorrodi, Islam, & Maranas, 2014; Zomorrodi & Maranas, 2012) proposed a multi-level and multi-objective framework, which is in theory adaptable for parasitic systems. To overpass the QSSA hypothesis, Baroukh, Muñoz-Tamayo, Steyer, and Bernard (2014), proposed to reduce metabolic networks into sub-networks at QSSA, allowing dynamics for the exchange metabolites between subnetworks, generating a smaller metabolic network that could be modeled without hypothesizing QSSA. These two methodologies were not applied to host-symbiont or host-pathogen interactions yet, but are encouraging directions to take.

3.3.3 Resource availability at the interaction site

To provide a reliable representation of biotic interactions, a critical point is the representation of the resources available at the interaction site between the two organisms. Metabolic modeling of an organism strongly depends on the calibration of the uptake rate of the substrates and the carbon/nitrogen sources modeled. Comparing the resource usage in the simulation with experimental data also provides a better understanding of the interaction (diCenzo et al., 2020). Upon biotic interaction, these aspects are dependent on the representation of the interaction site. Thus, it is necessary to: (i) define which resources are made available by the plant/the interacting organism, (ii) quantify at which amount they are present, (iii) indicate if these resources could be assimilated by the plant/the interacting organism, and (iv) quantify the assimilation capacity of the resources by the plant/the interacting organism. These details are not always available, in particular the quantity available of the resource and the organisms' assimilation capacities, and would require substantial data acquisition.

Currently, the generation of pathogens and symbionts metabolic models is growing fast. It provides large resources to understand how microorganisms interact with plants. The next step, as the number of plant metabolic models are also growing rapidly, would be to integrate pathogen/symbiont and plant in a global metabolic model. Even though three studies were published and provided some insights (diCenzo et al., 2020; Pfau et al., 2018; Rodenburg et al., 2019), in particular when quantitative aspects were addressed, this approach is still poorly used. Indeed, quantitative and dynamic metabolic models of biotic interactions are arduous, because they require both the application of non-standard metabolic modeling approaches and the integration of several experimental data, some of them difficult to acquire.

4. Visualizing metabolism

Data related to metabolism often makes little sense if not put into context, i.e. confronted with the knowledge that has been accumulated about the organism metabolism. The sheer number of known metabolites and biochemical reactions linking them require computational solutions to be exploited. Yet, it is sometimes hard to formulate the interpretation requirements as a problem that can be solved programmatically.

Visual inspection of data and knowledge side by side, highlighting discriminant metabolites on a pathway map, for example, became a handy tool to formulate hypotheses and design subsequent analysis. Knowledge related to metabolism can be intuitively represented as a network of metabolites connected by biochemical transformations, and the practical aspects of the visual exploration of such networks will be briefly introduced hereafter.

While the reconstruction of a metabolic network is a difficult task, as mentioned in the first section of this chapter, it is not to be confused with the creation of its visual representation, the network drawing, yet another far from trivial task on its own. A single network may be represented by several drawings of various usability and purposes, and this section aims at providing some insight about choosing the right one and avoiding common pitfalls.

The science of information visualization has provided numerous methods and tools that help to deal with complex and abstract data, including large networks (Herman, Melancon, & Marshall, 2000; von Landesberger et al., 2011). Many algorithms have been produced to create network layouts that maximize some aesthetic criterion, such as reducing cluttering, edge crossing or node overlaps (Ware, Purchase, Colpoys, & McGill, 2002; Yoghourdjian et al., 2018).

Such techniques have found an abundance of applications in network-related sciences, such as sociology or physics. Yet, their popularity has not pervaded much in the metabolism research community. One of the main reasons may be the discrepancy between the drawing produced automatically and the visual semantics expected of metabolic representations. The most striking example is the drawing of whole networks obtained using force-directed algorithms, which earned in the field the tongue-in-cheek title of "hairballs" or "ridiculograms," referring to their little use beyond advertising the complexity of a research question (Fig. 2).

The "quality" of visualization depends on a given task to be performed or a particular property to be shown. Many layouts are prone to reveal the global or meso-scale organization of a system represented as a network. Force-directed layouts, for example, can give a glimpse of a community organization in a network, with visible groups of nodes that seem to be more connected with each other than with those of the other groups.

While understanding the global properties of metabolism as a system is a relevant and still an ongoing research question to this day, metabolic network visualizations are mainly used to decipher the role of specific elements (metabolite or reaction), or group of elements, in the system, rather than properties of the system itself. We will hence focus on these applications for the rest of this section.

The complex chains of interactions that can relate the concentrations of two metabolites, involving numerous branching, usually makes visual interpretation challenging. One approach consists in dividing the metabolic

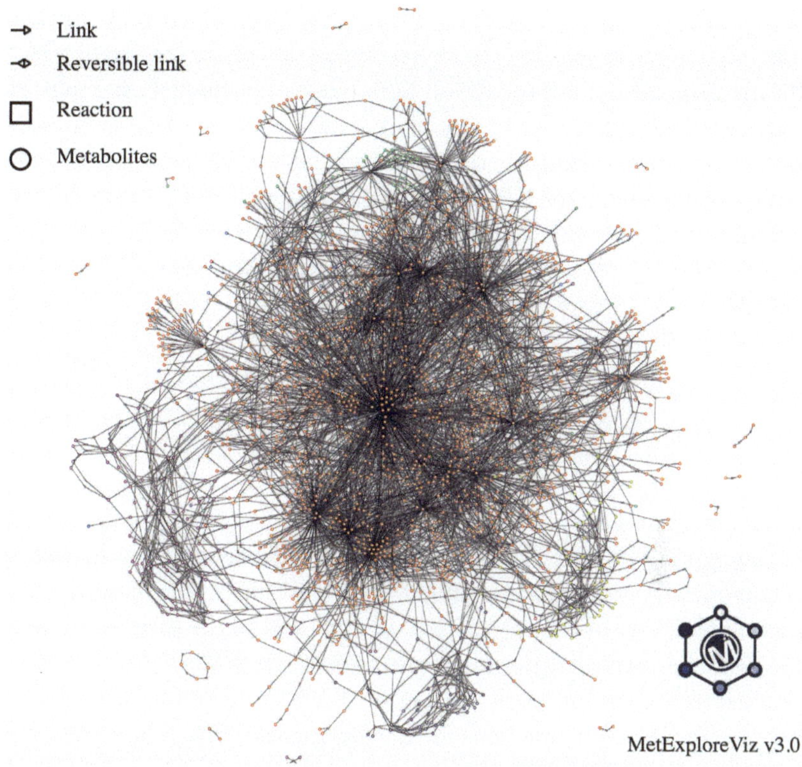

Fig. 2 ARAGEM metabolic network drawn using MetExploreViz with a force-directed layout.

network into smaller sub-parts that can be drawn manually and inspected more easily (Reimand et al., 2019; Saraiya, North, & Duca, 2005; Schuster, Pfeiffer, Moldenhauer, Koch, & Dandekar, 2002).

The definition of these sub-parts had historically been driven by their functional purposes (e.g. glycolysis, amino acid biosynthesis). Reasoning over individual metabolic pathways is still the most used strategy, following the aforementioned "divide and conquer" approach. However, the rise of global and untargeted experimental approaches (metabolomics, transcriptomics …) tends to dampen such practice, since observing entities of interest spanning multiple pathways is not rare (not to mention the subjective definition of pathways and consequently the lack of consensus between pathway databases). Thus, the call for global visualization of metabolism has grown strong (Dinkla & Westenberg, 2012; Gehlenborg et al., 2010).

The first global visualizations were proposed in the late 1960s, with the representation of G. Michal (Michal, 1999) and around the same time D. Nicholson (Nicholson, 2005). Not oriented toward the identification of large-scale structures, they opted for a planar representation dedicated to "knowledge navigation," and are often referred to as maps or atlases.[a]

Many proposed views today still look like those original drawings, with the addition of a layer of interactivity made possible by technological advances since then. Arguably the most peculiar and yet popular descendant of those maps has been the KEGG atlas (Okuda et al., 2008). Using colors for the related sub-parts, straight lines and smooth bends, and evenly spaced nodes, the KEGG atlas (Fig. 3) share a striking resemblance with Harry Beck's design of subway maps (Degani, 2013), which itself share similarities with electric diagrams.

The key of the worldwide application of such design to subway maps lies in the fact that it is task-oriented, it is focused solely on what those maps are used for: finding a path from station A to station B. Prior to Beck's design, the London underground map was drawn as an overlay on the classical London city map. Despite being familiar, it had several drawbacks regarding path search: stations from the city center were cluttered due to a higher density than country-side stations, tunnel twists and turns were hard to follow, and a lot of the displayed information were related to the surface and thus of little relevance for underground navigation (Tufte, 1997).

Path search is undoubtedly a relevant task in a metabolic network context, allowing finding biochemical transformations that could relate two compounds, which might explain the appeal of the KEGG Atlas. However, the analogy falls short when considering the number of "lines" and "stations" on a metabolic network, which makes performing this task far from an easy ride.

Another subtle difference with impairing consequences is related to topology. While not strictly planar, an underground railroad network can be projected on a two-dimensional map with little distortion, due to a small depth compared to the surface spanning. A well-known and more difficult case is one of the world maps. Projecting the surface of a near-spherical planet on a 2D map is at the cost of some distortions (Mulcahy & Clarke, 2001), which can have consequences on the way we perceive the represented object. The relative difference of areas between the African continent and Antarctica or Greenland is commonly biased by the classical Mercator representation (representation that has the benefits of representing parallels and meridians as straight lines, which have made it a fair visualization for nautical related tasks).

[a] http://biochemical-pathways.com/#/map/1.

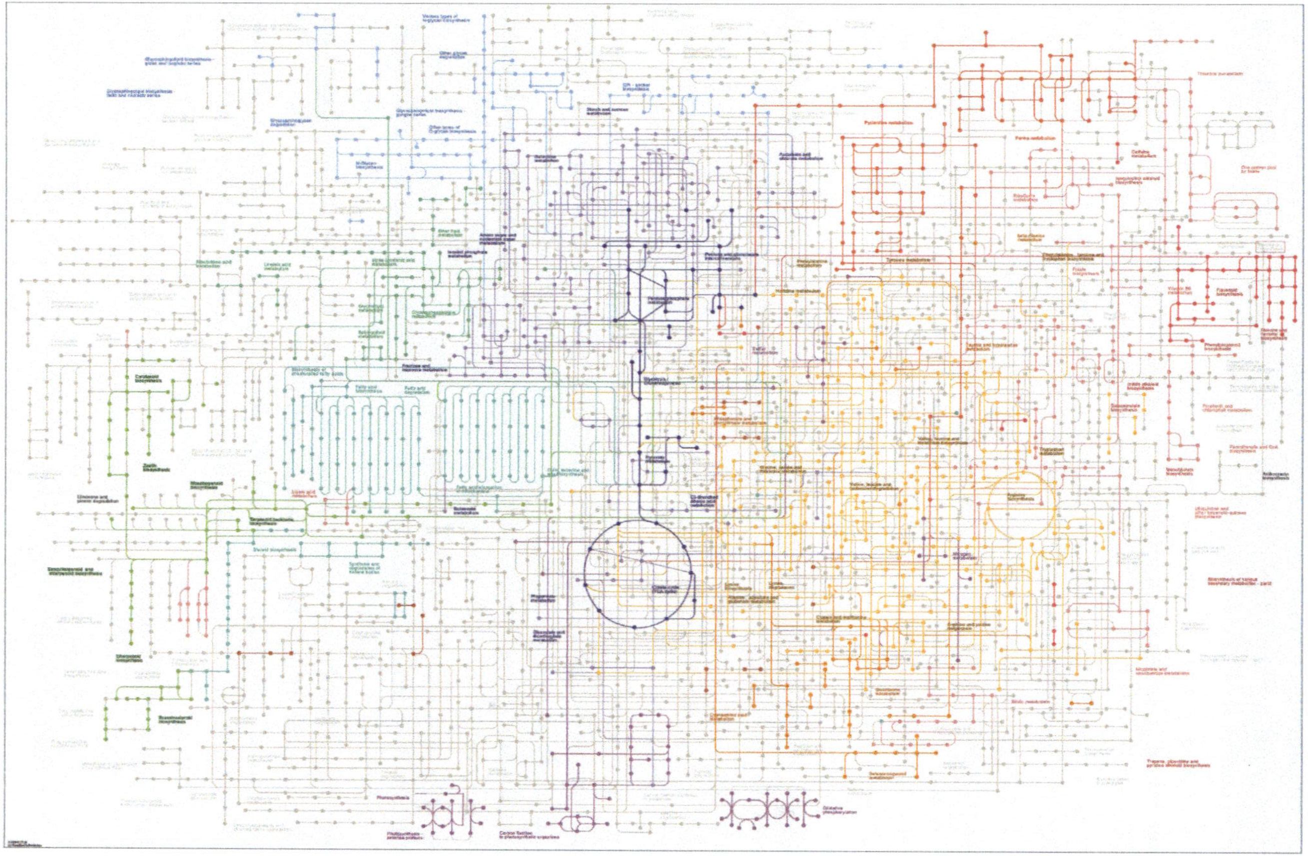

Fig. 3 KEGG overall representation of *A. thaliana* metabolic pathways.

Metabolic networks have a far more complex "shape," which makes the representation in a low dimensional space tedious. One should be aware that metabolic maps usually rely on some tricks to flatten the view, duplicating nodes in several places, or removing some edges, which is far from facilitating path search tasks or neighborhood exploration. Scratching the surface often reveals that most "hand-drawn" global visualizations of metabolism tend indeed to be "patchworks" of local pathways, where the necessary distortions are oriented to conserve their integrity, thus bearing the same bias as individual pathway views.

The old saying that "the map is not the territory" thus applies very well to metabolic networks, and their representations are still an open challenge.

While existing solutions can still be useful for navigating the network from an initial focal point to extract knowledge, one must be cautious when reasoning from sets of nodes. As extrapolating walk distance between stations from a subway map would be ill-advised, extending navigational "metabolic atlas" to other tasks can be biased. Therefore, from previously mentioned points, the perceived distances in metabolic drawings do not necessarily represent the "true" metabolic distances (usually estimated from path-length distances in the network). While it is tempting to draw biological hypotheses from a group of metabolites that would appear grouped or spread on the map, their relative positions bear little information and can be misleading. A drawing dedicated to this specific task must be considered, using, for example, multidimensional scaling, at the cost of less readable paths.

As the size of metabolic models has been subject to tremendous growth (Table 1), the manual drawing of their underlying network has become a fastidious task. Beyond scaling issues, hand-drawn maps' main weakness is their rigidity toward a constantly expanding understanding of metabolism, as small changes might lead to re-drawn large portions of the map. The future of metabolic network visualization may thus lie in the adaptation of drawing algorithms toward better-defined tasks.

The most commonly used graph-drawing algorithm to embed a network in a two- or three-dimensional space is the force-directed layout. There exist many variations of this method but, even if there was in past decades a clear improvement of the algorithms in terms of complexity and reproducibility, the underlying concept remains the same. These algorithms consider the network as a physical system made of physical bodies connected by a system of forces (like springs). The assumption is that such a system will reach an equilibrium status which will position nodes and edges in the representation space in such a way that aesthetic criteria will be satisfied (e.g. even spread of nodes, balanced length of edges). These algorithms are broadly used since they generate a representation for any kind of network, as they do not require any topological prerequisite. The main drawback, in the context of metabolic network studies, is that metabolic networks have topological features that usually lead to cluttered drawings. More precisely, dense parts of the network (nodes with high degrees and strong clustering index) tend to be overloaded in the center of the representation while, in the periphery, long isolated chains radiate in every direction, giving far more room to less central parts of metabolism. This representation captures well the overall organization of the metabolic network, but when it comes to navigating the network to glean some information about the role of a particular element, it is far from ideal.

At the other hand of the spectrum of graph drawing methods, lies planar representations (Didimo, Liotta, & Patrignani, 2019). This class of methods aims to provide a drawing with a good spread of nodes in the representation space while minimizing edge crossings by using edge bending. These drawings are quite efficient for instance for physical network representations such as electrical diagrams or software development schemas such as UML representations of program libraries. To achieve these goals the strategy is first to extract a subgraph which is planar (which can be drawn in two dimensions with no edge crossing) and then define an algorithmic solution to add the remaining edges to the initial drawing while minimizing the number of edge crossings. The advantage of this class of methods in the context of metabolic networks is the overall resemblance with atlas and maps, sharing the comfort of evenly spaced nodes for navigation. However, the main drawback is that, since metabolic networks are far from being planar, the algorithm will tend to create overly long edges that make visual path search quite challenging.

Based on the pros and cons of the various methods, some algorithms were proposed to combine the benefit of the existing approaches and try to fit biochemists' drawing conventions. The shared principle is to divide the network into sub-parts and apply the best-suited drawing algorithms to each part. For example, a circular layout could be applied to a sub-part containing a cycle, and a hierarchical layout could be applied to an acyclic sub-part. It was first designed to draw metabolic pathways (Fig. 4A) (Becker & Rojas, 2001). However, as acknowledged by authors, the approach is hardly scaling up to the network level. In fact, the challenge is to combine these separated drawings since subnetworks are usually highly overlapping. It thus requires additional operations (rotations, translation) to assemble sub-drawings in a harmonious arrangement.

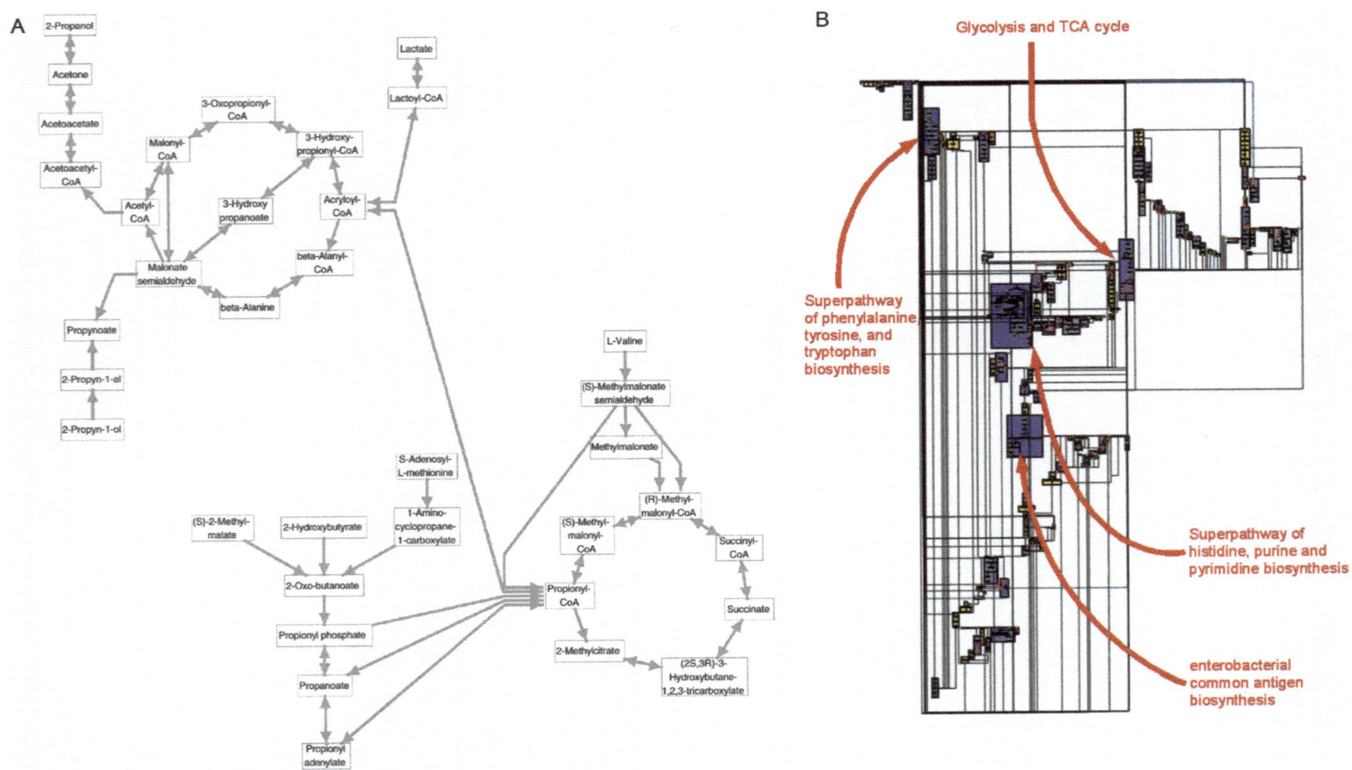

Fig. 4 Representation of metabolic pathways. (A) Automatic drawing of propanoate metabolism from (Becker & Rojas, 2001). (B) Automatic drawing of *E. coli* metabolic network using a combination of cyclic and hierarchical drawing algorithms for pathways and a planar graph drawing for the edges connecting these pathways. *From Bourqui et al. (2007).*

In their article, Bourqui et al. (2007) propose to take advantage of the decomposition of the metabolic network into pathways. To limit the challenge of overlapping pathways, they use a coloration algorithm to detect the maximum number of pathways, which can be drawn without influencing others. Then, pathways are arranged using a planar drawing with cycle and hierarchical substructure search applied internally to each pathway (Fig. 4B).

Even if automatic drawings provide a good first sketch of the drawing of metabolic networks, it still requires some manual curation and modifications. For example, a handcrafted list of "side compounds" is usually removed from the view, since those compounds, including water or ATP, appear overly connected (they are hubs, to speak in network terms) and cause drastic cluttering. Nevertheless, one should note that some highly connected nodes, such as Pyruvate, make the network harder to draw, yet their removal would cause crucial loss of information. The automation of the hub removal is thus far from trivial as careful expert curation is required. Hence, more and more tools provide interactive edition features on top of initial automatic drawings. This is, for instance, the case of Cytoscape (Shannon et al., 2003), or MetExploreViz (Chazalviel et al., 2018) which is designed especially for metabolic network representations (bipartite default style, edges bundling for reactions, etc).

Overall, moving from a segmented view using pathways to a more holistic, system-level representation is at cost of time-consuming manual curation, despite some progress regarding automatic drawing to generate "seeds". Beyond the creation of these multi-purpose global views aiming at knowledge exploration, it is possible to imagine task-oriented representations (see example in Fig. 5, adapted from (Frainay et al., 2018)).

In that case, it is of interest to apply more data-driven drawings of metabolic networks such as attribute layouts or hive plots (Krzywinski, Birol, Jones, & Marra, 2012) as support to convey a particular finding and identify intriguing patterns. Finally, it is important to notice that the quality of representation has not only to be assessed in terms of aesthetic criteria but also based on their ability to improve the user capacity to perform a search task (e.g. looking for a metabolic path between two metabolites of interest).

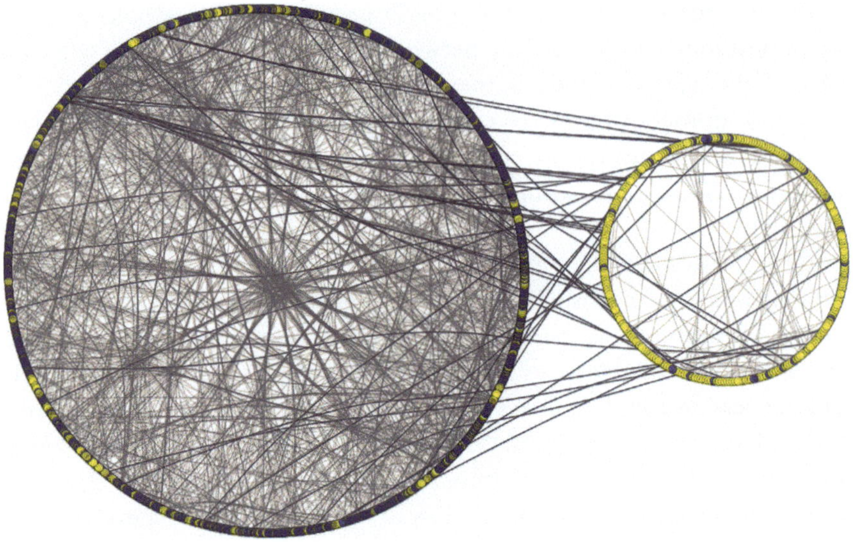

Fig. 5 An unconventional and task-oriented view of a metabolic network. The task here is the comparison of two groups of nodes partitioning the network. A circular layout has been applied to the two groups. This view emphasizes the difference between the two groups in terms of size and inner density, shows the relative scarcity of links between the two groups, and the overrepresentation of a given property in one of them, using color mapping. Drawn with Cytoscape using "group attributes layout" method.

7. Conclusion

As shown in this chapter, the development of omics-related techniques in plants have allowed metabolic modeling to become widespread. Since the acquisition of plant complete genomes remains complicated, compared to obtaining such data for bacteria, for example, plants GEMs reconstruction and analysis have benefited from all the methods that have been developed for metabolic modeling of bacteria and other eukaryotes.

Even if reconstructing and analyzing plant GEMs remain difficult (the number of existing plant GEMs speaks for itself), the number of GEMs for "non-Arabidopsis" species is growing, especially in the past few years. The next two big steps to improve modeling are to better take into account the compartmentalization of the models, and metabolomics will play a huge role here, and to better take into account the interaction among models, either among different tissues inside a plant or with the microbiome interacting with the plant. Once again, metabolomics will have a major role here, in particular by providing lists of metabolites shared among the different organs and/or species. Complementary to metabolomics data acquisition, developing biologically relevant models will also require careful considerations and adaptation of existing methodologies. Indeed, constraint-based modeling relies on mathematical hypotheses, such as QSSA, which may not match the complexity of compartmentalized, multi-tissue or multi-organism systems.

Finally, modeling plant metabolism will face the same challenges as modeling all other kinds of species: the amount of omics data to integrate into the models will keep growing, and modeling techniques will have to cope with such data. Improvements in the visualization of metabolic networks will undoubtedly help to provide context to the various data that may be associated with GEMs, and hence improve the understanding of such big networks.

References

Arkin, A. P., Cottingham, R. W., Henry, C. S., Harris, N. L., Stevens, R. L., Maslov, S., et al. (2018). KBase: The United States Department of energy systems biology knowledgebase. *Nature Biotechnology*, *36*(7), 566–569. https://doi.org/10.1038/nbt.4163.

Baroukh, C., Muñoz-Tamayo, R., Steyer, J.-P., & Bernard, O. (2014). DRUM: A new framework for metabolic modeling under non-balanced growth. Application to the carbon metabolism of unicellular microalgae. *PLoS One*, *9*(8), e104499. https://doi.org/10.1371/journal.pone.0104499.

Becker, M. Y., & Rojas, I. (2001). A graph layout algorithm for drawing metabolic pathways. *Bioinformatics*, *17*(5), 461–467. https://doi.org/10.1093/bioinformatics/17.5.461.

Bekaert, M., Edger, P. P., Hudson, C. M., Pires, J. C., & Conant, G. C. (2012). Metabolic and evolutionary costs of herbivory defense: Systems biology of glucosinolate synthesis. *New Phytologist*, *196*(2), 596–605. https://doi.org/10.1111/j.1469-8137.2012.04302.x.

Bogart, E., & Myers, C. R. (2016). Multiscale metabolic modeling of C4 plants: Connecting nonlinear genome-scale models to leaf-scale metabolism in developing maize leaves. *PLoS One*, *11*(3), e0151722. https://doi.org/10.1371/journal.pone.0151722.

Bordbar, A., Lewis, N. E., Schellenberger, J., Palsson, B.Ø., & Jamshidi, N. (2010). Insight into human alveolar macrophage and *M. tuberculosis* interactions via metabolic reconstructions. *Molecular Systems Biology*, *6*(1), 422. https://doi.org/10.1038/msb.2010.68.

Botero, K., Restrepo, S., & Pinzón, A. (2018). A genome-scale metabolic model of potato late blight suggests a photosynthesis suppression mechanism. *BMC Genomics*, *19*(S8), 863. https://doi.org/10.1186/s12864-018-5192-x.

Bourqui, R., Cottret, L., Lacroix, V., Auber, D., Mary, P., Sagot, M.-F., et al. (2007). Metabolic network visualization eliminating node redundance and preserving metabolic pathways. *BMC Systems Biology*, *1*(1), 29. https://doi.org/10.1186/1752-0509-1-29.

Caspi, R., Billington, R., Ferrer, L., Foerster, H., Fulcher, C. A., Keseler, I. M., et al. (2016). The MetaCyc database of metabolic pathways and enzymes and the BioCyc collection of pathway/genome databases. *Nucleic Acids Research*, *44*(D1), D471–D480. https://doi.org/10.1093/nar/gkv1164.

Chatterjee, A., Huma, B., Shaw, R., & Kundu, S. (2017). Reconstruction of Oryza sativa indica genome scale metabolic model and its responses to varying RuBisCO activity, light intensity, and enzymatic cost conditions. *Frontiers in Plant Science*, *8*, 2060. https://doi.org/10.3389/fpls.2017.02060.

Chatterjee, A., & Kundu, S. (2015). Revisiting the chlorophyll biosynthesis pathway using genome scale metabolic model of Oryza sativa japonica. *Scientific Reports*, *5*(1), 14975. https://doi.org/10.1038/srep14975.

Chazalviel, M., Frainay, C., Poupin, N., Vinson, F., Merlet, B., Gloaguen, Y., et al. (2018). MetExploreViz: Web component for interactive metabolic network visualization. *Bioinformatics*, *34*(2), 312–313. https://doi.org/10.1093/bioinformatics/btx588.

Cheung, C. Y. M., Poolman, M. G., Fell, D. A., Ratcliffe, R. G., & Sweetlove, L. J. (2014). A diel flux balance model captures interactions between light and dark metabolism during day-night cycles in C_3 and Crassulacean acid metabolism leaves. *Plant Physiology*, *165*(2), 917–929. https://doi.org/10.1104/pp.113.234468.

Cheung, C. Y. M., Williams, T. C. R., Poolman, M. G., Fell, D. A., Ratcliffe, R. G., & Sweetlove, L. J. (2013). A method for accounting for maintenance costs in flux balance analysis improves the prediction of plant cell metabolic phenotypes under stress conditions. *The Plant Journal*, *75*(6), 1050–1061. https://doi.org/10.1111/tpj.12252.

Contador, C. A., Lo, S.-K., Chan, S. H. J., & Lam, H.-M. (2020). Metabolic analyses of nitrogen fixation in the soybean microsymbiont *Sinorhizobium fredii* using constraint-based modeling. *MSystems*, *5*(1), e00516-19. /msystems/5/1/msys.00516-19.atom https://doi.org/10.1128/mSystems.00516-19.

de Oliveira Dal'Molin, C. G., Orellana, C., Gebbie, L., Steen, J., Hodson, M. P., Chrysanthopoulos, P., et al. (2016). Metabolic reconstruction of Setaria italica: A systems biology approach for integrating tissue-specific omics and pathway analysis of bioenergy grasses. *Frontiers in Plant Science*, 7. https://doi.org/10.3389/fpls.2016.01138.

de Oliveira Dal'Molin, C. G., Quek, L.-E., Palfreyman, R. W., Brumbley, S. M., & Nielsen, L. K. (2010a). AraGEM, a genome-scale reconstruction of the primary metabolic network in Arabidopsis. *Plant Physiology*, *152*(2), 579–589. https://doi.org/10.1104/pp.109.148817.

de Oliveira Dal'Molin, C. G., Quek, L.-E., Palfreyman, R. W., Brumbley, S. M., & Nielsen, L. K. (2010b). C4GEM, a genome-scale metabolic model to study C $_4$ plant metabolism. *Plant Physiology*, *154*(4), 1871–1885. https://doi.org/10.1104/pp.110.166488.

de Oliveira Dal'Molin, C. G., Quek, L.-E., Saa, P. A., & Nielsen, L. K. (2015). A multi-tissue genome-scale metabolic modeling framework for the analysis of whole plant systems. *Frontiers in Plant Science*, *6*. https://doi.org/10.3389/fpls.2015.00004.

Degani, A. (2013). A tale of two maps: Analysis of the London underground "Diagram". *Ergonomics in Design: The Quarterly of Human Factors Applications*, *21*(3), 7–16. https://doi.org/10.1177/1064804613489125.

diCenzo, G. C., Benedict, A. B., Fondi, M., Walker, G. C., Finan, T. M., Mengoni, A., et al. (2018). Robustness encoded across essential and accessory replicons of the ecologically versatile bacterium Sinorhizobium meliloti. *PLoS Genetics*, *14*(4), e1007357. https://doi.org/10.1371/journal.pgen.1007357.

diCenzo, G. C., Checcucci, A., Bazzicalupo, M., Mengoni, A., Viti, C., Dziewit, L., et al. (2016). Metabolic modelling reveals the specialization of secondary replicons for niche adaptation in Sinorhizobium meliloti. *Nature Communications*, 7(1), 12219. https://doi.org/10.1038/ncomms12219.

diCenzo, G. C., Tesi, M., Pfau, T., Mengoni, A., & Fondi, M. (2020). Genome-scale metabolic reconstruction of the symbiosis between a leguminous plant and a nitrogen-fixing bacterium. *Nature Communications*, *11*(1), 2574. https://doi.org/10.1038/s41467-020-16484-2.

Didimo, W., Liotta, G., & Patrignani, M. (2019). HV-planarity: Algorithms and complexity. *Journal of Computer and System Sciences*, *99*, 72–90. https://doi.org/10.1016/j.jcss.2018.08.003.

Dinkla, K., & Westenberg, M. A. (2012). Network visualization in cell biology. *Tsinghua Science and Technology*, *17*(4), 365–382. https://doi.org/10.1109/TST.2012.6297584.

Frainay, C., Schymanski, E., Neumann, S., Merlet, B., Salek, R., Jourdan, F., et al. (2018). Mind the gap: Mapping mass spectral databases in genome-scale metabolic networks reveals poorly covered areas. *Metabolites*, *8*(3), 51. https://doi.org/10.3390/metabo8030051.

Gehlenborg, N., O'Donoghue, S. I., Baliga, N. S., Goesmann, A., Hibbs, M. A., Kitano, H., et al. (2010). Visualization of omics data for systems biology. *Nature Methods*, 7(S3), S56–S68. https://doi.org/10.1038/nmeth.1436.

Gerlin, L., Cottret, L., Cesbron, S., Taghouti, G., Jacques, M.-A., Genin, S., et al. (2020). Genome-scale investigation of the metabolic determinants generating bacterial fastidious growth. *MSystems*, *5*(2), e00698-19. /msystems/5/2/msys.00698-19.atom https://doi.org/10.1128/mSystems.00698-19.

Hay, J. O., Shi, H., Heinzel, N., Hebbelmann, I., Rolletschek, H., & Schwender, J. (2014). Integration of a constraint-based metabolic model of Brassica napus developing seeds with 13C-metabolic flux analysis. *Frontiers in Plant Science*, *5*. https://doi.org/10.3389/fpls.2014.00724.

Herman, I., Melancon, G., & Marshall, M. S. (2000). Graph visualization and navigation in information visualization: A survey. *IEEE Transactions on Visualization and Computer Graphics*, *6*(1), 24–43. https://doi.org/10.1109/2945.841119.

Johnson, S. R., Lange, I., Srividya, N., & Lange, B. M. (2017). Bioenergetics of monoterpenoid essential oil biosynthesis in nonphotosynthetic glandular trichomes. *Plant Physiology*, *175*(2), 681–695. https://doi.org/10.1104/pp.17.00551.

Kanehisa, M. (2000). KEGG: Kyoto encyclopedia of genes and genomes. *Nucleic Acids Research*, *28*(1), 27–30. https://doi.org/10.1093/nar/28.1.27.

Kim, T. Y., Sohn, S. B., Kim, Y. B., Kim, W. J., & Lee, S. Y. (2012). Recent advances in reconstruction and applications of genome-scale metabolic models. *Current Opinion in Biotechnology*, *23*(4), 617–623. https://doi.org/10.1016/j.copbio.2011.10.007.

Koduru, L., Kim, H. Y., Lakshmanan, M., Mohanty, B., Lee, Y. Q., Lee, C. H., et al. (2020). Genome-scale metabolic reconstruction and in silico analysis of the rice leaf blight pathogen, *Xanthomonas oryzae*. *Molecular Plant Pathology*, *21*(4), 527–540. https://doi.org/10.1111/mpp.12914.

Krzywinski, M., Birol, I., Jones, S. J., & Marra, M. A. (2012). Hive plots—Rational approach to visualizing networks. *Briefings in Bioinformatics*, *13*(5), 627–644. https://doi.org/10.1093/bib/bbr069.

Lakshmanan, M., Lim, S.-H., Mohanty, B., Kim, J. K., Ha, S.-H., & Lee, D.-Y. (2015). Unraveling the light-specific metabolic and regulatory signatures of rice through combined in silico modeling and multi-omics analysis. *Plant Physiology*, 01379.2015. https://doi.org/10.1104/pp.15.01379.

Liu, L., Mei, Q., Yu, Z., Sun, T., Zhang, Z., & Chen, M. (2013). *An integrative bioinformatics framework for genome-scale multiple level network reconstruction of rice*. https://doi.org/10.2390/BIECOLL-JIB-2013-223.

Michal, G. (1999). *Biochemical pathways*. John Wiley & Sons, Inc.

Mintz-Oron, S., Meir, S., Malitsky, S., Ruppin, E., Aharoni, A., & Shlomi, T. (2012). Reconstruction of Arabidopsis metabolic network models accounting for subcellular compartmentalization and tissue-specificity. *Proceedings of the National Academy of Sciences of the United States of America*, *109*(1), 339–344. https://doi.org/10.1073/pnas.1100358109.

Moreira, T. B., Shaw, R., Luo, X., Ganguly, O., Kim, H.-S., Coelho, L. G. F., et al. (2019). A genome-scale metabolic model of soybean (*Glycine max*) highlights metabolic fluxes in seedlings. *Plant Physiology*, *180*(4), 1912–1929. https://doi.org/10.1104/pp.19.00122.

Mulcahy, K. A., & Clarke, K. C. (2001). Symbolization of map projection distortion: A review. *Cartography and Geographic Information Science*, *28*(3), 167–182. https://doi.org/10.1559/152304001782153044.

Nicholson, D. (2005). From metabolic pathways charts to animaps in 50 years. *Biochemistry and Molecular Biology Education*, *33*(3), 156–158. https://doi.org/10.1002/bmb.2005.494033032480.

Okuda, S., Yamada, T., Hamajima, M., Itoh, M., Katayama, T., Bork, P., et al. (2008). KEGG atlas mapping for global analysis of metabolic pathways. *Nucleic Acids Research*, *36*, W423–W426 (web server) https://doi.org/10.1093/nar/gkn282.

Orth, J. D., Thiele, I., & Palsson, B.Ø. (2010). What is flux balance analysis? *Nature Biotechnology*, *28*(3), 245–248. https://doi.org/10.1038/nbt.1614.

Peyraud, R., Cottret, L., Marmiesse, L., Gouzy, J., & Genin, S. (2016). A resource allocation trade-off between virulence and proliferation drives metabolic versatility in the plant pathogen Ralstonia solanacearum. *PLoS Pathogens*, *12*(10), e1005939. https://doi.org/10.1371/journal.ppat.1005939.

Peyraud, R., Mbengue, M., Barbacci, A., & Raffaele, S. (2019). Intercellular cooperation in a fungal plant pathogen facilitates host colonization. *Proceedings of the National Academy of Sciences of the United States of America*, *116*(8), 3193–3201. https://doi.org/10.1073/pnas.1811267116.

Pfau, T., Christian, N., Masakapalli, S. K., Sweetlove, L. J., Poolman, M. G., & Ebenhöh, O. (2018). The intertwined metabolism during symbiotic nitrogen fixation elucidated by metabolic modelling. *Scientific Reports*, *8*(1), 12504. https://doi.org/10.1038/s41598-018-30884-x.

Poolman, M. G., Kundu, S., Shaw, R., & Fell, D. A. (2013). Responses to light intensity in a genome-scale model of rice metabolism. *Plant Physiology*, *162*(2), 1060–1072. https://doi.org/10.1104/pp.113.216762.

Poolman, M. G., Miguet, L., Sweetlove, L. J., & Fell, D. A. (2009). A genome-scale metabolic model of Arabidopsis and some of its properties. *Plant Physiology*, *151*(3), 1570–1581. https://doi.org/10.1104/pp.109.141267.

Prigent, S., Frioux, C., Dittami, S. M., Thiele, S., Larhlimi, A., Collet, G., et al. (2017). Meneco, a topology-based gap-filling tool applicable to degraded genome-wide metabolic networks. *PLoS Computational Biology*, *13*(1), e1005276. https://doi.org/10.1371/journal.pcbi.1005276.

Reimand, J., Isserlin, R., Voisin, V., Kucera, M., Tannus-Lopes, C., Rostamianfar, A., et al. (2019). Pathway enrichment analysis and visualization of omics data using g:Profiler, GSEA, Cytoscape and EnrichmentMap. *Nature Protocols*, *14*(2), 482–517. https://doi.org/10.1038/s41596-018-0103-9.

Resendis-Antonio, O., Reed, J. L., Encarnación, S., Collado-Vides, J., & Palsson, B.Ø. (2007). Metabolic reconstruction and modeling of nitrogen fixation in Rhizobium etli. *PLoS Computational Biology*, *3*(10), e192. https://doi.org/10.1371/journal.pcbi.0030192.

Rodenburg, S. Y. A., Seidl, M. F., de Ridder, D., & Govers, F. (2018). Genome-wide characterization of *Phytophthora infestans* metabolism: A systems biology approach: A metabolic model for *Phytophthora infestans*. *Molecular Plant Pathology*, *19*(6), 1403–1413. https://doi.org/10.1111/mpp.12623.

Rodenburg, S. Y. A., Seidl, M. F., Judelson, H. S., Vu, A. L., Govers, F., & de Ridder, D. (2019). Metabolic model of the *Phytophthora infestans*-tomato interaction reveals metabolic switches during host colonization. *MBio*, *10*(4). e00454-19, /mbio/10/4/mBio.00454-19.atom https://doi.org/10.1128/mBio.00454-19.

Saha, R., Suthers, P. F., & Maranas, C. D. (2011). Zea mays iRS1563: A comprehensive genome-scale metabolic reconstruction of maize metabolism. *PLoS One*, *6*(7), e21784. https://doi.org/10.1371/journal.pone.0021784.

Salimi, F., Zhuang, K., & Mahadevan, R. (2010). Genome-scale metabolic modeling of a clostridial co-culture for consolidated bioprocessing. *Biotechnology Journal, 5*(7), 726–738. https://doi.org/10.1002/biot.201000159.

Saraiya, P., North, C., & Duca, K. (2005). Visualizing biological pathways: Requirements analysis, systems evaluation and research agenda. *Information Visualization, 4*(3), 191–205. https://doi.org/10.1057/palgrave.ivs.9500102.

Sarkar, D., & Maranas, C. D. (2020). SNPeffect: Identifying functional roles of SNPs using metabolic networks. *The Plant Journal.* tpj.14746 https://doi.org/10.1111/tpj.14746.

Scheunemann, M., Brady, S. M., & Nikoloski, Z. (2018). Integration of large-scale data for extraction of integrated Arabidopsis root cell-type specific models. *Scientific Reports, 8*(1), 7919. https://doi.org/10.1038/s41598-018-26232-8.

Schläpfer, P., Zhang, P., Wang, C., Kim, T., Banf, M., Chae, L., et al. (2017). Genome-wide prediction of metabolic enzymes, pathways, and gene clusters in plants. *Plant Physiology, 173*(4), 2041–2059. https://doi.org/10.1104/pp.16.01942.

Schuster, S., Pfeiffer, T., Moldenhauer, F., Koch, I., & Dandekar, T. (2002). Exploring the pathway structure of metabolism: Decomposition into subnetworks and application to mycoplasma pneumoniae. *Bioinformatics, 18*(2), 351–361. https://doi.org/10.1093/bioinformatics/18.2.351.

Seaver, S. M. D., Bradbury, L. M. T., Frelin, O., Zarecki, R., Ruppin, E., Hanson, A. D., et al. (2015). Improved evidence-based genome-scale metabolic models for maize leaf, embryo, and endosperm. *Frontiers in Plant Science, 6.* https://doi.org/10.3389/fpls.2015.00142.

Shannon, P., Markiel, A., Ozier, O., Baliga, N. S., Wang, J. T., Ramage, D., et al. (2003). Cytoscape: A software environment for integrated models of biomolecular interaction networks. *Genome Research, 13*(11), 2498–2504. https://doi.org/10.1101/gr.1239303.

Shaw, R., & Cheung, C. Y. M. (2018). A dynamic multi-tissue flux balance model captures carbon and nitrogen metabolism and optimal resource partitioning during Arabidopsis growth. *Frontiers in Plant Science, 9*, 884. https://doi.org/10.3389/fpls.2018.00884.

Simons, M., Saha, R., Amiour, N., Kumar, A., Guillard, L., Clement, G., et al. (2014). Assessing the metabolic impact of nitrogen availability using a compartmentalized maize leaf genome-scale model. *Plant Physiology, 166*(3), 1659–1674. https://doi.org/10.1104/pp.114.245787.

Stolyar, S., Van Dien, S., Hillesland, K. L., Pinel, N., Lie, T. J., Leigh, J. A., et al. (2007). Metabolic modeling of a mutualistic microbial community. *Molecular Systems Biology, 3*(1), 92. https://doi.org/10.1038/msb4100131.

The Arabidopsis Genome Initiative. (2000). Analysis of the genome sequence of the flowering plant Arabidopsis thaliana. *Nature, 408*(6814), 796–815. https://doi.org/10.1038/35048692.

Thiele, I., & Palsson, B.Ø. (2010). A protocol for generating a high-quality genome-scale metabolic reconstruction. *Nature Protocols, 5*(1), 93–121. https://doi.org/10.1038/nprot.2009.203.

Thiele, I., Vlassis, N., & Fleming, R. M. T. (2014). fastGapFill: Efficient gap filling in metabolic networks. *Bioinformatics, 30*(17), 2529–2531. https://doi.org/10.1093/bioinformatics/btu321.

Tufte, E. R. (1997). *Visual explanations: Images and quantities, evidence and narrative.* Graphics Press.

von Landesberger, T., Kuijper, A., Schreck, T., Kohlhammer, J. J., van Wijk, J., Fekete, J.-D., et al. (2011). Visual analysis of large graphs: State-of-the-art and future research challenges. *Computer Graphics Forum, 30*(6), 1719–1749. https://doi.org/10.1111/j.1467-8659.2011.01898.x.

Wang, C., Deng, Z.-L., Xie, Z.-M., Chu, X.-Y., Chang, J.-W., Kong, D.-X., et al. (2015). Construction of a genome-scale metabolic network of the plant pathogen *Pectobacterium carotovorum* provides new strategies for bactericide discovery. *FEBS Letters*, *589*(3), 285–294. https://doi.org/10.1016/j.febslet.2014.12.010.

Ware, C., Purchase, H., Colpoys, L., & McGill, M. (2002). Cognitive measurements of graph aesthetics. *Information Visualization*, *1*(2), 103–110. https://doi.org/10.1057/palgrave.ivs.9500013.

Wiechert, W. (2001). 13C metabolic flux analysis. *Metabolic Engineering*, *3*(3), 195–206. https://doi.org/10.1006/mben.2001.0187.

Yoghourdjian, V., Archambault, D., Diehl, S., Dwyer, T., Klein, K., Purchase, H. C., et al. (2018). Exploring the limits of complexity: A survey of empirical studies on graph visualisation. *Visual Informatics*, *2*(4), 264–282. https://doi.org/10.1016/j.visinf.2018.12.006.

Yuan, H., Cheung, C. Y. M., Poolman, M. G., Hilbers, P. A. J., & van Riel, N. A. W. (2016). A genome-scale metabolic network reconstruction of tomato (*Solanum lycopersicum* L.) and its application to photorespiratory metabolism. *The Plant Journal*, *85*(2), 289–304. https://doi.org/10.1111/tpj.13075.

Zoledowska, S., Presta, L., Fondi, M., Decorosi, F., Giovannetti, L., Mengoni, A., et al. (2019). Metabolic modeling of Pectobacterium parmentieri SCC3193 provides insights into metabolic pathways of plant pathogenic Bacteria. *Microorganisms*, 7(4), 101. https://doi.org/10.3390/microorganisms7040101.

Zomorrodi, A. R., Islam, M. M., & Maranas, C. D. (2014). D-OptCom: Dynamic multi-level and multi-objective metabolic modeling of microbial communities. *ACS Synthetic Biology*, *3*(4), 247–257. https://doi.org/10.1021/sb4001307.

Zomorrodi, A. R., & Maranas, C. D. (2012). OptCom: A multi-level optimization framework for the metabolic modeling and analysis of microbial communities. *PLoS Computational Biology*, *8*(2), e1002363. https://doi.org/10.1371/journal.pcbi.1002363.

CHAPTER NINE

Metabolite imaging by mass spectrometry: A new discovery tool

Heather J. Walker*
biOMICS Facility, Department of Animal and Plant Sciences, University of Sheffield, Sheffield, United Kingdom
*Corresponding author: e-mail address: h.j.walker@sheffield.ac.uk

Contents

Abstract

Mass spectrometry imaging is a rapidly emerging, powerful technique which not only allows the analysis of molecules such as lipids or peptides but also allows the spatial distribution of those metabolites to be mapped within flat biological tissue sections. It allows both targeted and untargeted imaging of thousands of metabolites with little or no preparation of the sample and negates the need for any complex chemical labeling. This makes it particularly useful for characterization of biological samples such as plant and mammalian tissues. Mass Spectrometry Imaging (MSI) utilizes several different sample introduction techniques however this review will concentrate mainly on matrix-assisted laser desorption/ionization mass spectrometry (MALDI) along with desorption electrospray ionization mass spectrometry (DESI) as these techniques are more widely used and well characterized.

Advances in Botanical Research, Volume 98
ISSN 0065-2296
https://doi.org/10.1016/bs.abr.2020.09.022

1. Introduction

Mass spectrometry imaging (MSI) is a technique used to visualize the spatial distribution of chemical compounds such as biomarkers, metabolites, peptides or proteins. It is fast, highly sensitive and specific with a wide dynamic range and allows the study of metabolites in situ without the need for complex sample preparation.

Plant imaging techniques encompass several different methods. This chapter however will concentrate on well characterized mass spectrometry based methods using both MALDI (matrix-assisted laser desorption/ionization) and DESI (desorption, electrospray ionization). It will aim to give an overview of the techniques and workflows along with some of the advantages and disadvantages each technique possesses in relation to the idiosyncrasies of plant tissues. It will focus on metabolite imaging, as opposed to protein imaging, and showcase some of the wide ranging examples of the analyses possible using MSI.

Early use of MALDI for imaging allowed the spatial distributions of several proteins to be determined (Caprioli, Farmer, & Gile, 1997), and this ground-breaking paper led to further applications mainly within biomedical and clinical settings. In recent years, MSI has significantly advanced, and can now provide high resolution both spatially and technologically using a variety of different ionization sources and approaches. These advances have rapidly increased the use of MSI in relation to the plant kingdom.

Although MALDI has been used as a sample introduction technique for imaging since the early 2000s, DESI, an atmospheric plant imaging technique, is only just starting to be popularized for plant imaging. Using the search term "plant imaging DESI" Google scholar found 378 papers published in 2019 as opposed to 2180 using the term "plant imaging MALDI" and 17,400 using the more general search term "plant imaging mass spectrometry." In fact the number of papers using Mass Spectrometry (MSI) imaging techniques overall has increased year upon year since the early 2000s as shown in Fig. 1. In particular over the past 10 years publications have at least doubled as the potential benefits of metabolite imaging for plants becomes apparent.

Many applications of imaging in medical or clinical settings have been published for a wide range of topics over the past 20 years for instance the analysis of collagen and elastin in formalin fixed tissues (Angel et al., 2018), understanding drug delivery systems using 3D cell culture models

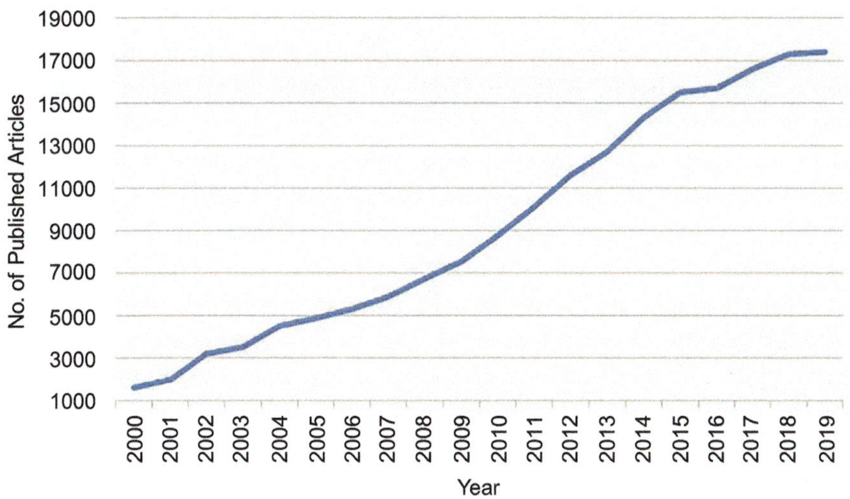

Fig. 1 Increase in papers using search term "Plant Imaging Mass Spectrometry" published from 2000 to 2019. *Data from Google Scholar.*

(Lukowski, Weaver, & Hummon, 2017), tracking Alzheimer's disease pathology biomarkers in the brain (Kakuda et al., 2017) and the molecular characterization of prostate cancer tissues (Randall et al., 2019) to name but a few. Many reviews have also been published on the use of mass spectrometry imaging techniques in clinical or biomedical settings (Buchberger, DeLaney, Johnson, & Li, 2018; Dilillo, Heijs, & McDonnell, 2018; Swales, Hamm, Clench, & Goodwin, 2019; Vaysse, Heeren, Porta, & Balluff, 2017). For the plant related community many reviews are also readily available as researchers become aware of this powerful technique (Boughton, Thinagaran, Sarabia, Bacic, & Roessner, 2016; Heyman & Dubery, 2016; Spraker, Luu, & Sanchez, 2020) and applications for plant MSI include areas such as disease resistance, primary and specialized metabolites using metabolomics, natural products, plant defenses, plant responses to abiotic and biotic stress, and the measurement of lipids. Some examples of these applications will be discussed in more detail later in the chapter in the examples section.

2. Basic concepts of imaging

Traditionally, biochemical mass spectrometry methods of analysis have tended to take whole tissues or organs and then grind them up to extract the various metabolites. Although very effective, this method loses valuable information on the potential differences in biochemistry which may occur, even between adjacent cells, as biological samples are not always homogenous and tissues are not made up of identical cells. Imaging metabolites, using mass spectrometry, within a tissue section negates the need for the destruction of the structure of the tissue and thus the loss of spatial information. It is therefore used to visualize the spatial distribution of chemical compositions within a tissue by using the molecular mass of their ions.

MSI relies upon powerful mass spectrometry equipment that is fast and has high resolution. The technique also has the potential to be a high throughput, screening tool particularly if a low spatial resolution is used. It is also highly sensitive and specific over a wide dynamic range and is often used as a qualitative tool but has also been shown to be quantitative (Sammour et al., 2019; Tang et al., 2019) and when combined with other more targeted techniques such as LC-MS it can become highly quantitative and confirmatory. For these reasons it is highly suitable for the analysis of metabolites in situ.

The main benefit of analyzing in situ is that it negates the need for complex sample preparation such as solvent or solid phase extractions of compounds. Any sample preparation steps could potentially alter metabolism or add contamination to the system so by removing the sample preparation step there is much less chance of this occurring. As with more traditional methods of mass spectrometry the analysis can either be untargeted where the need is to look for general changes or targeted where the need is to search for a specific metabolite.

The MALDI or DESI sample introduction system is used to raster across a sample whereby the source of ionization follows a linear, rectangular pattern of scanning lines across a set area. In this way the whole section area is sampled, with a mass spectrum collected at each point. Spectra are accumulated with many individual spectra collected for each area sampled.

The individual mass spectrums are collated to give the "image." By choosing a particular m/z ion an image of the distribution of that ion can be seen and can also be mapped onto a photographic image of the sample section. Usually a red, green, blue (RGB) scale is used whereby red indicates the highest levels of the particular ion. The following sections will give more detail about each of the workflow steps.

3. Ionization and sample introduction techniques

MALDI has long been used as an ionization technique for mass spectrometry imaging however it does require the sample to be under high vacuum and the laser is destructive to the tissue so the sample cannot be re-analyzed. DESI on the other hand is a technique used at atmospheric pressure and is non-destructive to the tissue which means the sample could be re-analyzed. Both MALDI and DESI are classified as soft ionization techniques which produce a low percentage of fragment ions which is particularly useful for imaging as there is little loss of information and this can aid identification. More technical information about each technique is given in the following sections.

3.1 MALDI

MALDI is a method where a laser is used with the help of a matrix to generate ions in a relatively gentle manner and is commonly used for metabolite profiling for small molecules or proteomics for high molecular weight species. For analysis by MALDI the sample is pre-mixed or coated with a highly absorbing matrix compound. Energy as light from the UV laser is absorbed by the matrix thus transforming the laser energy into excitation energy for the sample. This leads to the sputtering of analyte and matrix ions from the surface of the mixture, and in this gaseous phase the matrix ions transfer their energy to the analyte, which in turn ionizes the analyte due to the high energy absorbed.

This method of energy transfer is efficient and the sample analyte molecules are spared excessive direct energy, which may otherwise cause decomposition. Resulting ions are then subjected to a high electrical field to extract and accelerate them into the mass spectrometer as shown in Fig. 2.

Most commercially available MALDI mass spectrometers are UV absorbing and use lasers such as pulsed nitrogen lasers (337 nm) and Nd: YAG lasers (355 and 266 nm). A range of compounds are suitable for use as matrices and they usually work best for a particular group of analytes. For example, sinapinic acid (SA) is commonly used for protein analysis while α-cyano-4-hydroxycinnamic acid (CHCA) is often used for peptide or small molecule analysis.

For MALDI target spot analysis a small aliquot of a sample: matrix mixed solution is pipetted onto a sample target plate and is then allowed to crystallize before being analyzed, however for imaging the matrix must be

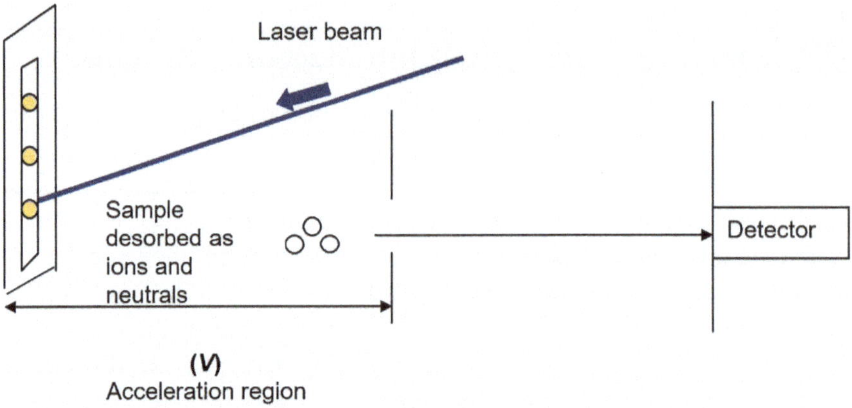

Fig. 2 Schematic of MALDI ionization/sample introduction technique.

applied on top of the sample, and this procedure is discussed in more detail in the section on matrix application later in the chapter. The plate is inserted into the mass spectrometer usually under vacuum (although atmospheric systems are available). The laser is fired at the sample in a standardized raster pattern over the sample, collecting a mass spectrum at many equidistant points across the sample area. The laser is pulsed hence the collection of data is not absolutely continuous. Using this technique samples can be screened very quickly, depending upon the chosen resolution, as there is very little "carry over" from spot to spot.

MALDI is classed as a soft ionization method and results predominantly in the generation of singly charged molecular-related ions regardless of the molecular mass, hence the spectra are relatively easy to interpret. Fragmentation of the sample ions does not usually occur during MALDI ionization, however the matrix can potentially interfere with the measurement of metabolite ions and it is also useful to note that intense peaks for the matrix can be seen in the spectrum.

3.2 DESI

DESI is a combination of electrospray (ESI) and desorption (DI) ionization methods. ESI is an atmospheric pressure ionization (API) technique and is particularly useful for the analysis of polar molecules ranging over a wide range of m/z, from less than 100 Da to more than 1,000,000 Da. Because it is a soft ionization technique, it allows the analysis of many biological molecules. John B. Fenn was awarded a Nobel Prize in Chemistry in the late 90s for its development (Yamashita & Fenn, 1984).

For standard ESI, the sample needs to be dissolved in or extracted into a polar, volatile solvent. The dissolved sample is then pumped at a flow rate between 1 μL/min and 1 mL/min through a narrow, stainless steel capillary. A high voltage of between 2 and 4 kV is applied to the tip of the stainless steel capillary, generating a strong electric field. The sample emerging from the tip is dispersed into an aerosol of highly charged droplets, which forms due to the electric field and the use of high pressure nitrogen gas flowing around the outside of the capillary. This high pressure nitrogen gas also helps to direct the spray emerging from the capillary tip toward the mass spectrometer.

For desorption ESI (DESI), the ionization occurs by directing this electrically charged solvent spray, formed using ESI, onto a sample surface that is only a few millimeters away. The spray is pneumatically directed at the sample and subsequently any splashed droplets that come up from the surface contain desorbed and ionized analytes. These ions are attracted through an inlet and accelerated into the mass spectrometer as seen in Fig. 3.

As the tissue is not destroyed by the ionized solvent spray, DESI-MSI has the ability to sequentially extract molecules from the surface of the tissue. By using a strongly polar solvent such as methanol only the strongly polar molecules such as amino or organic acids will be desorbed from the surface of the tissue. By using a less polar solvent such as chloroform other less polar classes of compounds can be desorbed such as lipids. In this way potentially several

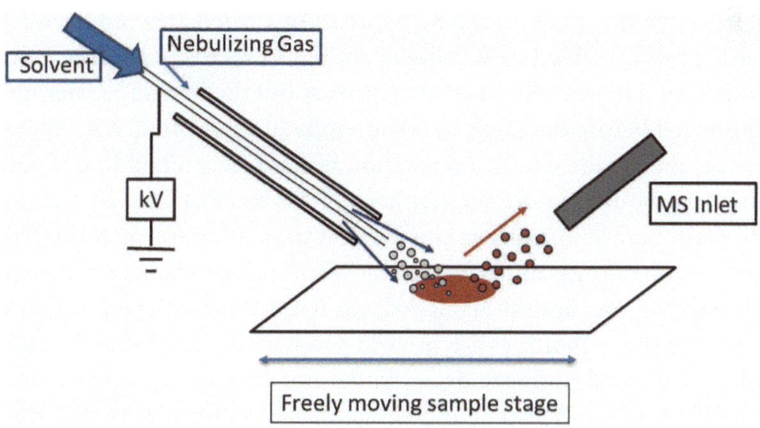

Fig. 3 Schematic of DESI ionization/sample introduction technique.

layers of image can be constructed and certain classes of compounds can be targeted.

DESI is a technique that allows for the ambient ionization of a range of compounds at atmospheric pressure, with little sample preparation. DESI, like MALDI, is classed as a relatively soft ionization technique so little fragmentation occurs allowing for easier interpretation of spectra. However unlike MALDI no matrix is needed for ionization to occur.

3.3 Sample preparation

Sample preparation for imaging techniques is highly critical to the success of the experiment. Although there are many protocols available for sample preparation for MSI, recent studies are still identifying the best methods for preparation of samples which can be highly specific to the species of interest and the type of tissue to be used (Dong et al., 2016; Goodwin, 2012; Nakabayashi, Hashimoto, Toyooka, & Saito, 2019). It is important to continue to optimize sample preparation procedures when undertaking new analyses in order to validate the analysis, as currently there is not an absolute standard method for imaging due to the wide variations in biological material.

The following sections will discuss some of the workflow steps involved in the preparation of samples with one of the key parameters being the need for samples to be flat otherwise it is possible to get variation in transmission from the laser ablation or solvent desorption.

Plant tissue should be prepared immediately after harvesting in order to avoid any degradation. This can be by freezing immediately upon collection or it could be embedded in gelatine or soaked in sucrose or another embedding medium. However it is useful to think about the imaging experiment to be performed before deciding upon an embedding technique as the experiment may dictate the sample preparation needed. By embedding in paraffin, for instance, would require the removal of this paraffin layer by the use of a xylene wash before imaging as the paraffin may interfere with the analysis. One of the other key points when using whole leaves is to freeze them flat at −80 °C so that they are not damaged and stay flat for imaging. Often cryogenic sectioning techniques are used to produce flat reproducible sections. Samples are also often freeze dried which not only stops compounds from being labile but also halts any metabolic processes from occurring thereby providing a reliable snapshot in time of metabolism.

Clinical samples are generally embedded in formalin for preservation which can negatively affect desorption of ions from the sample section, however with plant material this is not usually necessary and samples are frozen immediately into liquid nitrogen. This preserves the sample and stops de-localization of any metabolites. These frozen samples can then be cryogenically sectioned or can be freeze dried or mounted in OCT (optimal cutting temperature compound) and then sectioned. If they are cryogenically sectioned they are then freeze dried directly onto microscope slides (or plates) ready for imaging. For the analysis of proteins, trypsin can be spotted directly onto the sections to digest the proteins to peptides. After the trypsin digest the samples are washed to remove any salts with ethanol before analysis.

Optimizing these sample preparation and storage stages are key to obtaining robust data when imaging.

3.4 Cryosectioning and slide preparation

MALDI-MSI experiments are performed using conductively-coated glass microscope slides or steel or gold-plates. The use of glass slides has the added advantage that it also allows for microscopy and staining of the same sample in order to identify regions of interest and to maximize the information to be gained.

The plant tissue is attached to the slide using a small amount of tissue freezing medium such as OCT (optimal cutting fluid) or carbon tape. Alternatively samples can be attached using a freeze thaw technique which works particularly well for DESI as it reduces the possibility of contamination and higher noise levels from any fixing medium.

Samples are often freeze dried before analysis to suppress any metabolism from occurring or to stop any curling of the sample as it heats up, particularly a problem when using MALDI, however imaging can be done on fresh samples. Freeze drying has the added advantage that it flattens the tissue as it dehydrates and thus can produce more reproducible samples.

The size and thickness of the tissue sample section to be imaged will depend on the sample itself, for instance a leaf will yield much fewer sections than a stem or root, but also the type of tissue itself may dictate the sample size. As an example, potato tuber tissue once freeze dried loses its structure due to its high water content. In this case a thicker sample may need to be sectioned in order to preserve the structure of the sample. Routinely sections are cut to depths of between 10 and 50 μm, however depths outside this range are also possible but are limited by the dimensions of the instrument sample introduction area and the type of tissue. The cross sectional size of the sample to be analyzed is limited by its original size but also by the size of the slide or plate used for mounting the tissue.

At this stage sectioned samples can be stored at −20 °C however before analysis it is important to allow the sections to come up to room temperature slowly, and ideally in a vacuum desiccator so as to minimize any absorption of water by the section. For comparison, when preparing sections for the analysis of proteins, the tissue is fixed by washing sections briefly in ice cold 70% 2-propanol, followed by a wash in 95% 2-propanol, this procedure can also help to remove salts that interfere with the MALDI process (Grassl, Taylor, & Millar, 2011). Comprehensive reviews of sample preparation procedures including discussions around sectioning and preparing slides for MSI have been published (Dong et al., 2016; Goodwin, 2012; Nakabayashi et al., 2019), and most general reviews of MSI also include a section on sample preparation as it is key to obtaining robust datasets.

3.5 Matrix application (MALDI only)

For MALDI ionization the sample needs to be coated with a thin homogenous layer of a matrix compound consisting of crystallized molecules. These compounds are of fairly low molecular weight (to allow easy vaporization), but large enough (with low enough vapor pressure) not to evaporate during sample preparation or while in the mass spectrometer. They are often acidic, therefore act as a proton source to encourage ionization of the analyte, and have strong optical absorption in either the UV or IR range, so that they rapidly and efficiently absorb the laser irradiation. DESI ionization does not need a matrix as ionization is achieved from the ionized solvent spray hence the sample preparation for DESI is much simpler.

The matrix is key to MALDI ionization as it absorbs energy from the laser to transfer to the analytes. During the ablation by the laser the matrix molecules are energetically ablated from the surface of the sample and absorb the laser energy. They then carry the analyte molecules into the gas phase where they are ionized by being protonated or deprotonated by the nearby matrix molecules. The choice of matrix can depend on the particular types of compounds to be analyzed. The key to applying the matrix is to coat the sample in an even and homogenous layer. So the best way to apply the matrix is by using a nebulization system whereby the matrix is sprayed onto the sample in a thin layer allowing it to crystallize evenly without forming large crystals. Often it is better to coat the sample twice with matrix to make sure that the section is evenly covered making sure that each coat is perfectly dry before applying another layer. Routinely for small metabolites (<1500 Da) CHCA is used (α-cyano-4-hydroxycinnamic acid) as this matrix is well characterized and easily forms a homogenous layer over the section. Other matrices can be used, of which DHB (2,5-dihydroxybenzoic acid) is a common matrix however this particular compound easily forms crystals which can make it harder to get an even, homogenous layer over the section. The choice of matrix also depends on the compounds of interest and the polarity of ionization needed as some matrices are better for ionizing certain classes of compounds.

Although the application of the matrix is a key step it can potentially add extra noise and contamination to the analysis however as the m/z ions for the matrix are known they can be used as internal standards to correct for any mass drift over long analysis times but also for lining up samples with their images after analysis and also for normalization purposes.

3.6 Mass spectrometry instrumentation

Often for imaging experiments time of flight (TOF) instrumentation is used as this gives a wide dynamic range over a large mass range thus allowing a large number of masses to be scanned at both high and low concentrations. TOF is also a rapid scanning method allowing a large number of masses to be scanned within a small time window maximizing the information which can be collected within a short space of time. The limiting factor to instrumentation is usually the resolution not of the mass analyzer itself but of the sample introduction method so the resolution of the laser or DESI system is key. Spatial resolution refers to the size of the "pixel" used to determine 2D distribution across the section. Resolution is reliant but not wholly dependent upon spot size of laser (MALDI) or desorbing liquid (DESI) with the lowest spatial resolution achieved by commercial instruments being around the 30–100 μm range although this size range is continuously getting smaller. Currently MALDI provides higher spatial resolution than DESI as the pixel size for DESI is heavily reliant upon the solvent choice. Very viscous solvents, with a high percentage of organic as opposed to aqueous solvent, can give a very disperse spray thus increasing the pixel size and reducing resolution (Spraker et al., 2020).

Resolution effectively means the area which is analyzed in one "shot." However the smaller the resolution then potentially the less sensitive the technique becomes because the smaller each pixel then the fewer metabolites will be present at each pixel. This means that potentially some information could be lost as some metabolites may be under the limit of detection. For this reason resolution must be carefully chosen. It is a key criteria of MSI to eventually get to single cell resolution however if this means that the levels of metabolites are so low that lots of information is missed then there is no gain to reducing the resolution. Often a compromise must be found, however as technology advances then improvements in both sensitivity and resolution will occur and this will become less of an issue. For MALDI one of the ways to get around this problem is by using a technique called oversampling whereby the step window, the distance the stage moves to the next sampling position, is reduced to a size which is less than the resolution of the laser. In this way oversampling can be used to increase coverage. This is not necessary when using DESI as this is a continuous technique, as the ionized spray is pumped continuously, as opposed to MALDI which is a pulsed technique. However optimization of the diameter of the MALDI spot or DESI ionized liquid spray spot can be useful to increase the resolution. To correctly optimize these parameters it is useful to look at a signal to noise ratio as a poor ratio indicates that these parameters are not set correctly.

Isobaric species are common in plant material, where compounds have the same mass but a different structure, in particular for carbohydrates. This is a challenge to MSI as unlike chromatography procedures there is no separation of species before they enter the mass spectrometer. This is where potentially ion mobility becomes highly sought after with regards to imaging as it separates species, within the mass spectrometer, according to their cross sectional area. Ion mobility is a rapidly developing area in mass spectrometry. It is based on the principle that ions migrate through an electric field at characteristic velocities according to their relative collisional cross sections (CCS). The CCS of a molecule is dependent upon its size and shape but also upon its mass and charge state.

This method is particularly useful for the separation of isobaric species for example lipids or carbohydrates and is a way of potentially minimizing overlap of species within an imaging run (Fu et al., 2020).

4. Advantages and limitations

Both MALDI and DESI have advantages and limitations and the nature of the analysis required may dictate which technique to use.

Table 1 Optimization parameters for MALDI/DESI.

MALDI optimization parameters	DESI optimization parameters
Matrix compound	Solvent choice
Matrix concentration	Composition of solvent, i.e. ratio of solvents
Laser power	Flow rate
Voltage on sample plate	Angle of inlet tube/solvent spray
Voltage on MS inlet	Voltage on capillary tube

Table 1 lists the common optimization parameters to take into consideration for both sample introduction techniques.

MALDI has a main advantage that it is a much more characterized technique than DESI. As discussed in the introduction DESI is only just emerging as a technique for mass spectrometry imaging therefore is much less characterized and well documented. Also the use of the matrix compound with MALDI allows the matrix peaks to be used as markers to easily line up sections and images and also potentially to normalize and standardize data. The main advantage DESI has over MALDI is that the sample is not destroyed by the technique and can therefore be analyzed multiple times, even allowing for desorption using different solvents on the same sample as discussed in the DESI section. Another advantage of DESI is that less sample preparation is needed than with MALDI as it is not necessary to use a matrix compound. As DESI is at atmospheric technique as opposed to under vacuum it also allows for easy sample changing.

Generally when using MALDI as the sample introduction technique ions are analyzed using the positive mode hence the use of MALDI for lipid and protein analysis. In plants many secondary metabolites are small polar molecules some of which ionize better in the negative mode. Although MALDI can be used to create negatively charged molecules often the matrix compounds used for this are not as well characterized as those used for positive mode. This can mean that sometimes the whole picture of metabolism is not seen. This can give DESI an advantage as the system can easily be set up for the generation of positive or negative ions.

One of the main disadvantages for any mass spectrometry imaging technique is where the sample is not homogenous as this can lead to ion suppression and ultimately to non-quantitative results over the section. This can be overcome by using normalization techniques to make sure there is robustness in the data and this will be discussed further in Section 5.

5. Data handling and analysis

MSI analyses produce very large data sets and this data needs to be robust and reproducible. Replication is an obvious way of validating data sets as is using a region of interest (ROI) approach. For a ROI analysis, several small homogenous areas across the tissue are selected and a summed mass spectrum for each area is produced. These spectra can then be manipulated for use in a standard multivariate method such as a principal component analysis (PCA) and used to validate certain types or different areas of tissue. This ROI approach allows for several areas within one tissue to be selected. This is very useful for a metabolomics analysis where differences between tissue types or areas within a tissue are required but it can also give information about the homogeneity of the tissue.

Specialist software is available directly via manufacturers of the MSI equipment or freely available on the web using programs such as BioMap (Novartis) or msIQuant (Källback, Nilsson, Shariatgorji, & Andrén, 2016) and these types of software easily modify the raw data spectrums to images which can be overlaid onto photographs of the sections.

Normalization of datasets is very important for MSI and can be thought of as a mathematical function which returns data to its normal state, removing all systematical errors and noise in order to show only the differences caused by biological variations. It needs to adjust individual mass spectra so that they are comparable across an image and effectively correct for what could be seen as experimental bias. Normalization includes both intra-sample normalization, where systematic differences in the total amount of desorption and ionization efficiency of molecules within a sample is observed as well as inter-sample normalization where variations in instrument parameter settings, environmental conditions and sample preparation can lead to non-biologically relevant variability in overall peak intensity within sample datasets. The main factors responsible for intra-sample differences include the inhomogeneity of the sample itself as well as the matrix application (for MALDI only), the variability of the thickness of the tissue, and variation in ionization and detector efficiencies. But could also be due to ion suppression due to the different compositions of samples, for instance high starch content in potato can effect ionization efficiency, and potential differences in analyte extraction from different regions. Intra-sample normalization is often overcome by using a total ion count correction, which is a simple correction however may be problematic in cases where one large peak dominates the spectrum. Inter-sample variations are often overcome by using a sample specific scaling factor. Scaling factors can be based upon on a "homogenized" tissue molecular ion profile which is usually calculated as an average profile across the whole tissue. Several papers have discussed in detail normalization techniques and data handling for both MALDI and DESI (Fonville et al., 2012; Veselkov et al., 2018) and these are very useful reference points.

The images themselves once produced using the RGB scale, as discussed earlier in the basic concepts of imaging section, are highly visual tools and provide almost instant recognition of the variation of particular metabolites across a sample section. The validation of the MSI technique itself along with the data handling and analysis should not be underestimated for confidence and robustness within the dataset.

6. Applications

Early papers on plant imaging mainly focused on proteomic analyses, measuring large molecules due to the ease of measuring these with MALDI-MSI, however within plant material often the small polar metabolites and in particular the secondary metabolites are key to understanding changes in metabolism. Both MALDI-MSI and DESI-MSI are capable of detecting these types of metabolites.

Metabolites are important because they are the final step in the cascade and therefore closer to the phenotype than proteins; effectively they are the limiting currency of the cell and show us exactly what is happening in the cell at a particular point in time. Due to this, small metabolites and in particular secondary metabolites (<1500 Da) are key to understanding metabolism as they give a snapshot of what is happening at an exact moment in time and recently this has been an expanding area of research.

Much work around imaging plants has been published including work on preparation of samples (Dong et al., 2016; Goodwin, 2012; Nakabayashi et al., 2019) but also on the interpretation of the results (Fonville et al., 2012; Liu, Xiong, & Ouyang, 2015; Zhang & Liu, 2018) which is key to getting robust results. Articles vary in content from technological advances in MSI (Clench, 2016; Couvillion et al., 2019) to understanding the biological mechanisms behind the findings (Caprioli, 2016) and many reviews are available for plant imaging by mass spectrometry (Bjarnholt, Li, D'Alvise, & Janfelt, 2014; Parrot, Papazian, Foil, & Tasdemir, 2018; Sturtevant, Lee, & Chapman, 2016). The following examples section will expand on the use of MSI in some key areas of plant research and describe several published case studies to illustrate the wide range of applications of imaging mass spectrometry.

7. Examples

The following section will outline some examples of imaging of plant material by mass spectrometry; it is not intended to be a complete review but an illustration of different examples showing the breadth of research undertaken using mass spectrometry imaging techniques.

An example from my own research (Walker et al., 2016) is related to understanding metabolism of arginine in biological systems. *Arginine is a* major storage and transport amino acid and is an essential source of organic nitrogen for plants, but it also has roles as an essential amino acid for protein synthesis, a precursor for polyamines and nitric oxide (NO) production as well as involvement in many other cellular processes. In wheat grains amino acids are used for protein synthesis but also as the major source of nitrogen for the developing grain. The phloem sap in wheat leaves contains all of the amino acids required for protein synthesis. The type of protein and its content in the wheat grain is an important quality trait and dictates the end use of the crop, for instance whether it is used for baking and bread making is determined by this quality. As storage protein synthesis in wheat grain occurs between around 8 and 20 days post anthesis (dpa) the arginine ion distribution was mapped during this phase of development. The work clearly showed that there was transport of arginine through the wheat seed over time as can be seen in Fig. 4. The wheat seed before imaging is included in Fig. 5 as a comparison.

Although arginine was targeted other related metabolites within the same pathway; citrulline, ornithine and aspartate were imaged at the same time and these images are shown in Fig. 6. As the images for the different metabolites showed very similar patterns it can be concluded that developing wheat grain do have the capacity to synthesize arginine however, as arginine does need to be transported between compartments in the grain it may also be imported from other parts of the plant.

Ecology applications for MSI are on the increase and a recent study (Gadea et al., 2020) imaged metabolites using MALDI-MSI within lichens and related these to snail-lichen interactions. Lichens are slow-growing organisms which are assumed to synthesize specialized metabolites as defense protection against lichenivores, such as snails which graze on them.

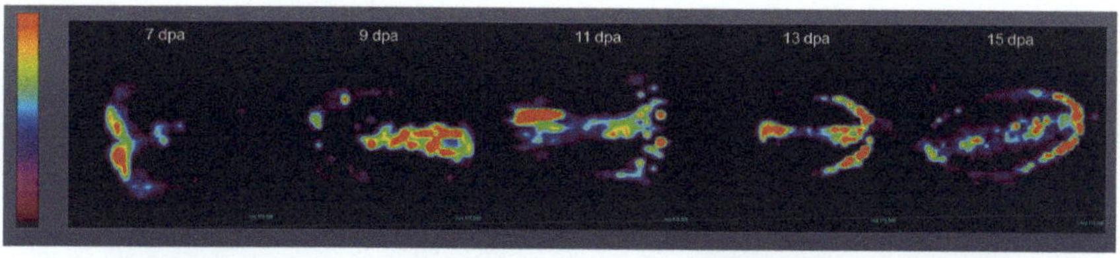

Fig. 4 Distribution of arginine during development. Sections were taken from the middle of the grain between 7 and 15 dpa, coated with α-CHCA and the ions of interest mapped with Biomap after normalization against the 172 *m/z* matrix mass. The images have all been scaled with using the RGB scale with red being the highest intensity peak. The images shown are for mid grain sections and are representative from three different grains at the same stage of development. The 4 dpa section is one-third of that of the 15 dpa section in size. The 15 dpa section is the same size as the representative section reproduced in Fig. 5. *Reproduced with permission from Walker, H. J., Steels, C., Bendell, L., Clench, M. R., Read, D. J., Cameron, D. D., & Burrell, M.M. (2016). Understanding metabolism of arginine in biological systems via MALDI imaging.* Proteomics, 16, *1690–1694. Copyright Wiley-VCH Verlag GmbH & Co. KGaA.*

Generally *lichenophage* species are found to prefer lichens with only a few specialized metabolites and as lichen, do not induce chemical defenses any specialized metabolites are likely to already be integral parts of the species. The research highlighted various examples of the accumulation of specialized metabolites, and in particular of phenolic compounds such as parietin, physodic acid, atranorin, pinastric acid, m-scrobiculin, and cytochalasin, in parts of lichens with high fitness value, such as the reproductive parts. However, the study also found that the accumulation of metabolites in reproductive parts is not the rule for all specialized metabolites. A dibenzofurane derivative, usnic acid, was found to be equally distributed within the external layers of the entire lichen, *Usnea taylorii*, and was found in both the branches and the apothecia. Accumulation of some specialized metabolites was thought to not only protect the lichen apothecia from lichenivores grazing but also from other external stresses such as UV radiation or oxidation. This study showed that the location of the metabolites is as important as identifying what the metabolites actually are and MALDI-MSI allowed this study to gain this spatial information about metabolites and link it to plant defense and ecology.

Another study used DESI-MS to focus on plant defenses in relation to pathogens and to map secondary metabolites produced in response to a plant-pathogen interaction. The complexity of interactions between plant and pathogen can be highly complex and in turn a distinct challenge for

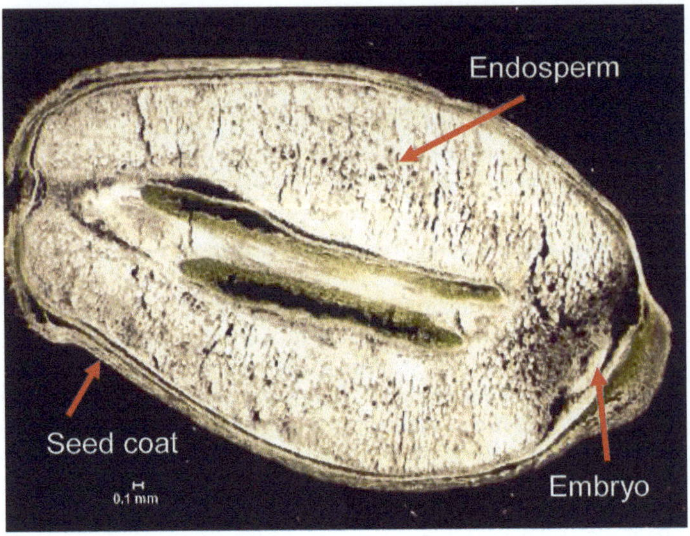

Fig. 5 For comparison the wheat seed before MALDI imaging. The 15 dpa section in Fig. 4 is the same size as the section shown here in Fig. 5. *Reproduced with permission from Walker, H. J., Steels, C., Bendell, L., Clench, M. R., Read, D. J., Cameron, D. D., & Burrell, M.M. (2016). Understanding metabolism of arginine in biological systems via MALDI imaging.* Proteomics, 16, *1690–1694. Copyright Wiley-VCH Verlag GmbH & Co. KGaA.*

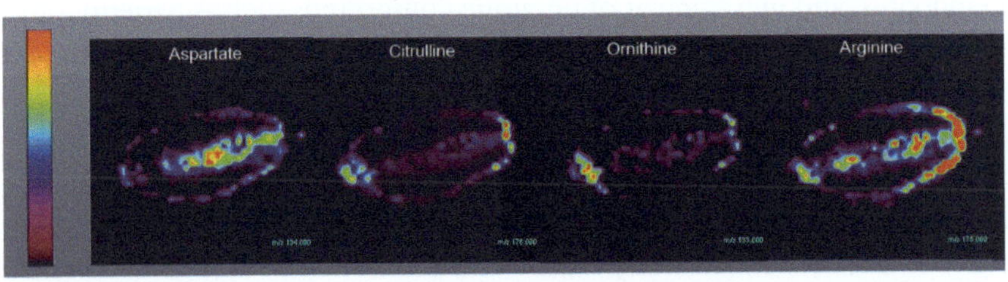

Fig. 6 Distribution of components of the arginine-ornithine cycle. Sections were taken from the middle of the grain at 15 dpa, coated with α-CHCA and the ions of interest mapped with Biomap. The images have all been scaled using the RGB scale with red being the highest intensity peak. The sections used are the same size as the section shown in Fig. 5. The identification of masses is, apart from arginine, based on accurate mass. *Reproduced with permission from Walker, H. J., Steels, C., Bendell, L., Clench, M. R., Read, D. J., Cameron, D. D., & Burrell, M.M. (2016). Understanding metabolism of arginine in biological systems via MALDI imaging.* Proteomics, 16, *1690–1694. Copyright Wiley-VCH Verlag GmbH & Co. KGaA.*

any analytical technique. DESI-MS offers the chance to get a real spatial snapshot in time to monitor the production of these secondary defense metabolites and to understand the mechanisms underlying the production of them. The study (Tata, Perez, Hamid, Bayfield, & Ifa, 2015) used a novel sampling technique which involved blotting or imprinting cut potato onto a flat surface. In this way the metabolites are transferred onto a flat surface such as tape, paper or even a microscope slide thus negating the need for cryosectioning. The slices of potato were inoculated with *Pythium ultimum*, a plant pathogen which causes damping-off and root rot disease in many plants including corn, soybean, potato, and wheat, and compared infected samples to untreated control samples. The images produced in the control samples were dominated by peaks at *m/z* 852 and 868 which corresponded to the glycoalkaloids: α-chaconine and α-solanine. These glycoalkaloids are the most abundant potato toxins and more than 95% of the total steroidal glycoalkaloid content of potatoes is comprised of these two. Both toxins are known to act as a defensive barrier for protection against pathogens. However after 8 days the metabolite profiles for the samples infected by the pathogen were found to be very different. Both α-chaconine and α-solanine, which were found to dominate the spectra of the control samples, were found to be much lower in the infected samples. The profiles of the infected samples showed elevated levels, as compared to the control samples, of the protonated forms of the glycoalkaloids: solasodiene (*m/z* 396), aglycon solanidine (*m/z* 398), solasodenone (*m/z* 412), solasodine (*m/z* 414), solanaviol (*m/z* 430), solaspiralide (*m/z* 428), γ-solanine/γ-chaconine (*m/z* 560), β-solanine (*m/z* 706), and β-chaconine (*m/z* 722), whose identities were confirmed by using MS/MS. The most intense peak of the infected potato profile was the ion which corresponded to the aglycon solanidine. Aglycons are the compounds remaining after the glycosyl group on a glycoside is replaced by a hydrogen atom. This matched expectations as fungal infections of potato sprouts are known to activate pathways which lead to the biosynthesis of alkaloids and in particular increase the levels of the aglycons: solanidine and solasodine leading to the de novo synthesis of several steroidal alkaloids that have either solasodine or solanidine as their common aglycon. By using this simple blotting sample preparation technique along with DESI-MSI the complex sample was profiled and imaged rapidly, using minimal amount of sample without the need for laborious extraction procedures. The findings matched expected conclusions however by having the added spatial information the synthesis of the alkaloids related to the infection could be easily traced. This type of study is vital in the pursuit of the kinetics of plant-pathogen systems as it gives a real visual aid and from this type of initial study quantification steps can be added to confirm mechanisms and rates.

Functional mechanisms within plants have also been studied using used MALDI-MSI (Sugahara, Kitao, Watanabe, & Yamagaki, 2019). This study focused on detecting flavonoids to understand the blue color expression mechanism of viola petals. The color expression of anthocyanin pigments in blue flowers is accurately controlled by their chemical and physical properties, including their pH and the presence of particular metal ions or colorless co-pigments. However despite the large number of known blue flowers, detail about the exact mechanism of their coloration remains limited.

By using a combination of MALDI-MSI along with LC-MS, for added identification, confirmation and quantification purposes the mechanism underpinning the coloration process has begun to be unpicked. The study used MALDI-MSI to image the petal of the viola and identified the structures and types of flavonols which co-localized with violanin, a key anthocyanin, in the discrete blue-colored region of the petal. This allowed the different areas of the petal to be further targeted for a quantitative analysis using LC-MS. The initial information from the imaging experiment led the authors onto the next step. Without the location information from the imaging step, the mechanism would not have been truly understood. The data collected from the MALDI-MSI and LC-MS experiments led onto the in vitro reconstruction of the color expression that was consistent with the visible spectrum of the viola petal. The results of this visible spectral analysis showed that the color development process was not just due to the increase in the solution pH inside the vacuole cells or the presence of metal ions. The authors found that the blue coloration of viola petals was actually expressed via the co-pigmentation of flavonol 3-*O*-glycosides and violanin. The study again demonstrated the power of MALDI-MSI and how the spatial information obtained can direct research and when combined with other techniques can become confirmatory and thus very powerful.

Natural product discovery is reliant upon the structural analysis of biomolecules especially for precious biomaterials. However the big challenge to the discovery and characterization of natural products is the huge cost in both time and manpower to the analysis of these highly diverse compounds. For this reason natural product research has begun to utilize both DESI-MSI and MALDI-MSI as screening tools for these highly sought after active biomolecules. Often natural products are polar molecules so they lend themselves in particular to analysis using DESI.

The plant *Rauvolfia tetraphylla* L. (family Apocynaceae), often referred to as the wild snakeroot plant, is native to Mexico, Central America, West Indies, and northern South America. It is an important wild plant for local medicinal use and also as the source of a dyestuff and ink. It has been introduced into India, China and Vietnam where it is experimentally cultivated for medicinal use. It is already used as a commercial source of the alkaloid reserpine for use in modern medicines and in particular for treatment of hypertension and diabetes, however it also produces several other indole alkaloids in its seeds and roots. To understand more the spatial localization of the various other indole alkaloids in developing fruits and plants of *R. tetraphylla* was studied using DESI-MSI (Kumara, Shaanker, & Pradeep, 2019). The imaging analysis allowed the distribution of several indole alkaloids to be mapped and it was found that ajmalicine, yohimbine, demethyl serpentine and mitoridine were largely localized in the fruit coat while ajmaline was restricted to the mesocarp of the fruit. The study also utilized a LC-MS based metabolomics approach, as a confirmatory experiment, to discriminate the different tissues, stems, leaves, roots and fruits, using the signatures of the indole alkaloids. Although the study did not discuss the mechanisms of the distribution of the indole alkaloids in depth it hypothesized that as ajmaline is a potent inhibitor of Na^+/K^+ channel it is therefore probable that the accumulation pattern in the mesocarp could help defend the fruit from insect herbivores. The study also proved useful as it identified concentrations of the alkaloids in the different tissues and DESI-MSI was vital to obtaining this information. Going forward these other alkaloids may become other important medicinal compounds.

MALDI-MSI can also be used for natural product studies and one study analyzed the distribution of natural products with Agarwood (Kuo, Huang, & Hsu, 2019). Agarwood is a fragrant, dark, resinous heartwood acquired from *Aquilaria* trees and is specifically used in Chinese medicines, perfume, and incense specifically in Asian religions. The formation of agarwood within the tree relies upon physical damage to the tree. This damage appears to stimulate biosynthesis of agarwood resin as the tree uses it to defend itself from biotic and abiotic stresses. As the agarwood is only produced when the tree is damaged this can lead to a deficit in the amount needed to meet demand. More than 150 agarwood natural products have been identified and these include chromones, sesquiterpenoids, and volatile aromatics and many of these compounds have been shown to possess promising bioactivities. However agarwood natural products are very diverse and are dependent upon species and also cultivating regions. The MALDI-MSI analysis identified a considerable number of agarwood specific ions and by using the metabolite ion patterns observed, the images could be categorized into three groups. The three groups of ions were categorized into those which were ubiquitous within the tissue section, those which were specific to the non-agarwood region and those which were specific to the agarwood region. In the agarwood region at least 36 different and distinct ions were found which would potentially represent different natural products. Again in this case an LC-MS/MS experiment was used as a confirmatory study alongside the imaging work and 14 agarwood natural products were putatively identified as being related to 2-(2-phenylethyl) chromone (PEC) including PEC monomers, bicyclic dimers, and *O*-linked dimers. Also 20 novel analogs of these PEC natural products were tentatively characterized using MS/MS spectral datasets. This study began to characterize agarwood compounds in *Aquilaria* trees using MALDI-MSI as a first step because it allowed the spatial identification of compounds specific to the agarwood to increase knowledge of wood biochemistry and natural products as well as adding to the Global Natural Product Social Molecular Networking database (GNPS).

As the previous examples have shown mass spectrometry based imaging has been used for many different types of analysis related to plants however more recently a new direction has begun to be taken using an in situ MALDI-MSI method to study the root-soil interface (Velickovic, Lin, Rivas, Anderton, & Moran, 2020). This research showed how imaging can still be used even with indirect samples. By using a specially adapted membrane for the rhizosphere it was possible to keep the structure of the root-soil interface and therefore keep the spatial information. This approach allowed the visualization and identification of compounds both on the root surface and those assumed to be root exudates within the whole rhizosphere. In fact, in only a 1.8 cm × 0.6 cm sampling area of a grass rhizosphere, at least four chemically distinct zones could be defined allowing a deeper understanding of rhizosphere processes on a sub-millimeter scale. Normally to do this type of work compounds would be extracted from the rhizosphere thus losing any spatial and location information.

This is a step into a new direction for MALDI-MSI as by imaging the whole rhizosphere it starts to open up new possibilities for the technique.

8. The future

This chapter was intended to give an overview of the MSI technique along with a flavor of the wide range of applications that it is used for. The technology is still evolving and an ever increasing spatial resolution and sensitivity could lead ultimately to routine single cell imaging by mass spectrometry. At the moment limitations to single cell imaging are due to the resolution of the systems available and the instrumentation sensitivity as there needs to be a high enough concentration of metabolites to record a response by the mass spectrometer detector. To get to the stage of single cell imaging improvements in sample preparation may also be necessary including the use of automated or robotic systems to guarantee consistency and reproducibility. On the data analysis side the files generated by MSI are invariably very large and these need to be handled routinely along with improvements in molecular identifications. These steps are necessary to move this technique onto the next level. With the absence of chromatography steps the identifications need to be reliable and reproducible. Currently LC-MS is used alongside imaging as a confirmatory analysis.

However on the positive side applications are reaching nearly every discipline where the position of molecules within a sample is key to understanding the underlying biology, and the plant related worlds of botany and agriculture are no different. The current greater emphasis upon food and in particular sustainability, security and yield create huge challenges across the globe. If MSI can start to address some of the questions related to these topics with its key advantages being sensitivity, spatial information and the ability to analyze without any major treatment of the sample almost in situ it will become a highly valuable tool. We are still a little way away from this ideal however with the rapid increases in technology and the ever increasing need for more detailed information we cannot be far away.

In final summary MSI brings the unique advantage of molecular specificity ensuring it a prominent role in both basic science and research and diagnostics.

References

Angel, P. M., Comte-Walters, S., Ball, L. E., Talbot, K., Mehta, A., Brockbank, K. G. M., et al. (2018). Mapping extracellular matrix proteins in formalin-fixed, paraffin-embedded tissues by MALDI imaging mass spectrometry. *Journal of Proteome Research, 17*(1), 635–646. https://doi:10.1021/acs.jproteome.7b00713.

Bjarnholt, N., Li, B., D'Alvise, J., & Janfelt, C. (2014). Mass spectrometry imaging of plant metabolites—Principles and possibilities. *Natural Product Reports*, *31*, 818–837. https://doi.org/10.1039/C3NP70100J.

Boughton, B. A., Thinagaran, D., Sarabia, D., Bacic, A., & Roessner, U. (2016). Mass spectrometry imaging for plant biology: A review. *Phytochemistry Reviews*, *15*, 445–488. https://doi:10.1007/s11101-015-9440-2.

Buchberger, A. R., DeLaney, K., Johnson, J., & Li, L. (2018). Mass spectrometry imaging: A review of emerging advancements and future insights. *Analytical Chemistry*, *90*(1), 240–265. https://doi:10.1021/acs.analchem.7b04733.

Caprioli, R. M. (2016). Imaging mass spectrometry: Molecular microscopy for the new age of biology and medicine. *Proteomics*, *16*, 1607–1612. https://doi:10.1002/pmic.201600133.

Caprioli, R. M., Farmer, T. B., & Gile, J. (1997). Molecular imaging of biological samples: Localization of peptides and proteins using MALDI-TOF MS. *Analytical Chemistry*, *69*, 4751–4760. https://doi:10.1021/ac970888i.

Clench, M. R. (2016). Advances in mass spectrometry imaging. *Proteomics*, *16*, 1605–1606. https://doi:10.1002/pmic.201600202.

Couvillion, S. P., Zhu, Y., Nagy, G., Adkins, J. N., Ansong, C., Renslow, R. S., et al. (2019). New mass spectrometry technologies contributing towards comprehensive and high throughput omics analyses of single cells. *Analyst*, *144*, 794–807.

Dilillo, M., Heijs, B., & McDonnell, L. A. (2018). Mass spectrometry imaging: How will it affect clinical research in the future? *Expert Review of Proteomics*, *15*(9), 709–716. https://doi:10.1080/14789450.2018.1521278.

Dong, Y., Li, B., Malitsky, S., Rogachev, I., Aharoni, A., Kaftan, F., et al. (2016). Sample preparation for mass spectrometry imaging of plant tissues: A review. *Frontiers in Plant Science*, 7, 60. https://doi.org/10.3389/fpls.2016.00060.

Fonville, J. M., Carter, C., Cloarec, O., Nicholson, J. K., Lindon, J. C., Bunch, J., et al. (2012). Robust data processing and normalization strategy for MALDI mass spectrometric imaging. *Analytical Chemistry*, *84*(3), 1310–1319. https://doi:10.1021/ac201767g.

Fu, T., Oetjen, J., Chapelle, M., Verdu, A., Szesny, M., Chaumot, A., et al. (2020). In situ isobaric lipid mapping by MALDI–ion mobility separation–mass spectrometry imaging. *Journal of Mass Spectrometry*, *55*(9), e4531. https://doi.org/10.1002/jms.4531.

Gadea, A., Fanuel, M., Le Lamer, A.-C., Boustie, J., Rogniaux, H., Charrier, M., et al. (2020). Mass spectrometry imaging of specialized metabolites for predicting lichen fitness and snail foraging. *Plants*, *9*, 70.

Goodwin, R. J. A. (2012). Sample preparation for mass spectrometry imaging: Small mistakes can lead to big consequences. *Journal of Proteomics*, *75*(16), 4893–4911. https://doi:10.1016/j.jprot.2012.04.012.

Grassl, J., Taylor, N. L., & Millar, A. (2011). Matrix-assisted laser desorption/ionisation mass spectrometry imaging and its development for plant protein imaging. *Plant Methods*, 7, 21. https://doi.org/10.1186/1746-4811-7-21.

Heyman, H., & Dubery, I. (2016). The potential of mass spectrometry imaging in plant metabolomics: A review. *Phytochemistry Reviews*, *15*, 297–316. https://doi:10.1007/s11101-015-9416-2.

Kakuda, N., Miyasaka, T., Iwasaki, N., Nirasawa, T., Wada-Kakuda, S., Takahashi-Fujigasaki, J., et al. (2017). Distinct deposition of amyloid-β species in brains with Alzheimer's disease pathology visualized with MALDI imaging mass spectrometry. *Acta Neuropathologica Communications*, *5*, 73. https://doi.org/10.1186/s40478-017-0477-x.

Källback, P., Nilsson, A., Shariatgorji, M., & Andrén, P. E. (2016). Quantitation software for mass spectrometry imaging enabling fast access, visualization, and analysis of large data sets. *Analytical Chemistry*, *88*(8), 4346–4353. https://doi:10.1021/acs.analchem.5b04603.

Kumara, P. M., Shaanker, R. U., & Pradeep, T. (2019). UPLC and ESI-MS analysis of metabolites of Rauvolfia tetraphylla L. and their spatial localization using desorption electrospray ionization (DESI) mass spectrometric imaging. *Phytochemistry*, *159*, 20–29. https://doi:10.1016/j.phytochem.2018.11.009.

Kuo, T.-H., Huang, H.-C., & Hsu, C.-C. (2019). Mass spectrometry imaging guided molecular networking to expedite discovery and structural analysis of agarwood natural products. *Analytica Chimica Acta*, *1080*, 95–103.

Liu, J., Xiong, X., & Ouyang, Z. (2015). Data processing and analysis for mass spectrometry imaging. *Methods in Molecular Biology*, *1203*, 195–209. https://doi:10.1007/978-1-4939-1357-2_19.

Lukowski, J. K., Weaver, E. M., & Hummon, A. B. (2017). Analyzing liposomal drug delivery systems in three-dimensional cell culture models using MALDI imaging mass spectrometry. *Analytical Chemistry*, *89*(16), 8453–8458. https://doi.org/10.1021/acs.analchem.7b02006.

Nakabayashi, R., Hashimoto, K., Toyooka, K., & Saito, K. (2019). Keeping the shape of plant tissue for visualizing metabolite features in segmentation and correlation analysis of imaging mass spectrometry in *Asparagus officinalis*. *Metabolomics*, *15*, 24. https://doi.org/10.1007/s11306-019-1486-5.

Parrot, D., Papazian, S., Foil, D., & Tasdemir, D. (2018). Imaging the unimaginable: Desorption electrospray ionization—Imaging mass spectrometry (DESI-IMS) in natural product research. *Planta Medica*, *84*(9–10), 584–593. https://doi:10.1055/s-0044-100188.

Randall, E. C., Zadra, G., Chetta, P., Lopez, B. G. C., Syamala, S., Basu, S. S., et al. (2019). Molecular characterization of prostate cancer with associated Gleason score using mass spectrometry imaging. *Molecular Cancer Research*, *17*(5), 1155–1165. https://doi:10.1158/1541-7786.MCR-18-1057.

Sammour, D. A., Marsching, C., Geisel, A., Erich, K., Schulz, S., Guevara, C. R., et al. (2019). Quantitative mass spectrometry imaging reveals mutation status-independent lack of Imatinib in liver metastases of gastrointestinal stromal tumors. *Scientific Reports*, *9*, 10698. https://doi.org/10.1038/s41598-019-47089-5.

Spraker, J. E., Luu, G. T., & Sanchez, L. M. (2020). Imaging mass spectrometry for natural products discovery: A review of ionization methods. *Natural Product Reports*, *37*, 150–162.

Sturtevant, D., Lee, Y.-J., & Chapman, K. (2016). Matrix assisted laser desorption/ionization-mass spectrometry imaging (MALDI-MSI) for direct visualization of plant metabolites in situ. *Current Opinion in Biotechnology*, *37*, 53–60. https://doi.org/10.1016/j.copbio.2015.10.004.

Sugahara, K., Kitao, K., Watanabe, T., & Yamagaki, T. (2019). Imaging mass spectrometry analysis of flavonoids in blue viola petals and their enclosure effects on violanin during color expression. *Analytical Chemistry*, *91*(1), 896–902. https://doi.org/10.1021/acs.analchem.8b03815.

Swales, J. G., Hamm, G., Clench, M. R., & Goodwin, R. J. A. (2019). Mass spectrometry imaging and its application in pharmaceutical research and development: A concise review. *International Journal of Mass Spectrometry*, *437*, 99–112. https://doi.org/10.1016/j.ijms.2018.02.007.

Tang, W., Chen, J., Zhou, J., Ge, J., Zhang, Y., Li, P., et al. (2019). Quantitative MALDI imaging of spatial distributions and dynamic changes of tetrandrine in multiple organs of rats. *Theranostics*, *9*(4), 932–944. https://doi.org/10.7150/thno.30408.

Tata, A., Perez, C. J., Hamid, T. S., Bayfield, M. A., & Ifa, D. R. (2015). Analysis of metabolic changes in plant pathosystems by imprint imaging DESI-MS. *Journal of the American Society for Mass Spectrometry*, *26*, 641–648. https://doi.org/10.1007/s13361-014-1039-0.

Vaysse, P.-M., Heeren, R. M. A., Porta, T., & Balluff, B. (2017). Mass spectrometry imaging for clinical research—Latest developments, applications, and current limitations. *Analyst, 142*, 2690–2712.

Velickovic, D., Lin, V. S., Rivas, A., Anderton, C. R., & Moran, J. J. (2020). An approach for broad molecular imaging of the root-soil interface via indirect matrix-assisted laser desorption/ionization mass spectrometry. *Soil Biology and Biochemistry, 146*, 107804. https://doi.org/10.1016/j.soilbio.2020.107804.

Veselkov, K., Sleeman, J., Claude, E., Vissers, J. P. C., Galea, D., Mroz, A., et al. (2018). BASIS: High-performance bioinformatics platform for processing of large-scale mass spectrometry imaging data in chemically augmented histology. *Scientific Reports, 8*, 4053. https://doi.org/10.1038/s41598-018-22499-z.

Walker, H. J., Steels, C., Bendell, L., Clench, M. R., Read, D. J., Cameron, D. D., et al. (2016). Understanding metabolism of arginine in biological systems via MALDI imaging. *Proteomics, 16*, 1690–1694.

Yamashita, M., & Fenn, J. B. (1984). Electrospray ion source. Another variation on the free-jet theme. *Journal of Physical Chemistry, 88*(20), 4451–4459. https://doi.org/10.1021/j150664a002.

Zhang, Y., & Liu, X. (2018). Machine learning techniques for mass spectrometry imaging data analysis and applications. *Bioanalysis, 10*(8), 519–522.

CHAPTER TEN

MS- and NMR-metabolomic tools for the discrimination of wines: Applications for authenticity

Josep Valls Fonayet[a,*], Grégoire Loupit[b], and Tristan Richard[a]
[a]Bordeaux Metabolome Facility, Unité de recherche Oenologie, Axe Molécules d'Intérêt Biologique, EA 4577, USC 1366 INRAE, Université de Bordeaux, ISVV, Villenave d'Ornon, France
[b]EGFV, Univ. Bordeaux, Bordeaux Sciences Agro, INRAE, ISVV, Villenave d'Ornon, France
*Corresponding author: e-mail address: josep.valls-fonayet@u-bordeaux.fr

Contents

Abstract

Wine is an extremely sophisticated chemical matrix in which the harmonious equilibria between many components (alcohols, acids, aromas, polyphenols, etc.) determine its final quality and value. The study of wine composition by metabolomic tools, referred as Wineomics, has emerged as a revolutionary approach to unravel the chemical space of wines and to associate wine composition with important viticultural and enological traits including grape variety, geographical origin, terroir or typicity.

Advances in Botanical Research, Volume 98
ISSN 0065-2296
https://doi.org/10.1016/bs.abr.2020.11.003

One of the most promising applications of metabolomic tools for wine analysis is the study of wine traceability and authenticity, a major issue for the wine industry and consumers. These studies are extremely complex, because counterfeits are becoming more sophisticated and they require powerful techniques capable to detect subtle differences between wines. In this chapter, we describe the targeted and untargeted approaches by Mass and NMR spectrometry-based metabolomics as well as some recent examples of their abilities in wine authenticity studies.

1. Introduction

Wine is one of the most extensively consumed beverages in the world and a comparatively expensive commodity in modern society. The global wine sector generates a great deal of wealth production: in 2018, it reached 292 million hL according to the International Organisation of Vine and Wine (International Organisation of Vine and Wine, 2019), with the top 10 countries exceeding 9 million hL of production: Italy (54.8 million), France (48.6), Spain (44.4), USA (23.9), Argentina (14.5), Chile (12.9), Australia (12.9), Germany (10.3), South Africa (9.5) and China (9.1). This solid implementation in all continents indicates the importance of the wine industry worldwide.

The commercial value of a wine brand is determined by parameters strongly related to its quality, history, provenance, or typicity, a term which refers to the extent to which a particular wine is associated to its style, variety, origin or vintage (Schueuermann, Khakimov, Engelsen, Bremer, & Silcock, 2016). Related to this concept of typicity, a French term, "terroir", is used to define the interaction between the physical and biological environment with the grapevines, in which participate several factors such as soil, climatic conditions, geology, geomorphology, biology of the vine and human activity (Canizo, Escudero, Pellerano, & Wuilloud, 2019). The quality and value of a specific wine will be then determined by the grape cultivars (either in monovarietal wines or in blendings), *terroir*, age or vintage year, and processing factors (Alañón, Pérez-Coello, & Marina, 2015; Pasvaska, Tzachristas, & Proestos, 2019; Pinu, 2018; Uttl et al., 2019; Versari, Laurie, Ricci, Laghi, & Parpinello, 2014).

Like other high-valued foods such as saffron or herb spices, wine is affected by different types of frauds and counterfeits, which represents a continuous increasing problem associated with economic issues and even with potential health risks for the consumers (Rubert, Zachariasova, & Hajslova, 2015). Wine counterfeit, mainly in terms of varieties and regions of geographical origin, is very widespread and it has been estimated that the share of counterfeit wine sales on the European market is more than 10% (Holmberg, 2010), reaching over 1.3 billion € in 2015. In China it has been described a persistent fraud for high-priced wines (Lecat, Brouard, & Chapuis, 2017), claiming that 50% of wine exceeding $35 per bottle in China is counterfeit (Muhammad & Countryman, 2019). Therefore, wine authenticity, closely related to the concept of traceability, is an issue of great importance nowadays.

Wine frauds include adulteration, namely, the unauthorized chaptalization (sugar addition) in musts; the dilution of wine with water; the addition of alcohol, coloring and flavor agents; or blending with unauthorized wines of inferior quality or different denomination. Moreover, counterfeit or incorrect mislabeling of wine is also considered a commercial fraud (Pasvaska et al., 2019; Versari et al., 2014): it is a legal requirement that the label accurately reflects the factors influencing the quality of wine, which can include the grape-growing region, grape variety or varieties, year of vintage, enological practices, etc. (Uttl et al., 2019). All statements on the label—representing the wine identity—should be verifiable by analytical data, which can be very challenging (Pasvaska et al., 2019).

Geographic origin is one of the major authenticity traits in wine due to its association with a particular quality or reputation (Kamiloglu, 2019). Most consumers associate geographical origin as a quality indicator before purchasing agricultural products (Wadood, Boli, Xiaowen, Hussain, & Yimin, 2020) and wine typicity is an important concept in oenology as it not only delineates geographical areas but also comprises recognizable sensory characters and composition (Kustos et al., 2020). In Europe, most wines with high economic value are linked to the origin area and to specific methods of production under Protected Designations of Origin (PDO), which set a specific regulation governing aspects such as grape varieties to be used, cultivation and vinification methods, aging, physical/chemical parameters and sensory characteristics (Villano et al., 2017). Moreover, PDO regulates also the geographical origin of the parcels from which the grape is harvested. Each European country has its own categories of PDO and Protected Geographical Indications (PGI, wine appellations). PGI production rules are less stringent than those bound to PDO (Perutka, Voženílek, & Šebela, 2020). Over 1500 PDO are registered in UE,[a] each with its specific rules. To illustrate some of these rules regarding varieties, Bordeaux PDO red wines can only be produced from Cabernet Sauvignon,

[a] https://ec.europa.eu/info/food-farming-fisheries/food-safety-and-quality/certification/quality-labels/geographical-indications-register/

Cabernet Franc, Merlot, Malbec (or "Cot"), Carmenere, and Petit Verdot varieties, while Burgundy red wines have to use Pinot noir as principal variety. Outside Europe, South Africa uses a similar Appellation Control system, while Australian legislation uses the term Geographical Indication which mimics the *Appellation Controllée* in France (Kustos et al., 2020). The appellations in the United States are based on Approved Viticultural Areas (Perutka et al., 2020). This heterogeneity makes the terms of authenticity and wine adulteration a very complicated subject since wine legislation differs depending on the country and even the region or type of wine. For example, chaptalization is prohibited in Argentina, California, Italy, Spain and South Africa, but it can be authorized in cold regions of France, Germany and United States which produce grapes with low sugar content. Water addition in minima quantities is accepted by the laws of California, but are not permitted in the UE (Versari et al., 2014). Blended wines are characteristic for some wine-producing areas (Bordeaux), but this procedure can be easily misused (Perutka et al., 2020).

There is a pressing need for verification of the wine authenticity and identity in the current context of growing market globalization (Pasvaska et al., 2019). This need for wine authentication is being increasingly demanded by the consumers and by the wineries to ensure the verification of the identity of the product, and preventing fraud and mislabeling is of high concern by regulation authorities in the UE and worldwide (Uttl et al., 2019). Control analysis for some common fraudulent practices has already been addressed, especially those that are easy to detect. For example, the OIV has officially established an isotopic analytical method for detecting the illegal addition of sugar and water (Kamiloglu, 2019), and the sugar addition in form of fructose-rich corn syrup may be detected by the presence of 5-(hydroxymethyl)-2-furaldehyde (Perutka et al., 2020). However, with counterfeit techniques becoming increasingly more sophisticated (Stupak, Goodall, Tomaniova, Pulkrabova, & Hajslova, 2018), more complex methodologies are being required for their detection. In this context, innovative applications of metabolomics tools can contribute to an exhaustive wine labeling (Ferri et al., 2015) by establishing chemical markers, fingerprints or "barcodes" for wine traceability (Dey & Montet, 2017; Medina, Pereira, Silva, Perestrelo, & Câmara, 2019).

1.1 Wine metabolomics

The analysis of food metabolome represents a new frontier in the evaluation of food quality (Ferri et al., 2015). Wine is one of the most complex foods in terms of metabolomic profile, because grapes, terroir, yeasts, bacteria, fining agents, exogenous oxidants and other enological practices and aging systems are involved in its elaboration (Arapitsas, Della Corte, et al., 2016). Besides, wine is a continuously evolving complex matrix, with thousands of different molecules that contribute to its characteristics (odor, taste, flavor, color, mouthfeel), determining the wine quality and the final acceptance by the consumer. Metabolomic analyses (whether targeted or untargeted) are specially fitted to cover such variability of chemical compounds. Wineomics is a foodomics term used to reference the study of the metabolites found through the winemaking process from grape to bottle (Alañón et al., 2015; Moyano, Serratosa, Marquez, & Zea, 2019).

Hundreds of metabolites have been detected in wines, but only a fraction of them have been identified. Targeted methods in wine metabolomics are mainly focused on the analysis of selected predefined groups of these known metabolites, normally belonging to specific classes (e.g., polyphenols, volatile compounds, sugars, amino acids, organic acids) (Gil et al., 2020) for which a quantitative measure is usually provided (or at least semi-quantitative in case of absence of standards) (Díaz et al., 2016). While targeted analyses have successfully been used for the analysis and discrimination of wines, they leave unconsidered a large part of the wine metabolome because of the restrictive nature in terms of specific metabolites analyzed. The whole chemical composition of wine should better reflect its history and traceability, including the grape variety and its environment (terroir), the yeast and bacteria strains that conducted the alcoholic and the malolactic fermentations, the containers used in the elaboration, the storage and aging, the enological practices, etc. (Arbulu, Sampedro, Gómez-Caballero, Goicolea, & Barrio, 2015). Therefore, recent developments by untargeted metabolomics can be much more appropriated for a better description of the chemical space associated with wine metabolome (Rocchetti, Gatti, Bavaresco, & Lucini, 2018) and the discrimination of wines with subtle differences and hence for the detection of sophisticated frauds (Arbulu et al., 2015).

Untargeted methods aim at characterizing the differential abundance (fingerprinting) of the metabolite responses among the samples (Díaz et al., 2016). An untargeted metabolite profile, based on *features* rather than compounds, usually contains information about both identified and unknown molecules (Pinu, 2018) and does not aim to provide absolute quantification of the metabolites. Untargeted metabolomics workflows usually consist of the following steps: sample selection and treatment, instrumental set-up, raw data acquisition, data analysis with chemometrics, marker annotation and database search, and hypothesis generation (Arapitsas, Ugliano, et al., 2016). Steps of annotation may sometimes not be indispensable to discriminate samples, but they should be mandatory for interlaboratory comparison and translating the methodologies into high throughput, more cost-effective methods. The annotation can be focused on selected *features* that are crucial for discriminating the samples, which can be assessed for example by the Variable Importance in the Projection (VIP) Score in chemometric analysis. Identification of features is based on comparisons with available pure standards, while tentative identification can be achieved using databases (Díaz et al., 2016). Non-targeted characterization of wine remains a challenge because of the enormous structural diversity of compounds, the large range of possible isomeric forms (especially for polyphenols), the lack of available standards for most wine metabolites and the lack of comprehensive databases of wine metabolites (Causon, Ivanova-Petropulos, Petrusheva, Bogeva, & Hann, 2019).

Multivariate data analysis and pattern recognition techniques are essential chemometric tools to conduct food authentication, especially in untargeted analysis. They include Principal Component Analysis (PCA), Partial Least Squares (PLS), Discriminant Analysis (DA), Hierarchical Clustering Analysis (HCA), and data mining/learning machines, among others. The utilization of these tools for food quality and authenticity has been reviewed elsewhere (Arvanitoyannis, 1999; Jiménez-Carvelo, González-Casado, Bagur-González, & Cuadros-Rodríguez, 2019; Kemsley, Defernez, & Marini, 2019).

Wine metabolomics can be assessed by many different spectroscopic techniques (Alañón et al., 2015; Arvanitoyannis, 2010; Canizo et al., 2019; Medina, Perestrelo, Silva, Pereira, & Câmara, 2019; Versari et al., 2014; Wadood et al., 2020). Isotopic analyses of $^{13}C/^{12}C$ ratios can indicate the addition of sugars, the $^{18}O/^{16}O$ ratio can indicate added water, and both could indicate geographic origin (Versari et al., 2014). The multi-elemental analysis of trace mineral composition by techniques such as ICP-MS has proven excellent capabilities for classification according to the wine origin (Azcarate, Martinez, Savio, Camiña, & Gil, 2015; Baxter, Crews, John Dennis, Goodall, & Anderson, 1997; Cabrita et al., 2018; Wu et al., 2019). Spectroscopic techniques based on UV–visible (Geană, Ciucure, Apetrei, & Artem, 2019), FT-IR (Basalekou, Pappas, Tarantilis, & Kallithraka, 2020; Geană et al., 2019) or Raman spectroscopy (Magdas, Guyon, Feher, &

Pinzaru, 2018; Zanuttin et al., 2019) have further been successfully applied for wine discrimination and reviewed recently (Chapman, Gangadoo, Truong, & Cozzolino, 2019). By combining several of these techniques the power of the analysis will be reinforced and will be able to address more complex studies and more subtle counterfeits.

In this chapter, we will focus on MS- and NMR-based profiling metabolomics, because they are versatile methodologies for describing a wide range of molecules including polyphenol and aroma compounds, which serve as criteria of quality in wines (Mayr, de Rosso, Dalla Vedova, & Flamini, 2018). Moreover, unlike other profiling spectroscopic techniques, where chemical information is harder to extract, NMR and MS can contribute to the identification or annotation of discriminating compounds, which can help to better understand the chemical space related to the different factors (whether varietal, geographical or practical) that determine wine authenticity. Fig. 1 aims to provide a graphical abstract of the several analytical methodologies which will be described in the chapter, which combined with chemometrics will serve to assess wine authenticity by discriminating wines according to different traits.

2. Liquid chromatography-mass spectrometry (LC-MS) in wine analysis: Phenolic compounds for authenticity studies

Because of its versatility, LC-MS is widely used for the analysis of wine components including amino acids, amines, sugars, organic acids, or polyphenols. Several of these compounds have been indicated as potential markers for wine authenticity studies because they can relate to variations from the grape cultivar, the viticulture practices or the winemaking technology. For example, the profile of free amino acids and ammonia (aminogram) has been described as a criterion for the authenticity of red and white wines in different countries (Duchowicz, Giraudo, Castro, & Pomilio, 2013) and would depend on fertilization of the soil, climatic conditions and duration of skin maceration. Organic acids can be indicators of the grape maturity (related to the climate), the yeasts used for fermentation, the winemaking processes and wine evolution (Robles, Fabjanowicz, Chmiel, & Płotka-Wasylka, 2019). However, the most commonly used compounds for wine authenticity studies remain phenolic compounds,

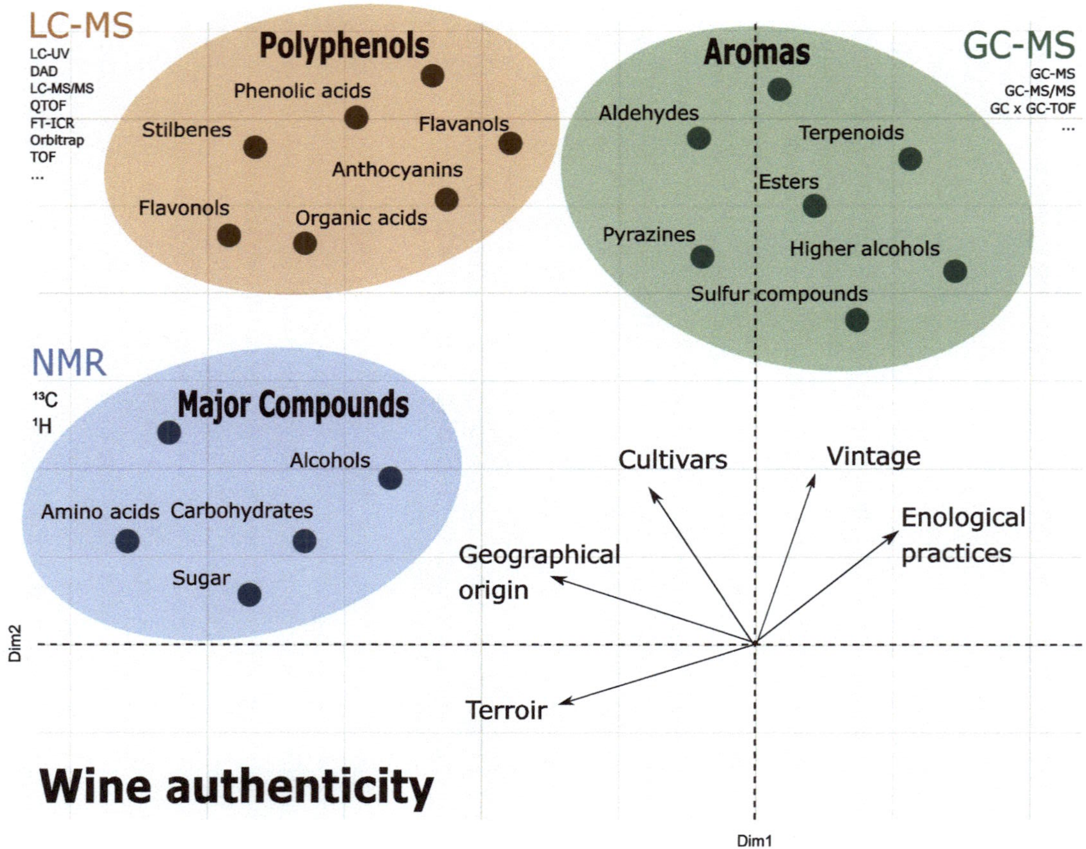

Fig. 1 Graphical abstract of the chapter, highlighthing the MS- and NMR-metabolomic methodologies applied for wine authenticity studies, the type of compounds analyzed with these techniques, and the need of chemometric tools for discriminating wines according to different factors (origin, grape varieties, enological practices, etc.).

because they consist of a large number of compounds which can better represent the variability among samples from different varieties, origins or vintage years.

Polyphenols in wine include a large diversity of structures (Fig. 2), mostly classified as flavonoids (including anthocyanins, flavan-3-ols and their oligomerised –proanthocyanidins- or polymerized –tannins- forms, flavonols or flavanones) and non-flavonoids (phenolic acids, stilbenes). Phenolic composition of wines depends on many factors. First of all, polyphenols are already present in grape, mainly in the skins and seeds. The content in grapes depends on the varietal potential but also the maturity, the climate, the soil, or the state of the grape. Many studies on the varietal characterization of grape polyphenols have indicated the ability of phenolic compounds to discriminate grape varieties and be used for authentication of wine varietal origin (Valls, Millán, Martí, Borràs, & Arola, 2009). Among them, anthocyanins have since long been established as a satisfactory criterion for the identification of red grape varieties, because anthocyanins profile is highly linked to the genetic source (Fraige, Pereira-Filho, & Carrilho, 2014).

Grape major anthocyanins, which are the molecules responsible for the red color in grapes and wine, are based on five glucosylated anthocyanidin skeletons. The glucose can be usually be found esterified with acetic acid and *p*-coumaric acid. These 15 major grape anthocyanins combinations can be easily analyzed by HPLC-DAD. In most red wine grapes, malvidin 3-glucoside is the main anthocyanin, but the profile among the rest of anthocyanins can be highly variable. For example, Pinot noir lacks the presence of acylated compounds. Early studies differentiated American *Vitis* species from *Vitis vinifera* L. because of the presence of diglucosides in the former. It is worth mentioning that the development of more sensitive methodologies based on LC-MS has revisited these profiles. Indeed, *Vitis vinifera* wine grapes also present minor amounts of pelargonidin derivatives, diglucosides, caffeoylated compounds, and coumaroyl forms *cis*- and *trans*- can be separated (Arapitsas, Perenzoni, Nicolini, & Mattivi, 2012). Sensitive LC-MS/MS analysis has been able to detect the presence of anthocyanins even in white grapes (Arapitsas, Oliveira, & Mattivi, 2015).

While the applicability of anthocyanins profile to discriminate grapes is well documented, their final composition in wine can differ enormously from the primary grape source depending on the extractability during the maceration process (Valls et al., 2017) and also the impact during wine evolution of multiple reactions and transformations of these polyphenols caused by fermentative processes and aging of wine. In aging wines, the

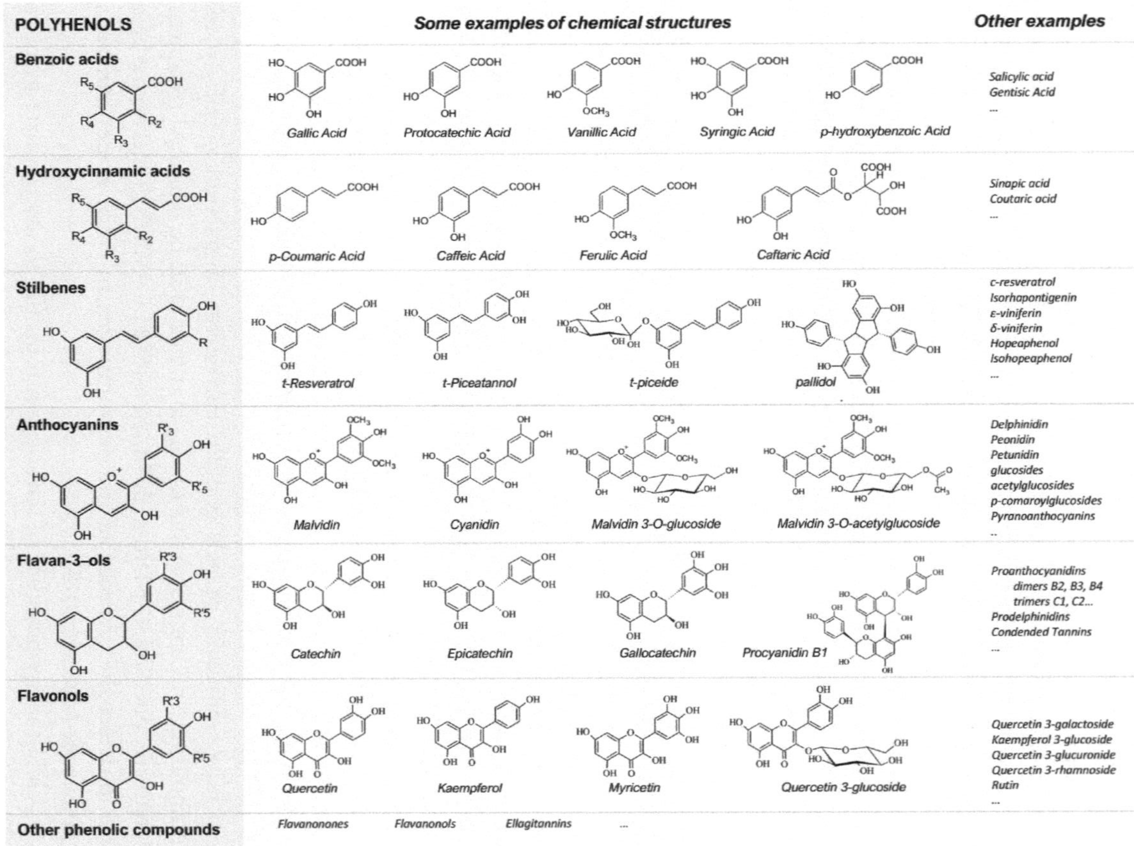

Fig. 2 Main polyphenol families of wine, including selected examples of their structure. Molecular structures designed with ACD/ChemSketch Freeware, version 2020.1.2, Advanced Chemistry Development, Inc., Toronto, ON, Canada, www.acdlabs.com, 2020.

contribution of native grape anthocyanins is highly reduced in favor of newly formed pyranoanthocyanins (including vitisins, carboxivitisins, vinylphenol-, vinylcatechol- and vinylguaiacol derivatives) and adducts with tannins. Pyranoanthocyanins are formed by reactions of anthocyanins with fermentation products. This implies that the anthocyanin chemistry becomes much more diverse and can be applied to detect aging markers in wine (Arapitsas, Perenzoni, et al., 2012; Zhang et al., 2020).

Together with the profile, the content of polyphenols (not only anthocyanins, but also other flavonoid and non-flavonoid forms) is a marker of authenticity. In this sense, stilbenes are an especially valuable family in terms of wine discrimination because of their known role as phytoalexins to be stimulated in the presence of pathogen attacks. This trait can contribute to differentiate among vintage years since the fungal pressure suffered by grapes (infections of mildew, *Botrytis*…) is highly dependent on the climatology and change between seasons (Guerrero, Valls-Fonayet, Richard, & Cantos-Villar, 2020).

2.1 Targeted wine polyphenols analysis

Table 1 compiles some relevant studies for the classification of wines based on the targeted analysis of phenolic compounds.

Simple targeted analysis of the 15 main anthocyanins profile obtained by LC-DAD at 520 nm has been applied to distinguish wines from different varieties and terroirs (Da Costa et al., 2018; de Andrade et al., 2013; Kumšta, et al., 2014), and good results can be achieved especially if the number of samples is limited, the conditions to be compared are quite differentiated, and the heterogeneity of the wines from the same category is limited. Other phenolic compounds can also be detected by UV spectroscopy: flavonols are detected at 360 nm, hydroxycinnamic acids at 320 nm, and all polyphenols (including flavan-3-ols) can be detected at 280 nm (Makris et al., 2006). The analysis of all these molecules can contribute to better explaining the discrimination among wines. In a large study including 111 monovarietal commercial bottled wines belonging to 11 red and white varieties from 4 different vintage years in Turkey (Sen & Tokatli, 2014), the PLS-DA analysis of 31 polyphenols analyzed by HPLC-DAD yielded clusters for all varieties except Merlot, who was confounded with the other cultivars.

Although the LC-UV methodologies can lead to satisfactory results in some studies, the development of LC-MS methodologies in recent years,

Table 1 Relevant studies of wine authenticity based on the targeted analysis of polyphenols by liquid chromatography, including LC-MS.

Methodology	Criteria	Samples	Analyzed compounds, (most discriminating compounds)	Chemometric method	References
HPLC-UV-Vis	Geographical origin	12 commercial Brazilian and Chilean dry red wines from Cabernet Sauvignon, 3 wine regions and 2 vintage years, 26 wines from other varieties (Syrah, Tannat, Merlot)	9 Anthocyanins (*levels of anthocyanins, Pt3glu, Pn3glu, Mv3glu, Pn ac glu, Pn pc glu*)	PCA	de Andrade, do Nascimento, Pereira, Hallwass, and Paim (2013)
HPLC-DAD	Varietal and geographical origin	17 monovarietal commercial Czech red wines, from 3 cultivars (Blaufränkrisch, Blauer Portugieser, Saint Laurent), 3 wine regions and different vintage years	15 Antocyanins (*variety: Mv3glu, Dp3glu, Dp ac glu*) (*terroir: Mv3glu, Dp3glu, Dp pc glu*)	PCA CDA (Canonical discriminant analysis)	Kumšta, Pavloušek, and Kárník, (2014)
HPLC-DAD	Geographical origin	26 Argentinian and 11 Chilean Monovarietal Syrah wines	15 Anthocyanins (*Mv3glu, Pn3glu, Pt ac glu, Pn ac glu, Mv ac glu*)	Extreme Learning Machine	Da Costa, Llobodanin, de Lima, Castro, and Barbosa (2018)
HPLC-DAD	Varietal and geographical origin	40 experimental monovarietal Greek young red wines, 6 cultivars (Merlot, Cabernet Sauvignon, Syrah, and 3 autochthonous varieties), 9 wine regions	19 polyphenols: 7 Anthocyanins, 4 Flavanols, 4 Hydroxycinnamates, 4 Flavonols (*variety: Procyanidin B1, B2, Mv 3glu, Mv ac glu*)	Discriminant analysis	Makris, Kallithraka, and Mamalos (2006)

HPLC-DAD	Varietal origin, vintage year	111 commercial Turkish monovarietal wines, of 11 cultivars (including autochtonous varieties and Cabernet sauvignon, Merlot, Syrah, Muscat, Chardonnay), and 9 wine regions	31 polyphenols: 13 Anthocyanins, 8 Flavonols, 6 phenolic acids, 3 flavanols, 1 stilbene (*red wines, vintage year: malvidin compounds*) (*white wines, vintage year: Q3gal, vanilic acid, o-coumaric acid*)	PLS-DA	Sen and Tokatli (2014)
UPLC-PDA-MS/MS	Varieties, Enological practices	2 experimental wines from 2 varieties (Zweigelt, Rondo) with 10 vinification process	55 polyphenols: 24 Anthocyanins, 7 Flavonols, 12 Flavanols, 8 Phenolic acids, 4 stibenes	PCA	Stój, Kapusta, and Domagała (2020)
HPLC-MS/MS	Vintage year	20 commercial red wines purchased in China, from 2 groups of vintage range (10 years between them)	37 Anthocyanin derivatives (*vitisins, flavanol-anthocyanin adducts*)	PLS-DA	Zhang et al. (2020)
UPLC-QqQ-MS/MS	Wine type (dry wine/ dessert wine)	6 experimental wines of 2 types (dessert wines or dry wines) made from 2 Croatian cultivars (red Plavac mail, white Posip) in 4 different vintege years	45 polyphenols: 12 Phenolic acids, 2 phenol aldehydes, 6 Stilbenes, 8 Flavan-3-ols, 15 Flavonols, 1 dehydroflavonol, 1 hydroquinone (*phenolic acids, phenolic aldehydes, flavan-3-ols and flavonols*)	PCA LDA	Budić-Leto, Zdunić, Gajdoš-Kljusurić, Mucalo, and Vrhovšek (2017)

Continued

Table 1 Relevant studies of wine authenticity based on the targeted analysis of polyphenols by liquid chromatography, including LC-MS.—cont'd

Methodology	Criteria	Samples	Analyzed compounds, (most discriminating compounds)	Chemometric method	References
UHPLC-QqQ-MS/MS	Geographical origin	12 commercial rosé wines, from 11 wine regions (of 6 different countries)	152 polyphenols: 95 Anthocyanins, 15 Phenolic acids, 5 Stilbenes, 11 Flavanols, 5 Flavonols, 2 dihydroflavonols, 5 others (*caftaric acid, coutaric acid, GRP (grape reaction product), B1; tyrosol, gallic acid, catechin mv3glu, Mv ac glu*)	PCA	Lambert et al. (2015)
UPLC-QqQ-MS/MS	Varietal origin	167 commercial wines from 2015 purchased in Croatian wineries: 67 monovarietal red wines from 4 cultivars (Cabernet Sauvignon, Merlot, Teran, Plavac mali); 105 monovarietal white wines from 6 cultivars (Chardonnay, Muscat Blanc, autochthonous varieties)	38 Polyphenols: 15 anthocyanins, 11 phenolic acids, 8 stilbenes, 7 flavan-3-ols, 17 flavonols, 3 others (*red varieties: pn ac glu, taxifolin*) (*White varieties: c-piceid*)	SLDA	Lukić, Radeka, Budić-Leto, Bubola, and Vrhovsek (2019)

Cy = cyanidin, Dp = delphinidin, Mv = malvidin, Pt = petunidin, Pn = peonidin, glu = glucoside, ac glu = acetylglucoside, pc glu = p-coumaroylglucoside.

especially for targeted analysis of phenolic compounds by triple quadrupole (TQD) spectrometers, has proven to be a more efficient strategy for the differentiation of wines. These instruments operate in Multiple Reaction Monitoring mode, which means that they only monitor selected targeted MS/MS transitions. The MS/MS provide more selectivity and better signal to noise ratios, characteristics that have contributed to establishing TQD as the paradigm of targeted LC-MS analysis. Of course, the potential developments based on targeted analysis rely on the improvement in the number of compounds to be detected. A recently described high-throughput UHPLC-QqQ-MS method for polyphenol profiling in rosé wines allowed the selective quantification of 152 phenolic compounds (including 95 anthocyanin derivatives, 15 phenolic acid derivatives, 5 stilbenes, 7 flavonols and dihydroflavonols, 11 flavan-3-ols, and 2 additional non-phenolic wine components (tryptophan and tryptophol)). This methodology was successfully applied to 12 rosé wines from different European origins (Lambert et al., 2015).

This applicability for varietal discrimination of targeted methods has also been proven in larger studies. For example, in a study with 173 young wines produced from 4 red and 6 white grape cultivars in different regions of Croatia, the targeted analysis of 58 polyphenols for red wines and 32 in white wines was successfully used for varietal differentiation (correct classification higher than 95% in all cases) (Lukić et al., 2019).

The effects of aging in wine can be easily detected by the targeted analysis of the pyranoanthocyanin derivatives formed from reactions with the wine matrix, especially between anthocyanins and fermentation products such as phenols and carbonyls (Zhang et al., 2020).

Targeted methods can also be able to discriminate between types of wines. For example, traditional sweet dessert wines are produced with dried grapes to naturally concentrate sugars. A targeted analysis of 45 compounds was applied to analyze 18 dessert wines and dry wines made from the same autochthonous cultivars in Croatia (Budić-Leto et al., 2017). The PCA analysis described groupings based on the type (dry or dessert wine) as well as on the variety for the dry wines. Recent studies have also indicated that targeted analysis of polyphenols can be able to discriminate among different types of maceration (Stój et al., 2020).

2.2 Untargeted wine LC-MS analysis

Non targeted methods by LC-MS are based on defining the profile of metabolites in terms of features, which can be defined as "a molecular entity with a unique m/z and retention time" (Arapitsas, Della Corte, et al., 2016). High-Resolution Mass Spectrometry (HRMS) by TOF, orbital ion traps and Fourier-transform ion cyclotron resonance (FT-ICR) have been employed for the non-targeted analysis of wines because of the high-level confidence in sum formula annotation (Causon et al., 2019).

In the case of FTICR-MS, it is not unusual a direct injection of the sample to the Mass Spectrometer, because of its slow scan rate (1 Hz), which means that coupling to LC implies reducing to second scale scans with consequent reduction in sensitivity and spectral resolution (Roullier-Gall et al., 2015). However, normally the HRMS instruments are coupled to UHPLC systems, which increases the range of detected metabolites, decreases putative ion suppression effects and achieves separation of isobaric substances (Roullier-Gall et al., 2015) depending on the selectivity of the columns used. Reverse phase C18 chromatography and Hydrophilic interaction chromatography (HILIC) are the best complementary approaches in untargeted metabolomic fingerprinting. HILIC offers a different selectivity with better retention and sensitivity for polar metabolites (Arapitsas, Della Corte, et al., 2016). In an experiment comparing HILIC and C18 columns, HILIC columns showed substantial, better suitability for sugars and organic acids, while phenolic compounds, amino acids, fatty acids and higher alcohols presented a similar number of identified compounds in both columns evaluated. Despite this, HILIC columns often present difficulties in terms of variation of the retention times of peaks detected, which, associated with the relatively new introduction of these types of columns and the traditional use of C18 columns for wine analysis could explain the wider implementation of C18 columns for untargeted wine analysis (Arbulu et al., 2015).

Another factor that influences the untargeted studies is the extractive process. While wine analysis could be spared from these extractive limitations because the matrix requires no extraction (except if a previous cleaning step by Solid Phase Extraction is performed), the sample processing affects highly the profiles obtained by the untargeted analysis. The profiles where PTFE or nylon membrane syringe filters were used resulted in fewer putative compounds detected in the MS profile compared to the use of centrifugation to decant large peptides and proteins (Arbulu et al., 2015).

The typology of MS analysis has also a great impact on the detection of features and, in consequence, in the results obtained from untargeted LC-MS studies. The positive ion mode generally produces a larger number of features than in negative mode (Arapitsas et al., 2020; Vaclavik, Lacina, Hajslova, & Zweigenbaum, 2011). The great number of initial raw features should be filtered by different criteria (removal of false positives and redundancies, frequency of presence among the samples, statistical significance, Variable Importance in the Projection (VIP) score, or by the fold-change) to determine the features which contribute the most to the separation of samples, which can then be tentatively identified (Vaclavik et al., 2011).

The different instrumental platforms used for the untargeted analysis can also impact on the results, normally not in terms of discriminating power between the samples groups but in terms of detection and annotation of the features (Díaz et al., 2016). The comparison between Orbitrap and Q-TOF platforms for discriminating 43 Spanish wines from 3 different regions (Penedes, Ribera del Duero, and Rioja) showed that both instruments were able to separate the wines in terms of the respective Protected Denomination of Origin, but surprisingly none of the 8 top discriminating compounds was common between both methodologies (Díaz et al., 2016). A similar observation was obtained from the comparison of Direct Injection FTICR-MS, HILIC/UPLC-qTOF-MS and RP/UPLC-qTOF-MS: only 30% of all of the annotations were detected in at least two of the three platforms, and masses common to all three instruments represented only 2.38% of the detected masses, highlighting the complementarity of FTICR-MS in relation to qTOF (Roullier-Gall et al., 2015). New technologies based on ion mobility (IM) can be extremely promising because IM can provide additional structural information not determined using LC-HRMS alone, which can be of particular value for the characterization of unknown isomeric compounds (Causon et al., 2019).

Table 2 compiles some relevant untargeted LC-MS studies for wine classification.

2.2.1 Profile analysis for varietal discrimination

The varietal analysis is the most common analysis of wine discrimination by untargeted LC-MS. This can be successfully applied for the traceability of the monovarietal wines and blendings. For example, Graciano is often used as a blending partner for Tempranillo wines in Rioja (Spain), an authorized practice that contributes to the overall quality of the final wine thanks to the increase of aromatic properties, red color and acidity provided by Graciano in the blending (Arbulu et al., 2015). The analysis by HPLC-qTOF of 18 Graciano and Tempranillo monovarietal wines from different wineries and vintage years in Rioja (Spain) showed a clear differentiation in both positive and negative mode for both varieties after PCA analysis.

Table 2 Relevant studies of wine authenticity based on the untargeted analysis by liquid chromatography-MS.

Methodology	Criteria	Samples: Number and types of wines	Chemometric method	Number of features Type of discriminant Compounds	References
HPLC-QTOF-MS	Varietal	18 Graciano and Tempranillo wines from one Spanish region	PCA	3102 features (positive), 2600 features (negative) Phenolic compounds, organic acids, lignans, terpenes	Arbulu et al. (2015)
HPLC-QTOFMS	Varietal	45 commercial red wines of 3 cultivars (Merlot, Cabernet Sauvignon, Pinot noir) from 12 different countries and different vintage years	PCA PLS-DA	35,000 initial features (positive mode); 7800 features (negative mode); filtered to 25 features Anthocyanins	Vaclavik et al. (2011)
UPLC-qTOF FT-ICR-MS	Geographical origin	152 commercial Burgundy wines (France), from 2 cultivars (white: Chardonnay; red: Pinot noir), 5 wine producers	PCA	Features related to secondary metabolites, amino acids, aromatic compounds	Roullier-Gall, Witting, Gougeon, and Schmitt-Kopplin (2014)
UPLC-FT-ICR-MS	Clone, harvest year	18 Experimental wines wines from 3 Pinot noir clones rape cluster types, in 2 vintage years	PCA PLS PLS-DA	26 features with VIP >1 for harvest years Anthocyanins, Flavanols, Organic acids	Cuadros-Inostroza, Verdugo-Alegría, Willmitzer, Moreno-Simunovic, and Vallarino (2020)

UPLC-QTOF-MS	Monovarietal red Wines, Terroirs	110 commercial Italian wines from 11 varieties (Aglianico, Cannonau, Corvina, Montepulciano, Nebbiolo, Primitivo, Raboso, Sagrantino, Sangiovese, Teroldego) and 12 wine regions in Italy	PCA	11,274 features (positive mode); 7397 features (negative mode) Anthocyanins, flavanols, amino acids, N-containing metabolites	Arapitsas et al. (2020)
HPLC-micrOTOF	Varietal and geographical origin	26 experimental wines from 8 varieties (Aspiran, Bonarda, Cabernet Sauvignon, Malbec, Merlot, Sangiovese, Syrah, Tempranillo) and 11 wineries from 3 regions in Argentina	PCA D-UPLS	Anthocyanins: Mv 3 glu, Mv ac glu, Mv vp glu or Mv pc glu, 3 unknowns	Pisano, Silva, and Olivieri (2015)
UPLC FT-ICR MS	Varietal, vintage year, wine quality	400 commercial Chilean wines, 4 cultivars (Carmenere, Cabernet Sauvignon, Merlot, Syrah), 6 vineyards, 3 vintage years	PCA (Varietal) LDA	5028 features (positive), 2804 features (negative)	(Cuadros-Inostroza et al., 2010)
UHPLC-QTOF-MS	Geographical origin	6 commercial Chardonnay wines from 6 wine regions in 4 countries (Italy, Israel, France, Australia)	O-PLS-DA	258 annotated polyphenols 24 VIP (Flavonols, Flavanols, phenolic compounds)	Rocchetti et al. (2018)

Continued

Table 2 Relevant studies of wine authenticity based on the untargeted analysis by liquid chromatography-MS.—cont'd

Methodology	Criteria	Samples: Number and types of wines	Chemometric method	Number of features Type of discriminant Compounds	References
HPLC-QTOF-MS/MS	Varietal	43 white wines from 3 cultivars (Riesling, Pinot Gris, Roter Traminer) and 3 vintage years from Czech Republic	PCA PLS-DA	2096 features (positive), 1045 features (negative) 14 final selected variables not annotated	Uttl et al. (2019)
HPLC-Orbitrap HPLC-QTOF	Geographical origin	42 commercial wines from 3 Spanish wine regions	PCA PLS-DA O-PLS-DA	Stilbenes, Flavanols	Díaz et al. (2016)
UPLC-QTOF-MS	Geographical origin	60 commercial rosé wines, from 3 french wine regions, different vintages	LDA, Random Forest	14,000 features 72 molecules (Phenolic acids, Anthocyanins, Flavanols)	Gil et al. (2020)
UPLC-QTOF	Cultivars, Oxygen	216 commercial wines from 6 different cultivars (pinot Gris, Inzolia, Muller Thurgau, Chardonnay, Traminer, Grillo), 2 exposures to oxygen	PCA (separation for cultivars, not for oxidation state) OPLS-DA (oxidation state)	8000 features (Organic acids, Sulfonated compounds)	Arapitsas, Ugliano, et al. (2016)
FT-ICR-MS	Concentration of SO_2	18 experimental french Chardonnay wines with3 concentrations of SO_2, 2 stoppers	PCA	11,181 features (Flavonols, Anthocyanins, Amino acids	Roullier-Gall et al. (2017)

HILIC UPLC-QTOF	Storage conditions	5 wines, 2 storage conditions	PCA O-PLS-DA	6311 features 29 storage markers metabolites (quercetin, catechin, malvidin 3-glucoside and pyranomalvidin 3-glucoside, Vitamins, One phenolic acid)	Arapitsas, Della Corte, et al. (2016)
UHPLC/QTOF-MS	Oxidation status	5 white wines	PCA OPLS-DA	Anthocyanins Flavonols Phenolic acids	Romanini, Colangelo, Lucini, and Lambri (2019)
UPLC-QTOF-MS	Wine micro-oxygenation	1 wine, 8 different oxygenations	PCA ICA	Amino acids Carbohydrates Amines Flavonoids Organic and phenolic acids Fatty acids	Arapitsas et al. (2012)
FTICR-MS and UPLC-QTOF-MS	Aging	1 young and 1 older wine		Flavonoids	Roullier-Gall et al. (2015)

In a recent study, 43 Riesling, Pinot Gris, and Roter Traminer wines were analyzed by an untargeted LC-qTOF approach after ethyl acetate extraction of the polyphenols. Multivariate date analysis by PLS-DA applied to the positive mode enabled a successful classification of 92% of the samples into the three varieties (Uttl et al., 2019). By setting a Variable Importance in Projection (VIP) threshold higher than 3, a total of 14 features were retained as the most significant ones to build the model, and the most important features for varietal classification in each variety could be established.

Similar results were obtained for the metabolome composition analysis of six red wines by microLC-MS/MS methodology of the polyphenolic fraction obtained after ethyl acetate extraction. A PCA was conducted for 264 annotated compounds, from which 165 were classified as polyphenols. Clear differences were observed for the six varieties used (Syrah, Zinfandel; Petit Syrah, Merlot, Cabernet Sauvignon) from different wineries in the United States (Ma, Tanaka, Vaniya, Kind, & Fiehn, 2016). This study highlights also the new developments in high throughput metabolomics by miniaturization (microfluidics) since the microLC-MS/MS method consisted of a 1 mm i.d. column at a flow of 0.05 mL/min with a 4 min gradient (Ma et al., 2016).

2.2.2 Profile analysis for geographical discrimination

The untargeted analysis by LC-qTOF has also proven to be able to discriminate monovarietal wines growing in different origins (Rocchetti et al., 2018). Chardonnay is a white variety that originated in the Burgundy region of France and widely expanded in other countries such as USA, Australia, China and South Africa. This variety is very versatile to produce from light and fruity to complex and full-bodied wines. An untargeted metabolomic approach used to profile phenolic compounds of 6 wines from different countries (France, Australia; Israel and 3 regions of Italy) resulted in a total of 258 annotated polyphenols. Unsupervised hierarchical cluster analysis differentiated wines from different origins, and supervised analysis by OPLS-DA could discriminate the aging environment between oak wood barrels (considered traditional winemaking for producing high-quality wines) vs stainless steel aging (Rocchetti et al., 2018).

The origin of wines has also been discriminated in the case of 30 rosé wines from 3 French regions by a UPLC-qTOF methodology that detected 14,000 features, which were further annotated with a database to assign 72 molecules. Among them, four phenolic compounds contributed the most to the discrimination (Gil et al., 2020).

2.2.3 Profile analysis for winemaking, enological practices, vintage year and wine conservation

A further level of separation of wines beyond the variety level has also been achieved by FT-ICR-MS to differentiate among wines obtained in microfermentations with three different Pinot noir clones in 2 vintage years. A total of 26 molecular features exhibited a significant contribution to explain the variation between harvest years (Cuadros-Inostroza et al., 2020). Moreover, the integration of the metabolite results with aroma and taste quality attributes evaluated by a sensory panel allowed to differentiate according to its quality and annotate some molecular features strongly associated with the specific sensory attributes for the clones. The results showed that the compounds responsible for the discrimination between the sweetness, bitterness, acidity and stringency belonged to the families of anthocyanins, flavonols and flavanols; and that specific polyphenols were strongly associated with aroma attributes (e.g., piceid with coffee/chocolate; pelargonidin 3-glucoside with animal attribute).

Wine can be stored for a long time and indeed aging is considered positively linked to quality and price. Premium quality wines are expected to last and even improve with age, and are frequently stored up for several years, while cheap young wines are consumed in a short time. Metabolomic analysis can also assess the effects of storage of wines (Arapitsas, Della Corte, et al., 2016), as it was stated in a study with two different storage conditions over 24 months: optimum storage conditions in a wine cellar (stable 16 °C) and typical domestic storage (variable Temperature between 20 °C and 27 °C). OPLS-DA analysis highlighted 29 tentative markers for storage conditions which could serve for authenticity studies.

Another enological practice than can induce differences in the metabolome composition and therefore be used for traceability and authenticity is the addition of sulfur dioxide (SO_2), which is commonly used as an additive in musts and wines for its antioxidant, antiseptic and antioxidative properties (Roullier-Gall et al., 2017). This agent is still considered as an essential tool for most winemakers, but because of the social concern of potential latent toxicities of sulfites; alternative strategies are being currently employed for replacing or at least reducing the doses of exogenous SO_2 added to the wine. Metabolic fingerprints obtained from FT-ICR MS data were able to discriminate 18 Chardonnay wines according to the added concentration of SO_2 (0, 4 and 8 g/hL). From the 11,181 features detected in the samples, a total of 686 were increased or decreased depending on the sulfite concentration. Among these compounds, the whole chemical space was represented (amino acids; polyphenols, carbohydrates, fatty acids, etc.), which illustrates the diversity of compounds which are involved in the sulfite treatment. A more detailed analysis indicated the presence of several sulfonated forms of polyphenols derived from gallic, caffeic and caftaric acids. Moreover, the same study also revealed discrimination of wines according to the type of bottle stopper (agglomerated cork or natural cork); the use of natural cork leads to a relative increase in the number of discriminant sulfur and nitrogen-containing compounds (Roullier-Gall et al., 2017).

Also related to the effects of small amounts of oxygen in bottled wines, a study including 216 bottles of 12 different white wines produced from 6 different cultivars showed the differentiation related to the varieties (with some overlapping between Chardonnay and Traminer) after PCA analysis. 35 metabolites were considered to be putatively induced by the different amounts of oxygen, including ascorbic acid, tartaric acid and some sulfonated compounds detected in wine for the first time (Arapitsas, Ugliano, et al., 2016).

3. Gas chromatography-mass spectrometry (GC–MS) in wine analysis: Volatile compounds for authenticity studies

The composition in volatiles (i.e., the volatolome) is one of the most important factors for determining wine characteristics, typicity and quality (Zhang et al., 2010). The volatolome of a particular wine can be composed by over 800 different compounds belonging to chemical families (Fig. 3) such as terpenes, higher alcohols, carbonyl compounds (aldehydes and ketones), acids, esters, lactones, vanillin derivatives, pyrazines, sulfur compounds, etc. (González-Barreiro, Rial-Otero, Cancho-Grande, & Simal-Gándara, 2015), and whose origin can be traced to the grapes, the microorganisms, the wood or the aging process (Villano et al., 2017).

One of the most widely used classifications of wine aromas distinguishes the varietal (or "primary") aromas of wine, which are already present in the berries and depend on the variety and its ecosystem (rootstock, climate, soil, cultural techniques of cultivation). However, the actual aroma potential from berries is not limited to the chemical "free aromas" structures directly responsible for the grape and must aroma. The potential aroma of winemaking grapes depends also on a bound aroma fraction, which consists of specific precursors (e.g., glycosidic forms) of relevant wine aroma molecules (Ferreira & Lopez, 2019). Secondary aromas are those formed during the fermentation processes

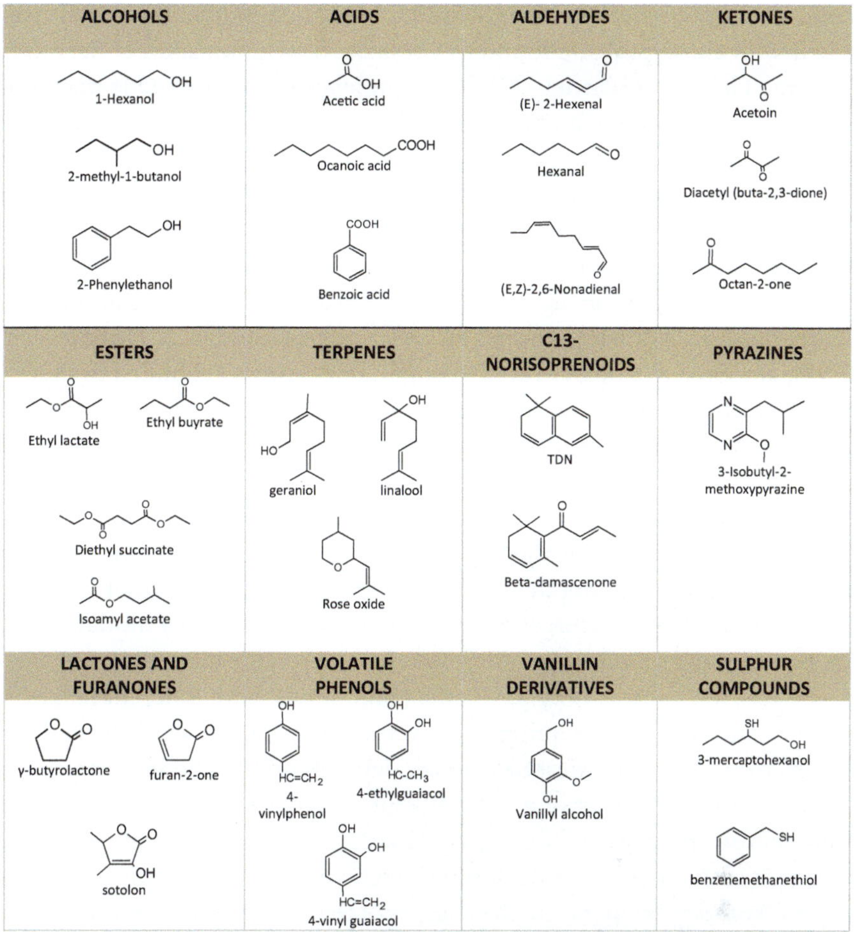

Fig. 3 Main volatile compounds families of wine, including selected examples of their structure. Molecular structures designed with ACD/ChemSketch Freeware, version 2020.1.2, Advanced Chemistry Development, Inc., Toronto, ON, Canada, www.acdlabs.com, 2020.

by the action of yeast and bacteria metabolism or hydrolysis reactions. Finally, tertiary aromas are those formed during wine evolution in bottles or barrels through aging and maturation, secondary fermentations and oxidation-reduction reactions (Ferreira & Lopez, 2019; Forleo, Zappi, Gottardi, & Melucci, 2020; Zhang et al., 2010).

Some peculiar aromas have been associated with specific compounds in several cultivars. The green pepper odor attributed to the pyrazines in

Cabernet sauvignon (but also present in other varieties), menthol notes associated with eucalyptol and found in Cabernet Sauvignon and Fer Servadou (Poitou, Thibon, & Darriet, 2017), the presence of *cis*-rose oxide that confers floral notes in Gewürztraminer (Villano et al., 2017), the association of linalool with Muscat floral aroma, the kerosene-like odor typical for Riesling wines caused by TDN (Springer et al., 2014), the mint nuances found in the typical aging bouquet of Bordeaux red wines associated with limonene-like derivated compounds (Picard, Franc, de Revel, & Marchand, 2018), or the higher content of volatile thiols in Sauvignon blanc (Carpentieri et al., 2019; Tominaga, Des Peyrot Gachons, & Dubourdieu, 1998), are some examples of well-recognizable odor notes linked to precise compounds and which in some cases have been associated to some specific cultivars. As a consequence, some of these molecules have been proposed as candidates for varietal authenticity studies (Villano et al., 2017). However, these simple linkages *molecule-aroma-variety* must be taken with caution, because in reality such specificity is mostly due to an average higher amount of these compounds rather than to a complete lack of them in other cultivars. Overall, the concentrations of some of these notes are strongly modulated by the maturity and vinification conditions. Moreover, the complex wine bouquet is due to a great variety of volatile compounds, with different ranges of concentrations. The perception of a specific compound in wine depends on its olfactory detection threshold, which depends on the chemical structure, but synergistic effects among the different molecules can enhance or limit their sensory perception. So, in accordance to this complexity, in most studies of authenticity for establishing criteria of genuineness (varieties, origin, etc.) or to prevent frauds, the analysis of the whole volatile profile rather than the determination of single molecules is used for wine differentiation (Zhang et al., 2010).

3.1 Gas chromatography coupled with mass spectrometry for the analysis of volatile compounds to assess the authenticity

Because of the relatively simple and early implementation of the hyphenation between GC and Mass spectrometry via Electron Ionization, together with the high peak capacity and robustness of the methodology and the development of reproducible databases such as the NIST, Gas Chromatography hyphenated with Mass Spectrometry (GC–MS) has been easily established as the technique of choice for the study of low-molecular-weight volatile compounds of wine aroma (Cajka, Showalter, Riddellova, & Fiehn, 2016).

Despite the sensitivity of the new generation of GC–MS for the detection and analysis of volatiles, sample enrichment is often required, because the contribution of these compounds to the wine aroma and its defaults can be expressed even at very low concentrations (ng/L–μg/L). The different methodologies used for the sample pretreatment will enormously impact the quality of the results in terms of representativity. Liquid-liquid extractions with an organic solvent such as dichloromethane and direct GC-injection of the extract were of common use but need large times of sample preparation and have been replaced by other techniques. In general, to avoid the introduction of an aqueous matrix in the GC instrument, the study of the headspace (HS) is preferred. Over the past years, there has been an increasing utilization of sorption methods for the extraction of the volatile aromas, including off-line solid-phase extraction (SPE), the use of online headspace solid-phase microextraction (HS-SPME) or stir-bar sorptive extraction (SBSE) (Schueuermann et al., 2016). The use of SPME fibers is the most widely applied methodology in aroma studies, with different available coatings from combinations of polyacrylamide (PA), divinylbenzene (DVB), polydimethylsiloxane (PDMS), or carboxen (CAR). The different selectivities of the microfibers will influence the results of the metabolomic study (Haggerty, Bowyer, Jiranek, & Taylor, 2016). The DVB/CAR/PDMS SPME fiber, with a large spectrul of adsorption, is the most commonly applied for wine aroma studies.

The MS ionization is also a source of influence in the results since the response will be highly dependent on the type of compound. For example, esters tend to give a higher signal by Electron Ionization than other compounds (Schueuermann et al., 2016). In some cases, the use of Chemical Ionization instead of Electron Ionization to provide less fragmented mass spectrum can be an interesting option (Thibon et al., 2015).

Afterward, results will be affected by the data treatment; which includes preprocessing methodologies such as noise filtering, baseline correction, peaks detection, alignment, identification and normalization (Castro, Martins, Teixeira, & Silva Ferreira, 2014).

Conceptually, GC–MS methodologies for the analysis of wines could be divided into targeted and untargeted methodologies. By its definition, targeted analysis should be based in the utilization of standards to determine a relatively limited number of individual compounds, often belonging to single families (Haggerty et al., 2016; Liu et al., 2020). The use of external standards for all the analytes present in a sample can be very expensive. Concerning most non-targeted analyses of wine aroma by GC, the metabolomics pipeline differs from the one used for LC-MS methodologies in the sense that the identification of volatile compounds is usually performed previously to the application of multivariate data analysis. The reason is the easiness to obtain tentative identifications by the retention index and by comparing the spectra with databases such as the NIST database (Robinson, Boss, Heymann, Solomon, & Trengove, 2011). Some precautions should be considered for the annotation of compounds since the NIST database can present several problems of mismatching because structurally related volatile compounds give similar spectra and thus ambiguous identifications can be obtained (Stój, Czernecki, Domagała, & Targoński, 2017a).

3.2 Targeted wine GC–MS analysis

The strict targeted analysis for a reduced number of selected compounds is rarely used for wine authenticity studies, but could find its place in particular cases were some specific markers can be directly linked to some wine characteristics or condition which have been previously described by wine profiling. For example, GC–MS has been applied to identify products linked to the aging process, which could be used for authenticity analysis in small studies. In Port wines, several dioxane and dioxolane isomers, furfural and 5-hydroxymethylfurfural were identified as potential biomarkers of the aging process (Castro et al., 2014) caused by the presence of oxygen and high temperatures, which impact the sensory perception and the final quality of these wines. Other compounds associated with oxidation in oaked Chardonnay wines were studied by HS-SPME-GC-MS in a targeted method which compared three types of wines (control, least oxidized and most oxidized wines). The study revealed several carbonyl compounds, including new oxidation biomarkers such as beta-damascenone, pentanal or vinyl decanoate (Pinto et al., 2018). Other oak-related volatile compounds such as volatile phenols and phenolic aldehydes have been successfully used for comparing wines aged in Spanish, French or American oak barrels (Fernández de Simón et al., 2008). Sotolon is also a flavor compound typical of oxidized white wines and is important in wines such as Port wines or Xeres, where oxidation during winemaking plays a role (Pons, Lavigne, Landais, Darriet, & Dubourdieu, 2008, 2010). Other markers can be linked to the maturity of the grape and therefore indicate climatic conditions of the year. For example, γ-nonalactone, β-damascenone and 3-methyl-2,4-nonanedione were identified by GC-Olfactimetry and MS as markers in Merlot and Cabernet wines of red wines made with overripe or dehydrated grapes (Allamy, Darriet, & Pons, 2018; Pons, Lavigne, Eric, Darriet, & Dubourdieu, 2008).

When some of these specific markers are established, more sensitive GC–MS/MS methods based on Triple Quadrupole instrumentation can enhance their detection. For example, the determination of volatile phenols associated with smoke taint could be obtained by GC–MS/MS analysis using specific MRM transitions for these derivatives (Liu et al., 2020). Similarly, an SPME-GC–MS/MS method (Green, Parr, Breitmeyer, Valentin, & Sherlock, 2011) or a liquid-liquid extraction followed by GC–MS/MS (Thibon et al., 2015) have been applied to reach low detection limits for the analysis of thiols in Sauvignon blanc wines.

3.3 GC–MS profile analysis for authenticity studies

Table 3 reflects some representative studies of wine traceability using the volatile profile obtained by GC–MS.

3.3.1 Profile analysis for varietal discrimination

Several studies using simple GC–MS instrumentation have attempted to use the volatile profile for discriminating wines according to the grape varieties. For example, 30 major peaks were chosen and enough to discriminate experimental Chinese wines from 3 different varieties produced in the same winery. After Principal Component Analysis and Soft Linear Discriminant Analysis (SLDA), a 100% recognition according to the variety was achieved, and the 11 most discriminant peaks were identified by the NIST database as esters (such as isobutyl acetate, ethyl hexanoate or diethyl butanedioate) and alcohols (1-pentanol; 2,3-butanediol). However, none of these compounds can really be considered variety-specific, because they are generated during the yeast metabolism. Indeed when this methodology was applied for commercial wines, the identification descended to 65% due to the presence of wines collected from different wineries and with different aging. These discrepancies between the results indicate that models for varietal discrimination will be highly influenced by the types of samples included in the study (Zhang et al., 2010), and that for complex studies a wide description of the volatile profile will be needed for increasing the discriminating power. An example of this complexity can be found in a chemometric study including 334 wines of 6 different cultivars (Sauvignon blanc, Chardonnay, Pinotage, Shiraz, Cabernet Sauvignon and Merlot) from 6 different South African regions and various producers. The analysis of 39 volatile components in these samples was not able to provide a satisfactory prediction for most red wine cultivars (except for Pinotage) (Weldegergis et al., 2011).

In such complex cases, an untargeted approach, so that no a priori GC peaks or mass features are selected for the statistical analysis, but the full scan

Table 3 Relevant studies of wine authenticity based on targeted and untargeted analysis of volatile compounds by GC–MS.

Methodology	Criteria	Samples	Number of features/ analyzed compounds *Most Discriminant Compounds*	Chemometrics	References
HS-SPME-EI-Ion Trap-GC–MS	Varietal	13 experimental Chinese wines and 36 commercial wines from 3 cultivars (Cabernet Sauvignon, Merlot, Cabernet Gernischt) and different vintages	30 features; *11 discriminating peaks (Esters and alcohols)*	ANOVA, PCA, SLDA	Zhang et al. (2010)
HS-SPME-EI-GC–MS	Varietal	272 german white wines from different origin and vinatge and from 5 cultivars (Riesling, Muller-Thurgau, Sivaner, Pinot Gris, Pinot Blanc)	6911 features 5638 feayures after signal treatment *Monoterpenoids, C-13 noirsoprenoids, esters*	PCA PLS-DA	Springer et al. (2014)
HS-SPME_ GCxGC-TOF	Varietal	43 Brazilian monovarietal wines from 5 cultivars (Chardonnay, Sauvignon Blanc, Merlot, Cabernet Sauvignon, Pinot noir) and 11 blends (Chardonnay/ Pinot noir)	480 analytes *12 discriminating compounds (2,3-butanediol; 4-carene, 3-Penten-2-one…)*	PCA LDA	Welke, Manfroi, Zanus, Lazzarotto, and Alcaraz Zini (2013)

HS-SPME-GCxGC-TOF	Geographical	70 nothern Italian sparkling wines from 2 regions (Trentodoc and Franciacorta) in 48 wineries and 6 vintage years	1695 features 196 biomarkers *Varietal terpenes, prefermentative compounds (C5-C6), sulfur compounds, ethyl esters of diprotic acids, cyclic acetals, norisoprenoids (safranal, alpha-isophorone)*	PCA Network analysis	Carlin et al. (2016)
HS-SPME-GC–MS	Geographical	32 Australian Chardonnay wines, 31 Shiraz wines from 2 subregions	*17 sensory attributes, 20 volatiles (esters, alcohols, isoprenoids, acids and carbonyls)*	PCA Discriminant analysis	Kustos et al. (2020)
HS-SPME-GC-EI-MS	Geographical	35 commercial wines from different countries (France, Italy, Spain and Poland), 4 different vintages and diverse (12) grape varieties	35 compounds (alcohols, acids, aldehydes, esters, ketones, volatile phenols) *3-methylbutan-1-ol, butane-2,3-diol, phenylmethanol, diethyl butanedioate, 4-ethylphenol…*	HCA	Stój, Czernecki, Domagała, and Targoński (2017b)
HS-SPME-GC-EI-MS (Targeted for pyrazynes and esters); SPME-GC–MS/MS (targeted for thiols)	Geographical	18 Sauvignon blanc wines from 3 different countries (France, Austria, New Zealand) and 2 vintage years	18 compounds analyzed and sensorial description *IBMP (New Zealand), 4MMP (Austria), benzaldehyde (France)*	PCA	Green et al. (2011)

Continued

Table 3 Relevant studies of wine authenticity based on targeted and untargeted analysis of volatile compounds by GC–MS.—cont'd

Methodology	Criteria	Samples	Number of features/ analyzed compounds *Most Discriminant Compounds*	Chemometrics	References
SPE_GC_MS	Geographical/ Yeast	30 experimental (Microferementation) and 2 commercial Negroamaro wines from 2 regions in South Italy	39 analyzed compounds *Fatty acids, ethyl octanoate, ethyl lactate, ethyl hexanoate, methyl vanillate, 2-methylbutanoic acid…*	PCA	Tufariello et al. (2014)
HS-SPME-GC-FID	Geographical	244 Spanish red wines from 5 different regions and 3 different varieties (Tempranillo, Garnacha and Airén)	12 major compounds (esters, higher alcohols, acids) *3-methyl-butyl acetate, 3-methyl-1-butanol, diethyl succinate, hexanoi acid, geraniol, octanoi acid*	PCA LDA	Sáenz, Cedrón, and Cabredo (2010)
Headspace stir bar sorptive extraction-GC–MS	Cultivar and Origin	334 young south African wines of six different cultivars (Sauvignon blanc, Chardonnay, Pinotage, Shiraz, Cabernet Sauvignon, Merlot) in 6 different regions	39 volatile compounds *Isoamyl acetate, Isoamyl alcohol, Ethyl octanoate (Pinitage wines)*	PCA LDA	Weldegergis, de Villiers, and Crouch (2011)
HS-SPME-GC–MS	Geographical	28 commercial Chilean Carignan wines from 6 different regions in two different vintage years	62 volatile compounds (esters, alcohols, terpenes, acids, aldehydes, lactones, norisoprenoids *Esters*	PCA LDA	Úbeda, del Barrio-Galán, Peña-Neira, Medel-Marabolí, and Durán-Guerrero (2017)

HS-GC–MS	Geographical, Method of production	33 Italian Prosecco wines from different Italy regions	45 volatile compounds *Alcohols, Esters, Norisoprenoid, Terpeness, etc.*	PCA	Forleo et al. (2020)
HS-SPME-GC–MS	Geographical, Winery	32 Italian wines (7 Valpolicella, 10 Amarone and 5 Reciotto)	54 volatile compounds *Ethyl-9-decanoate, benzyl alcohol (winery)* *Nonanal, Isobutyl alcohol, Ethyl phenol, Isoamyl acetate (wine tipology)*	PCA HCA	Dall'Asta, Cirlini, Morini, and Galaverna (2011)
SPE-GC-EI-MS (Untargeted)	Geographical Vintage effect, Barrel maturation effect, Tipicity/ clones	26 New Zealand Pinot Noir wines from 6 regions and different clones in 2 vintage years	99 compounds after PARAFAC 2 modeling *Vintage effect:esters,a cids, alcohols, acetamides* *Barrel effect: esters, lactones, vanillin derivatives* *Tipicity effect: terpenes, acetamide, alcohols*	Parallel Factor Analysis (PARAFAC 2) PCA	Schueuermann et al. (2016)
LLE-GC-FID	Physiology: Vigor of the plant	25 experimental (microvinification, industrial) Grenache wines from Somontano	29 major compounds (carbonyl compounds, esters, alcohols, acids, lactones) *Ethyl esters, acetaldehyde*	PCA	Arias, Lacau, Astraín, Barón, and Fernandez-Zurbano (2019)
LLE-GC-FID (major compounds) SPE-GC–MS (minor compounds)	Wine aging: Barreled wines vs oak chips	231 Garnacha Tinta wines from 6 Spanish DOs	75 compounds *11 dsicriminating compounds (lactones, vanillin derivatives, furfural derivatives)*	PCA	Hernández-Orte et al. (2014)

MS spectra, has proven to be highly efficient for variety discrimination. In a study with 198 wine samples focused on the 5 most frequent German white varieties (Riesling, Müller-Thurgau, Silvaner, Pinot Gris, Pinot blanc), the full information on 7290 scans was first preprocessed and cleaned to provide 5638 features. After PCA analysis, clear clusters were obtained except for differentiating between the two Pinot varieties. A PLS-DA model was validated and correctly classified Silvaner (97% correct classification), Riesling (93%), Müller-Thurgau (80%, probably due to a too reduced number of samples in the set), and Pinot (91%). After matching with MS databases and some available standards, the substances contributing to the classification were identified and belonged to the families of monoterpenoids, C13-norisoprenoids, and some esters (Springer et al., 2014). However, this study could not discriminate between vintages or locations of the samples. In these cases, a rebuilding of the model reducing its complexity could be necessary. To illustrate these effects, in the same study a better variety classification could be achieved by simplifying the model to a two-class comparison between Riesling and Pinot, for which a 100% correct classification was achieved (Springer et al., 2014).

It becomes more difficult to distinguish clones of the same variety. Some attempts have proven to discriminate clones that confer quite different characteristics to the wine. An Australian study with six clones of Pinot Noir was unable to distinguish all the six clones, but was able to discriminate between the clones that produced high and low site-typicity wines. The PCA analysis presented 30 volatile compounds that could be good candidates as markers for the typicity effect. By analyzing the consistency of these markers between 2 consecutive vintage years, a reduced number of metabolites seemed to be consistently associated with the typicity, including beta-citronellol, homovanillyl alcohol, *N*-(3-methylbutyl)acetamide and *N*-(2-phenylethyl)acetamide. In conclusion, the discrimination of clones remains extremely complex, since the pattern of discrimination for clone groups seems to be depending on other factors such as the vintage year (Schueuermann et al., 2016).

3.3.2 Profile analysis for origin discrimination

Geographical origin is one of the factors that have also been tested for wine discrimination and classification, which is a crucial factor for Protected Denominations of Origin (PDO). In reality, the "origin" includes factors as the type of grape variety used in the PDO, climate factors (temperature, rainfall), soil (limestone, calcareous, etc.), yielding of the plant, etc.

The variability in these factors will influence the complexity of the data. For example, the classification of wines from 5 Spanish origin denominations was achieved with high correct classification (98%) and good prediction ability (75%) using just 12 volatile compounds (Sáenz et al., 2010). Sometimes the separation is mostly associated with the different varieties used in each country rather than to the terroir or climate effect (Stój et al., 2017b). If considering monovarietal wines from 1 variety, a targeted method focused on the analysis of 15 major compounds (mostly esters and higher alchohols), 3 thiols and 1 pyrazine was successfully used for studying the influence of the source and the origin in New Zealand Sauvignon blanc wines compared with French and Austrian origins (Green et al., 2011). Despite satisfactory results can be found for discriminating wines from different countries, small geographical distances between countries difficult the discrimination. However, good results can also be achieved for closer geographic areas. In a study with Carignan wines in six different regions of Chile, the PCA showed clear differences according to the geographical origin, with lower contents of esters and acids in production areas closer to the mountains compared to the ones closer to the ocean. Linear Discriminant Analysis correctly classified the 100% of the wines to one of six regions (Úbeda et al., 2017). A good separation was also found in Negroamaro wines cultivated in two close zones of Salento, Puglia (Italy) through the analysis of major and minor volatiles. This discrimination due to the origin can also be influenced by the endogenous native yeasts that contribute to the must fermentation and which can be different in both areas (Tufariello et al., 2014). As a consequence, such studies can contribute to the selection of autochthonous yeast strains that can define the typicity of a wine.

The GC–MS analysis was also used for determining the volatile fingerprinting in Valpolicella, Amarone and Reciotto wines from the Italian Valpolicella region. These wines are produced from the same autochthonous grape cultivars (Corvina, Corvinone and Rondinella) aged in wood or oak barrels but differ mainly in the withering treatment time before vinification. The results allowed to discriminate between these three products and even were able to discriminate, for the same brand, between two different wineries (Dall'Asta et al., 2011).

3.3.3 Profile analysis for vintage discrimination

Besides grape variety and origin, it is well-known that vintage can affect the sensory perception and the chemical composition of wine aroma. Variations on weather condition, especially at certain stages of the grapes, are known to contribute to its chemical changes. A vintage effect was identified in Pinot blanc wines in New Zealand in 2 consecutive vintage years, with differences in the composition of esters, volatile acids, acetamides, alcohols and terpenes (Schueuermann et al., 2016). Other parameters that can contribute to discriminating between wines are the type of vineyard, or more concretely, its "potential" in terms of the vigor of the plant (Arias et al., 2019).

3.3.4 *Profile analysis for enological practices discrimination*

Other factors that have proven to be dominating on the discrimination of wines by the aroma profile are the winemaking conditions. Barreled wines showed increased amounts of oak- and aging-related compounds such as furanic compounds, phenolic aldehydes and oak lactones; while pre-barreled wines were higher in a range of higher alcohols and esters (Schueuermann et al., 2016). Vintage climate can also affect the aroma composition. The analysis of the volatile profile can also distinguish wines macerated with oak chips compared to wines aged in oak barrels. This can be highly usefully for avoiding possible frauds related to the accepted practice of using small oak fragments (chips) for a fast maceration of wines to mimic the barrel aging but with a faster and less expensive process (Hernández-Orte et al., 2014).

Recent studies have also highlighted the ability of the GC–MS volatile profiling in differentiating wines from other enological practices. In a study with Prosecco wines in Italy obtained from the same grape (Glera), several categories could be studied: the denomination/origin of the wines, but also discrimination of biological wines from the other samples, and even discrimination between wines produced by hand-picked harvesting and mechanical harvesting (Forleo et al., 2020).

3.4 Comprehensive GCxGC–MS for authenticity studies

In overall, the utilization of simple GC–MS has proven its applicability for authenticity studies but can be limited if the dataset is too complex and the categories to be discriminated include a high variance. For these cases, an increase in the coverage of the metabolome can lead to better results. Recent developments in GC–MS have greatly extended the analytical power of the technique.

Comprehensive GCxGC (two-dimensional gas chromatography) coupled to High-Resolution MS has emerged as a powerful analytical technique with excellent qualities to unravel the complex volatolome of wine (Welke et al., 2013). One-single dimensional GC–MS is limited in terms of co-elution of peaks, and can identify at most 10% of the known volatile compounds in wine. GCxGC improves the resolving power of the methodology and allows a higher number of tentatively identified compounds (Robinson et al., 2011). The technique is based on the application of two GC columns of different retention characteristics, so that the sample undergoes two different separation process, improving the analytes separation and enhancing the sensitivity of the analysis compared to the traditional monodimensional 1D-GC (Welke et al., 2013).

This GCxGC-TOF methodology has been used to study the varietal differentiation by analyzing the volatiles in 54 wines elaborated with 5 grape varieties: Cabernet Sauvignon, Merlot, Chardonnay, Sauvignon Blanc and Pinot Noir. 480 compounds could be detected by this two-dimensional chromatography, improving the high number of co-elutions which would have happened in 1D-GC. A PCA was performed on the 77 analytes which presented the highest variation among the 5 varieties. The PCA showed a clear separation by variety among the monovarietal wines and also for blendings 50% Pinot noir/50% Chardonnay. 12 volatile compounds were selected as markers of differentiation and classification of wines according to these grape cultivars. Chardonnay/Pinot noir blendings were differentiated from other wines by their contents in terpenes, Chardonnay wines were discriminated by two esters, and cabernet sauvignon wines were differentiated by two furanones (Welke et al., 2013).

The GCxGC methodology can also be combined with untargeted metabolomics, enhancing the power of the technique. GCxGC was applied for the untargeted study of regional origin in northern Italian premium sparkling wines of Trentodoc and Franciacorta (from Trento and Lombardy, respectively), elaborated with the same major varieties (Chardonnay, Pinot noir and Pinot blanc). After spectral deconvolution, 1695 features were detected, which were further shortlisted to 969 features present in at least 44 samples. Masses were matched with the NIST 2.0, Wiley 8 and the FFNSC2 mass spectral libraries. The 2D strategy allowed for the detection of many new compounds that would have probably coeluted and left unanalyzed in 1D chromatography. Some of these newly identified (or tentatively identified) compounds helped for the discrimination of the sparkling wines among the two Italian areas (Carlin et al., 2016). A higher level of monoterpenes, probably due to a difference in the microclimate of both regions, and several C6 and C5 alcohols were found in Trentino wines, while Franciacorta wines showed to be richer in diethyl esters and sulfur compounds.

3.5 Electronic nose for authenticity studies

Another alternative methodology used for the differentiation and classification of wines based on the analysis of volatile compound is the use of electronic noses (e-noses) based on chemical sensors (for example, metal oxide semiconductors or polymers) or mass spectrometers (Berna, Trowell, Clifford, Cynkar, & Cozzolino, 2009; Ziółkowska, Wąsowicz, & Jeleń, 2016). The principle relies on injecting the headspace of the wine without separation of its various constituents, and the sensor response (based in the physical or chemical interactions with the compounds) or the MS detected ions constitute a fingerprinting which is further analyzed by chemometric techniques. The speed and simplicity of the direct analysis is the main advantage of the technique compared to the time-demanding separation by GC–MS. However, several sample pretreatments must also be considered before analysis including SPME, purge and trap, or pre-evaporation to solve the interference of ethanol in the analysis (Beltran, Duarte-Mermoud, & Munoz, 2009).

The main disadvantage for electronic noses based on chemical or physical sensors was the lack of translation of the recorder signal into useful information about the specific compounds which could explain the discrimination among categories. However, the development of MS sensors, where the mass-to-charge ratio of the ions and their fragments can be obtained, improves this aspect. Besides, ethanol signal interference can be easily reduced in MS by not including the corresponding fragmentation ions in the analysis (Berna et al., 2009). Similar in conception is the Proton Transfer Reaction–MS (PTR-MS) instruments, frequently built with TOF spectrometers and which provide a rapid, non-invasive fingerprinting of the volatile organic compounds of the samples without chromatographic separation (Schueuermann, Bremer, & Silcock, 2017).

Such devices have been successfully applied for wine classification according to variety and geographical origin. For example, an SPME-MS e-nose profiling of 79 monovarietal wines from 5 different white and red grape varieties and from 9 different countries allowed a 100% varietal classification and prediction ability in LDA models among several types of red and white wines. For geographical origin, whose models were built specifically for each variety, results could be divergent (100% for Cabernet Sauvignon, 90% for Melot) (Ziółkowska et al., 2016). The geographical origin was also studied for 34 Sauvignon blanc samples from 3 different countries (Berna et al., 2009) but showed greater discrepancies, with wines clustered by classes rather than by regions. Better results could be observed in a study for discriminating 35 Tempranillo Australian wines and 25 Tempranillo Spanish wines by MS-enose fingerprinting. The SLDA correctly classified the origin for 86% of the samples (Cynkar, Dambergs, Smith, & Cozzolino, 2010) The analysis of Pinot noir wines by PTR-MS for investigating differences based on vineyard site in Australia allowed the discrimination between wines coming from high typicity sites from those originated in low typicity regions (Schueuermann et al., 2017).

4. Nuclear magnetic resonance (NMR) for wine authenticity studies

NMR spectroscopy has been increasingly applied in food science and agriculture for analysis, control and authentication purposes (Amargianitaki & Spyros, 2017; Schwarzinger, 2018; Sobolev et al., 2019). Despite the low sensitivity of the methodology in relation to MS, wine NMR analysis treasures multiple advantages for high-throughput studies: the ease of the sample preparation, which reduces the risk of manipulation artifacts and shortens the overall analysis time (Picone et al., 2016), its high reproducibility, and the fact that the spectra contain a rich amount of information allowing the identification of major components in complex mixtures in a single analysis (Gougeon et al., 2018).

Several targeted methods based on NMR have been established for the analysis of wine. Alcohol content can be easily determined by benchtop MHz NMR spectrometers (Isaac-Lam & Zuiderweg, 2016). The application of Site-specific Natural Isotopic Fractionation nuclear magnetic resonance (SNIF-NMR) for determining the isotope Deuterium/Hydrogen ratio of ethanol has been used for geographical origin detection (International Organisation of Vine and Wine, 2019), and a ^{1}H NMR technique for the quantification of glucose, malic acid, acetic acid, fumaric acid, shikimic acid and sorbic acid has been adopted within the OIV scientific chair as a promising quantitative analytical technique (Herbert-Pucheta, Mejía-Fonseca, Zepeda-Vallejo, Milmo-Brittingham, & Maya, 2019). Other studies have developed a quantitative ^{13}C NMR analysis that could be used to detect the presence of added sugar in sweet wines (Colombo, Aupic, Lewis, & Pinto, 2015).

Recent developments in quantitative ^{1}H NMR (qNMR) targeted metabolomics aim at identifying and quantifying concrete compounds from selected recognized signals in the spectra (Gougeon et al., 2018). The quantitative approach benefits of the high dynamic concentration range of the technique and the universal response factor for every hydrogen in the sample, which is independent of the molecule where it is found (Picone et al., 2016). That implies that the signal can be easily translated in terms of concentration by integrating the peak areas of the compounds in the sample and referring them to an internal standard added in the mixture at a known concentration. However, untargeted methods are much more interesting for metabolomic NMR studies in wines. The whole NMR spectrum of a wine sample can be considered as a fingerprint with inherent information about variety, origin, vintage, or enological treatments, among others (Godelmann et al., 2013). This information hidden in the spectra fingerprint can be used for comparison and identification purposes (Amargianitaki & Spyros, 2017) by applying chemometric techniques.

Most NMR wine studies for authenticity testing are based on ^{1}H NMR acquisition (Amargianitaki & Spyros, 2017; Sobolev, Circi, & Mannina, 2016) because of the greater sensitivity of the technique over ^{13}C NMR approaches. Remarkably, sensitivity is the main issue for NMR studies. Since NMR sensitivity and resolution is dependent on the field strength, moving to higher fields improves sensitivity (Schwarzinger, 2018). Nevertheless, instrumental costs and the difficulty to establish reference data to be included in homogeneous spectral databases have lead to a *de facto* industry standard of 400 MHz instrument, which is the most widely used NMR spectrometer for food analysis (Schwarzinger, 2018). It is worth noting that ^{13}C can also find a place for some metabolomics studies, because the results can be complementary to those obtained from ^{1}H NMR (Geana et al., 2016), reinforcing the annotation of compounds by applying both techniques. The main advantage of ^{13}C NMR is that carbon frequencies are spread over a larger spectral window (between 0 to about 220 ppm, while in the case of ^{1}H NMR the spectra range between 0 to about 12 ppm). The larger spectral window of ^{13}C can enhance the resolution of peaks and reduce overlaps (Mannu et al., 2020), which can be of interest in the case of complex studies. However, the experiments with ^{13}C will compromise the high throughput qualities of the technique in terms of the expended time of acquisition to achieve an acceptable sensitivity. Analogously, the use of two-dimensional experiments could serve to assign novel resonances in crowded regions of the spectra involving relevant metabolites associated to enological practices or quality control parameters (Herbert-Pucheta, Pino-Villar, et al., 2019), but their use is limited because of the long acquisition times required for the analysis.

4.1 Sample and data treatment in NMR analysis of wines

Despite the easiness of the NMR technique for fast and simple wine analysis, minimum sample pretreatment is required. Because NMR is a structural-dependent technique, it must be taken into account that many wine components (organic acids, amino acids, phenolic compounds) can adopt different conformations depending on the pH. As a consequence, wine pH can be adjusted in order to minimize chemical shift variations and harmonize all the samples of a set before pursuing comparative studies (Gougeon, da Costa, Richard, & Guyon, 2019). Individual pH adjustment could, however, result in destroying meaningful physical-chemical information that is contained in signal shifts, for which a single addition of a buffered solution to wine samples, allowing to reduce the pH variation range without a complete adjustment at the same pH for all the samples has been proposed as an alternative approach (Magdas, Pirnau, Feher, Guyon, & Cozar, 2019).

Another major concern is that water and ethanol, as the major constituents in wine, provide too intense overlapping peaks in the wine spectra, hindering the capability of the technique for obtaining valuable information of other compounds of the sample. Both compounds can be eliminated physically by convenient sample pretreatment (evaporation, freeze-drying). However, the development of spectroscopically NMR suppression techniques (Amargianitaki & Spyros, 2017) has enormously simplified the acquisition of informative spectra of wines. The suppression of water signal by optimized pulse sequences is common in many metabolomic studies (Giraudeau, Silvestre, & Akoka, 2015). The challenge in the wine matrix is going beyond the standard water presaturation pulse program to obtain the simultaneous suppression of water and ethanol resonances, which has been achieved by inserting a shaped pulse sequence during the relaxation delay of the 1D 1H NOESYGPPS NMR pulse sequence to suppress the eight ^{1}H NMR frequencies of water and ethanol peaks (Monakhova et al., 2011). The WET1D pulse sequence can suppress the methyl and ethyl peaks of ethanol as well as the residual water (Fotakis & Zervou, 2016). The choice of the solvent suppression scheme and the associated parameters has a high impact on the accuracy of the resulting spectra (Fotakis & Zervou, 2016) but a good compromise must be reached between the sensitivity limitations and the reduced experimental time, which is a critical issue in metabolomics studies (Giraudeau et al., 2015).

The usual spectra preprocessing in untargeted approaches includes phase correction, baseline correction applied over the entire spectral range, segment-wise peak alignment over the whole chemical shift range of the ^{1}H NMR spectra to be analyzed (Fan et al., 2018), spectra normalization and data reduction by bucketing of the spectra into sequential segments (bins) obtaining an integral for each of them (Godelmann et al., 2013). Afterward, some selected bins containing uninteresting compounds (residual water and ethanol regions after suppression) are usually removed to provide a dataset of buckets that will be further analyzed by multivariate data analysis.

Concerning the identification of wine major components excluding water and ethanol, the most common major compounds identified by NMR include (Fig. 4) alcohols (methanol, 3-methylbutanediol, 2,3-methylbutanediol, glycerol, and ethanol); organic acids (lactic, malic, acid, citric, succinic, acetic, fumaric, tartaric, shikimic, caftaric), carbonyl compounds (ethanal), amino acids (proline, alanine, valine), sugars (glucose, fructose, arabinose, xylose) and polyphenols (gallic acid, caffeic acid, catechin) (Ali, Maltese, Toepfer, Choi, & Verpoorte, 2011; Cassino, Tsolakis, Bonello, Gianotti, & Osella, 2017; Gougeon, da Costa, Guyon, & Richard, 2019).

4.2 Application of NMR for authenticity studies

Table 4 resumes some studies that have applied NMR techniques for the discrimination of wines.

4.2.1 NMR for varietal, geographical and vintage year discrimination

Several studies have shown a discrimination of wines by ^{1}H NMR according to the variety and type of wine. Results depend on the heterogeneity of the samples in terms of other confounding factors (geographical origin, vintage year) and also on the genetic homology between cultivars. In a study with 20 experimental red wines from 5 varieties, the HCA dendrogram showed the closeness between Cabernet Sauvignon, Cabernet Franc, and Merlot clustered in one node, which was expected for its shared progeny, while Syrah and Zinfadel, were clustered in another group. The most discriminating compounds were proline, valine, gallic acid, tartaric acid, glucose, glycerol, 2,3-butanediol, succinic acid, lactic acid and acetic acid (Hu et al., 2015), showing that compounds found in grape and compounds formed during fermentation can contribute to the discrimination of wines.

When a larger amount of samples is included, the ^{1}H NMR spectroscopy has also given satisfactory results despite the heterogeneity of the samples. One of the first large studies was conducted by Godelmann et al. (Godelmann et al., 2013) with 600 wines from 5 wine-growing areas of Germany. Correct predictions of 95% were obtained for Pinot noir,

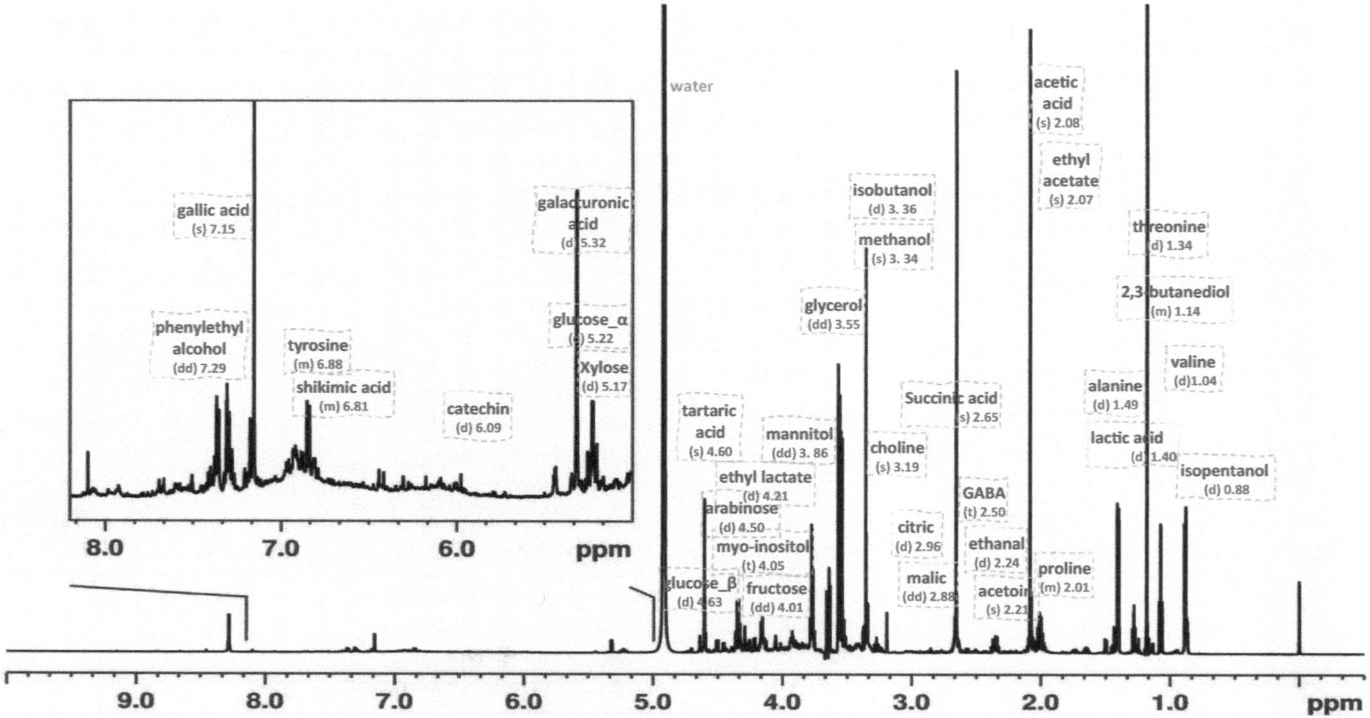

Fig. 4 Example of a ^{1}H NMR spectrum of a commercial wine, including assignation of compounds to the signals of the spectrum. s: singlet, d: doublet, dd: double doublet, t: triplet, m: multiplet.

Table 4 Relevant studies of wine authenticity based on NMR spectroscopy.

Experiment/ Instrumentation	Criteria	Samples and pretreatment	Range of study Identified compounds	Chemometrics	References
^{1}H NMR, 600 MHz	Varietal	20 experimental wines, same vintage year, same winery, 5 cultivars (Merlot, Syrah, Zinfander, Ruby Cabernet, Cabernet sauvignon) *Lyophylised wines*	17 identified compounds	PCA PLS-DA	Hu et al. (2015)
^{1}H NMR, 400 MHz	Varietal	170 chinese wine samples, from different regions and 5 vintages.: 99 red wines (Cabernet Sauvignon, Rose Honey, Cabernet Gernitsh); 71 white samples (29 Ugni Blanc, 21 Chardonnay, 21 Long Yan *pH adjustment 3.00*	0.8–9,7 ppm (0.01 ppm buckets)	PCA/LDA models	Fan et al. (2018)
^{1}H NMR, 500 MHz	Wine type, Variety	917 samples of wines from different regions of Czech Republic, different types of wine (white wines, sweet white wines, rosés, red wines) and different varieties *pH adjustment to 4.00*	226 bins of 0.04 ppm 38 identified metabolites	PCA PLS-DA	Mascellani et al. (2020)
^{1}H NMR, 500 MHz	Variety, Vintage year, Wine quality	59 German white wines from 2 varieties (Riesling and Mueller-Thurgau) and 2 vintage years *Ethyl acetate extraction*	0–10 ppm, bins of 0.04 ppm 41 identified compounds	PCA PLS-DA	Ali et al. (2011)

^{1}H NMR, ^{13}C NMR, 400 MHz LC-UV (anthocyanins, organic acids), Isotopic analysis	Varietal, Vintage year	56 experimental wines from Romania, of 5 different varieties (Cabernet Sauvignon, Merlot, Feneasca Neagra, Pinot Noir, Mamaia), from one vineyard, and different vintage years *pH adjustment 3.10*	19 ^{1}H NMR significant variables, 3 ^{13}C NMR significant variables	LDA	Geana et al. (2016)
^{1}H NMR, 500 MHz	Geographical origin, Vintage year	6 Barbera red wines from 3 Italy Piedmont region in 2 vintage years *pH adjustment 3.10*	0.5–10 ppm (bins of 0.04 ppm) 65 identified metabolites	PCA	Cassino et al. (2017)
^{1}H NMR, 600 MHz	Geographical origin	19 Chinese wines from 2 different regions *Wine at pH 3.10*	33 identified compounds	PCA	Gougeon et al. (2018)
^{1}H NMR, 600 MHz	Geographical origin	127 Bordeaux wines from 6 different appellations *Addition of buffer solution*	40 identified compounds	PCA OCS-PLS-DA	Gougeon, da Costa, Guyon, and Richard (2019)
^{1}H NMR, 600 MHz	Geographical origin	22 experimental Rioja wines from 9 wineries in 3 vintage years *Lyophylised wines*	Three different regions of the spectra: 1.3–3.2 ppm (organic acids), 3.2–5.5 ppm (sugars), 6.4–10 ppm (aromatic compounds) 31 identified compounds	PCA	López-Rituerto et al. (2012)
^{13}C NMR 500 MHz	Geographical origin	36 wine samples of several varieties from 7 greek regions	0–220 ppm (bins of 0.1 ppm)	PCA MANOVA k-NN PLS-DA	Mannu et al. (2020)

Continued

Table 4 Relevant studies of wine authenticity based on NMR spectroscopy.—cont'd

Experiment/ Instrumentation	Criteria	Samples and pretreatment	Range of study Identified compounds	Chemometrics	References
^{1}H NMR, 400 MHz LC-DAD-MS	Varietal, Geographical origin	22 southern Italy wines from 22 varieties, 5 regions, one vintage year *pH adjustment to 3.10*	0.5–10 ppm (bins of 0.04 ppm)	PCA PLS-DA	Ragone et al. (2015)
^{1}H NMR, 500 MHz	Varietal, Geographical	107 wine samples from 2 countries (Romania and France), from 4 cultivars (Sauvignon Blanc, Riesling, Chardonnay and Pinot Gris), 5 different vintage years *pH adjustment to 3.10*	Range: spectra divided in 4 distinct areas Identification of discriminating compounds	LDA	Magdas et al. (2019)
^{1}H NMR, 400 MHz	Variety, Region and Vintage	67 Greek wines from 4 varieties (Agiorgitiko, Mandilaria, Moschofilero, Asyrtiko), from 3 different wineries and in 2 different years *Fractionation of the polyphenols in XAD resins*	Aromatic region: 5.40–10.48 ppm (bins of 0.04 ppm) 10 identified phenolic compounds	PCA PLS-DA	Anastasiadi et al. (2009)
^{1}H NMR, 400 MHz	Varietal, Geographic origin, Vintage year	574 german wines from 28 varieties, 5 regions and 2 vintage years *pH adjustment 3.10*	0.5–9.5 ppm 16 identified compounds	PCA LDA MANOVA Monte-Carlo cross validation	Godelmann et al. (2013)

^{1}H NMR, 600 MHz	Varietal, Geographical origin	28 commercial wines from 3 varieties (Cabernet Sauvignon, Campbell Early, Shiraz), different countries (France, USA, Australia, Korea) and 4 vintage years *Lyophilisation*	bins of 0.001 ppm	PCA PLS-DA	Son et al. (2008)
^{1}H NMR, LC-MS, 600 MHz	Viticultural practices, vintage year	88 experimental wine samples from 2 varieties (Syrah and Chenin Blanc) in two different years, two different training systems (espalier and lyre) and 6 different rootstocks *Vaccum dried*	0.7–9.9 ppm, Bins of 0.04 ppm 23 identified compounds	PCA iPLS	Alves Filho et al. (2019)
^{1}H NMR, 600 MHz	Viticultural practice and winemaking effect	32 experimental Negroamaro red wines from Italy (Apulia) obtained with 3 vinification technologies (traditional, ultrasounds, cryomaceration) and 2 different soil management practices (soil tillage, cover crop) and two training systems (monolateral or bilateral Guyot) *Evaporated wines*	0.5–9.5 ppm, bins of 0.04 ppm 19 identified compounds	PCA OPLS-DA	de Pascali et al. (2014)

Continued

Table 4 Relevant studies of wine authenticity based on NMR spectroscopy.—cont'd

Experiment/ Instrumentation	Criteria	Samples and pretreatment	Range of study Identified compounds	Chemometrics	References
^{1}H NMR, 600 MHz	Viticultural practice	16 experimental wines from organic and biodynamic grapes	Two regions: polyphenols and carbohydrates 11 identified compounds	PCA HCA	Laghi, Versari, Marcolini, and Parpinello (2014)
^{1}H NMR, 400 MHz	Winemaking effect	40 experimental wines with autochthonous or commercial yeasts	0.4–7.8 ppm, 115 buckets with variable lengths 5 discriminating metabolites identified	PCA HCA	Mazzei, Spaccini, Francesca, Moschetti, and Piccolo (2013)
^{1}H NMR, 800 MHz	Variety, Viticultural practices	Over 50 experimental wines of 2 varieties (Cabernet Sauvignon, Shiraz) with 4 diffeent canopy exposure treatments (control, exposed, highly exposed, shaded)	0.2–9 ppm, 28,373 data points	PCA PLS-DA	Rochfort, Ezernieks, Bastian, and Downey (2010)

Lemberger, Pinot blanc/Pinot gris, Müller-Thurgau, Riesling and Gewürztraminer, although another variety (Sylvaner) was classified only 64% correctly. Following a similar NMR approach, the correct classification in a set of 170 Chinese red wines were 68% for Cabernet Sauvignon, 100% for red Honey, 81% for Cabernet Gernisch, 93% for Ugni Blanc, 95% for Long Yan, and 94% for Chardonnay (Fan et al., 2018). The predictive models built by the NMR profile of 917 Czech wines classified correctly Pinot noir (96%), Blaufrankisch (96%), Riesling (92%), Cabernet Sauvignon (77%), Chardonnay (76%), Gewurztraminer (60%), Hibernal (60%), Gruner Veltliner (52%), Pinot gris (48%), Sauvignon blanc (45%) and Palava (40%) (Mascellani et al., 2020).

In many studies, the analysis of the varieties is linked with the analysis of the geographical origin or the vintage year, which are other discriminating factors that can be revealed by NMR. The differentiation of Cabernet Sauvignon wines from Australia, France and California was observed by PLS-DA (Son et al., 2008), together with the differentiation from other cultivars (Australian Shiraz and Korean Campbell early). The methodology can also be applied to close regions. For example, a 100% of correct geographical classification was achieved for long distances (Romanian vs French samples) and short distances (three Romanian regions) of Sauvignon, Riesling and Chardonnay wines (Magdas et al., 2019). A singular imprint of Bordeaux wines with classification rates ranging from 71% to 100% was found in comparison to other five major French regions, in agreement with the distribution of grape varieties (Gougeon, da Costa, Guyon, & Richard, 2019). The aforementioned study of Godelmann et al. showed that the geographic origin of the 600 studied wines could be correctly predicted with an average correction of 89% for the 5 German regions of the study (Godelmann et al., 2013).

NMR studies can also be applied for discrimination according to the vintage year. Wine evolution was followed during bottle aging by ^{1}H NMR and the comparison of profiles between 2 and 4 years storage established clear differences by PCA, associated to a decrease in organic acids, catechin and epicatechin, and an increase in esters and gallic acid (Cassino, Tsolakis, Bonello, Gianotti, & Osella, 2019).

4.2.2 NMR for viticultural and winemaking practices

NMR can also reveal differences in cultural practices. Training systems influence strongly the chemical composition of the grape due to the different sunlight exposure level of the vine canopy (Alves Filho et al., 2019), and

PCA showed clear discrimination for Cabernet Sauvignon and Syrah wines obtained from shaded grapes and highly exposed grapes (Rochfort et al., 2010). Soil management practices (soil tillage and cover crop) also impacted the NMR profile of the samples and could be differentiated by PCA (de Pascali et al., 2014). Interesting results have also been found for wines obtained for different cultivating methods (organic and biodynamic) grapes of Sangiovese (Laghi et al., 2014) The NMR analysis followed by PCA discriminated among both types of wine, and the organic wine showed higher levels of resveratrol and caffeic acid. Resveratrol is a known phytoalexin that is stimulated in fungal attacks, which could explain its higher presence in organic wines.

NMR-based metabolomics was also able to discriminate according to the microorganisms used in the fermentation. For example, PCA of the ^{1}H NMR spectra showed a significant diversity in the metabolome of a wine obtained through fermentation by a commercial yeast compared to the wine obtained through fermentation with a selected autochthonous *Saccharomyces cerevisiae* starter (Mazzei et al., 2013). A significant difference assessed by ^{1}H NMR was also found in the metabolome between control wines and wines added with different inactive yeasts (Hu, Cao, Zhu, Xu, & Wu, 2019).

The comparison of the NMR fingerprinting for alternative winemaking techniques has also been proven. The NMR profile of wines obtained with ultrasonic-assisted extraction or cryomaceration could be differentiated from traditional maceration wines (de Pascali et al., 2014).

5. Conclusions and future perspectives

Wine metabolomics analyses by MS and NMR approaches have proven a powerful tool for the classification of wines according to the variety, the geographical origin, the vintage year or the winemaking conditions, which constitute the main factors to be addressed for authenticity studies to combat frauds and counterfeits. Wine authenticity studies can be highly complex because wine is a very complex matrix and usually there are no specific markers, but rather a relative difference in levels of wine metabolites that are dependent on all those factors and their interactions (Gil et al., 2020). As a consequence, wine samples can be highly heterogeneous and include a lot of statistical noise caused by the intrinsic variability within the samples. Moreover, since metabolomics is a relatively young technique—expanding very fast—the workflows are not as robust and well-designed as for targeted analysis, and there is still debate about experimental design and method validation (Arapitsas, Ugliano, et al., 2016).

The new generation of metabolomics instruments, aiming at covering a higher density of the metabolome, will certainly contribute to a better characterization of biomarkers in the field of wineomics. However, the advances on this research area do not solely depend on more performant acquisition instruments leading to an increase of the number of compounds or features to be included in the analysis, but also on the development of powerful approaches for data analysis (such as machine learning techniques) and automatisation that maximize the actual treatment of the complex data arising from metabolomic studies.

The success of wine metabolomics will depend also on the application of multiplatform methodologies rather than a technique alone. Data integration of complementary non-redundant information from NMR, LC-MS, GC–MS and other spectroscopic techniques will improve the discriminating power of the metabolomics approach, increase the robustness of the studies and allow for revealing subtle differences between wines. In order to maximize the impact of wine metabolomic studies related to authenticity and typicity, data fusion should be able to incorporate also wine sensorial assessments by trained experts, which will contribute to a deeper knowledge of the relationship between the chemical composition and wine intrinsic quality and typicity.

Acknowledgments

The authors want to thank P. Pétriacq for his continuous assistance in the manuscript, A. Pons and C. Thibon for their kind and valuable comments related to the GC methodologies, and G. da Costa for his assistance in the NMR subchapter. This work was supported by the Bordeaux Metabolome Facility and MetaboHUB (ANR-11-INBS-0010) project. G.L. acknowledges the grant provided by the French Ministry of Higher Education and Research.

References

Alañón, M. E., Pérez-Coello, M. S., & Marina, M. L. (2015). Wine science in the metabolomics era. *TrAC Trends in Analytical Chemistry*, *74*, 1–20. https://doi.org/10.1016/j.trac.2015.05.006.

Ali, K., Maltese, F., Toepfer, R., Choi, Y. H., & Verpoorte, R. (2011). Metabolic characterization of Palatinate German white wines according to sensory attributes, varieties, and vintages using NMR spectroscopy and multivariate data analyses. *Journal of Biomolecular NMR*, *49*(3–4), 255–266. https://doi.org/10.1007/s10858-011-9487-3.

Allamy, L., Darriet, P., & Pons, A. (2018). Molecular interpretation of dried-fruit aromas in merlot and cabernet sauvignon musts and young wines: Impact of over-ripening. *Food Chemistry*, *266*, 245–253. https://doi.org/10.1016/j.foodchem.2018.06.022.

Alves Filho, E. G., Silva, L. M. A., Ribeiro, P. R. V., Brito, E. S. de, Zocolo, G. J., Souza-Leão, P. C., ... Canuto, K. M. (2019). 1h NMR and LC-MS-based metabolomic approach for evaluation of the seasonality and viticultural practices in wines from São Francisco River valley, a Brazilian semi-arid region. Food Chemistry, 289, 558–567. https://doi.org/10.1016/j.foodchem.2019.03.103.

Amargianitaki, M., & Spyros, A. (2017). NMR-based metabolomics in wine quality control and authentication. *Chemical and Biological Technologies in Agriculture*, *4*(1), 126. https://doi.org/10.1186/s40538-017-0092-x.

Anastasiadi, M., Zira, A., Magiatis, P., Haroutounian, S. A., Skaltsounis, A. L., & Mikros, E. (2009). 1H NMR-based metabonomics for the classification of Greek wines according to variety, region, and vintage. Comparison with HPLC data. *Journal of Agricultural and Food Chemistry*, *57*(23), 11067–11074. https://doi.org/10.1021/jf902137e.

de Andrade, R. H. S., do Nascimento, L. S., Pereira, G. E., Hallwass, F., & Paim, A. P. S. (2013). Anthocyanic composition of Brazilian red wines and use of HPLC-UV–vis associated to chemometrics to distinguish wines from different regions. *Microchemical Journal*, *110*, 256–262. https://doi.org/10.1016/j.microc.2013.04.003.

Arapitsas, P., Della Corte, A., Gika, H., Narduzzi, L., Mattivi, F., & Theodoridis, G. (2016). Studying the effect of storage conditions on the metabolite content of red wine using HILIC LC-MS based metabolomics. *Food Chemistry*, *197*(Pt B), 1331–1340. https://doi.org/10.1016/j.foodchem.2015.09.084.

Arapitsas, P., Oliveira, J., & Mattivi, F. (2015). Do white grapes really exist? *Food Research International*, *69*, 21–25. https://doi.org/10.1016/j.foodres.2014.12.002.

Arapitsas, P., Perenzoni, D., Nicolini, G., & Mattivi, F. (2012). Study of Sangiovese wines pigment profile by UHPLC-MS/MS. *Journal of Agricultural and Food Chemistry*, *60*(42), 10461–10471. https://doi.org/10.1021/jf302617e.

Arapitsas, P., Scholz, M., Vrhovsek, U., Di Blasi, S., Biondi Bartolini, A., Masuero, D., et al. (2012). A metabolomic approach to the study of wine micro-oxygenation. *PLoS One*, 7(5), e37783. https://doi.org/10.1371/journal.pone.0037783.

Arapitsas, P., Ugliano, M., Marangon, M., Piombino, P., Rolle, L., Gerbi, V., et al. (2020). Use of untargeted liquid chromatography-mass spectrometry metabolome to discriminate Italian Monovarietal red wines, produced in their different terroirs. *Journal of Agricultural and Food Chemistry*, *68*(47), 13353–13366. https://doi.org/10.1021/acs.jafc.0c00879.

Arapitsas, P., Ugliano, M., Perenzoni, D., Angeli, A., Pangrazzi, P., & Mattivi, F. (2016). Wine metabolomics reveals new sulfonated products in bottled white wines, promoted by small amounts of oxygen. *Journal of Chromatography. A*, *1429*, 155–165. https://doi.org/10.1016/j.chroma.2015.12.010.

Arbulu, M., Sampedro, M. C., Gómez-Caballero, A., Goicolea, M. A., & Barrio, R. J. (2015). Untargeted metabolomic analysis using liquid chromatography quadrupole time-of-flight mass spectrometry for non-volatile profiling of wines. *Analytica Chimica Acta*, *858*, 32–41. https://doi.org/10.1016/j.aca.2014.12.028.

Arias, I., Lacau, B., Astraín, J., Barón, C., Fernandez-Zurbano, P., & Ferreira, Vicente, Escudero, Ana (2019). Effects of vineyard 'potential' and grape maturation on the aroma-volatile profile of Grenache wines. OENO One, 53(4). https://doi.org/10.20870/oeno-one.2019.53.4.2381.

Arvanitoyannis, I. (1999). Application of quality control methods for assessing wine authenticity: Use of multivariate analysis (chemometrics). *Trends in Food Science & Technology*, *10*(10), 321–336. https://doi.org/10.1016/S0924-2244(99)00053-9.

Arvanitoyannis, I. S. (2010). Wine authenticity, traceability and safety monitoring. In *Managing Wine Quality* (pp. 218–270). Elsevier. https://doi.org/10.1533/9781845699284.2.218.

Azcarate, S. M., Martinez, L. D., Savio, M., Camiña, J. M., & Gil, R. A. (2015). Classification of monovarietal Argentinean white wines by their elemental profile. *Food Control*, *57*, 268–274. https://doi.org/10.1016/j.foodcont.2015.04.025.

Basalekou, M., Pappas, C., Tarantilis, P. A., & Kallithraka, S. (2020). Wine authenticity and traceability with the use of FT-IR. *Beverages*, *6*(2), 30. https://doi.org/10.3390/beverages6020030.

Beltran, N. H., Duarte-Mermoud, M., & Munoz, R. E. (2009). Geographical classification of Chilean wines by an electronic nose. *International Journal of Wine Research*, *2009*, 209. https://doi.org/10.2147/IJWR.S4609.

Berna, A. Z., Trowell, S., Clifford, D., Cynkar, W., & Cozzolino, D. (2009). Geographical origin of sauvignon blanc wines predicted by mass spectrometry and metal oxide based electronic nose. *Analytica Chimica Acta*, *648*(2), 146–152. https://doi.org/10.1016/J.aca.2009.06.056.

Budić-Leto, I., Zdunić, G., Gajdoš-Kljusurić, J., Mucalo, A., & Vrhovšek, U. (2017). Differentiation between Croatian dessert wine Prošek and dry wines based on phenolic composition. *Journal of Food Composition and Analysis*, *62*, 211–216. https://doi.org/10.1016/j.jfca.2017.05.015.

Cabrita, M. J., Martins, N., Barrulas, P., Garcia, R., Dias, C. B., Pérez-Álvarez, E. P., et al. (2018). Multi-element composition of red, white and palhete amphora wines from Alentejo by ICPMS. *Food Control*, *92*, 80–85. https://doi.org/10.1016/j.foodcont.2018.04.041.

Cajka, T., Showalter, M. R., Riddellova, K., & Fiehn, O. (2016). Advances in mass spectrometry for food authenticity testing. In *Advances in food authenticity testing* (pp. 171–200). Elsevier. https://doi.org/10.1016/B978-0-08-100220-9.00007-2.

Canizo, B. V., Escudero, L. B., Pellerano, R. G., & Wuilloud, R. G. (2019). Quality monitoring and authenticity assessment of wines: Analytical and Chemometric methods. In *Quality control in the beverage industry* (pp. 335–384). Elsevier. https://doi.org/10.1016/B978-0-12-816681-9.00010-2.

Carlin, S., Vrhovsek, U., Franceschi, P., Lotti, C., Bontempo, L., Camin, F., et al. (2016). Regional features of northern Italian sparkling wines, identified using solid-phase micro extraction and comprehensive two-dimensional gas chromatography coupled with time-of-flight mass spectrometry. *Food Chemistry*, *208*, 68–80. https://doi.org/10.1016/j.foodchem.2016.03.112.

Carpentieri, A., Sebastianelli, A., Melchiorre, C., Pinto, G., Trifuoggi, M., Lettera, V., et al. (2019). Fiano, Greco and Falanghina grape cultivars differentiation by volatiles fingerprinting, a case study. *Heliyon*, *5*(8), e02287. https://doi.org/10.1016/j.heliyon.2019.e02287.

Cassino, C., Tsolakis, C., Bonello, F., Gianotti, V., & Osella, D. (2017). Effects of area, year and climatic factors on Barbera wine characteristics studied by the combination of 1 H-NMR metabolomics and chemometrics. *Journal of Wine Research*, *28*(4), 259–277. https://doi.org/10.1080/09571264.2017.1388225.

Cassino, C., Tsolakis, C., Bonello, F., Gianotti, V., & Osella, D. (2019). Wine evolution during bottle aging, studied by 1H NMR spectroscopy and multivariate statistical analysis. *Food Research International (Ottawa, Ont.)*, *116*, 566–577. https://doi.org/10.1016/j.foodres.2018.08.075.

Castro, C. C., Martins, R. C., Teixeira, J. A., & Silva Ferreira, A. C. (2014). Application of a high-throughput process analytical technology metabolomics pipeline to port wine forced ageing process. *Food Chemistry*, *143*, 384–391. https://doi.org/10.1016/j.foodchem.2013.07.138.

Causon, T. J., Ivanova-Petropulos, V., Petrusheva, D., Bogeva, E., & Hann, S. (2019). Fingerprinting of traditionally produced red wines using liquid chromatography combined with drift tube ion mobility-mass spectrometry. *Analytica Chimica Acta*, *1052*, 179–189. https://doi.org/10.1016/j.aca.2018.11.040.

Chapman, J., Gangadoo, S., Truong, V. K., & Cozzolino, D. (2019). Spectroscopic approaches for rapid beer and wine analysis. *Current Opinion in Food Science*, *28*, 67–73. https://doi.org/10.1016/j.cofs.2019.09.001.

Colombo, C., Aupic, C., Lewis, A. R., & Pinto, B. M. (2015). In situ determination of fructose isomer concentrations in wine using (13)C quantitative nuclear magnetic resonance spectroscopy. *Journal of Agricultural and Food Chemistry, 63*(38), 8551–8559. https://doi.org/10.1021/acs.jafc.5b03641.

Cuadros-Inostroza, A., Giavalisco, P., Hummel, J., Eckardt, A., Willmitzer, L., & Peña-Cortés, H. (2010). Discrimination of wine attributes by metabolome analysis. *Analytical Chemistry, 82*(9), 3573–3580. https://doi.org/10.1021/ac902678t.

Cuadros-Inostroza, Á., Verdugo-Alegría, C., Willmitzer, L., Moreno-Simunovic, Y., & Vallarino, J. G. (2020). Non-targeted metabolite profiles and sensory properties elucidate commonalities and differences of wines made with the same variety but different cultivar clones. *Metabolites, 10*(6), 220. https://doi.org/10.3390/metabo10060220.

Cynkar, W., Dambergs, R., Smith, P., & Cozzolino, D. (2010). Classification of Tempranillo wines according to geographic origin: Combination of mass spectrometry based electronic nose and chemometrics. *Analytica Chimica Acta, 660*(1–2), 227–231. https://doi.org/10.1016/j.aca.2009.09.030.

Da Costa, N. L., Llobodanin, L. A. G., de Lima, M. D., Castro, I. A., & Barbosa, R. (2018). Geographical recognition of Syrah wines by combining feature selection with extreme learning machine. *Measurement, 120*, 92–99. https://doi.org/10.1016/j.measurement.2018.01.052.

Dall'Asta, C., Cirlini, M., Morini, E., & Galaverna, G. (2011). Brand-dependent volatile fingerprinting of Italian wines from Valpolicella. *Journal of Chromatography. A, 1218*(42), 7557–7565. https://doi.org/10.1016/J.chroma.2011.08.042.

Dey, G., & Montet, D. (2017). Chemical and biological 'barcodes' or markers in food traceability a case study on wines. In D. Montet, & R. C. Ray (Eds.), *Food traceability and authenticity* (pp. 90–115). Boca Raton, FL: CRC Press. 2017 | Series: Food biology series | "A science publishers book.": CRC Press. https://doi.org/10.1201/9781351228435-5.

Díaz, R., Gallart-Ayala, H., Sancho, J. V., Nuñez, O., Zamora, T., Martins, C. P. B., et al. (2016). Told through the wine: A liquid chromatography-mass spectrometry interplatform comparison reveals the influence of the global approach on the final annotated metabolites in non-targeted metabolomics. *Journal of Chromatography. A, 1433*, 90–97. https://doi.org/10.1016/j.chroma.2016.01.010.

Duchowicz, P. R., Giraudo, M. A., Castro, E. A., & Pomilio, A. B. (2013). Amino acid profiles and quantitative structure–property relationship models as markers for Merlot and Torrontés wines. *Food Chemistry, 140*(1–2), 210–216. https://doi.org/10.1016/j.foodchem.2013.02.064.

Fan, S., Zhong, Q., Fauhl-Hassek, C., Pfister, M. K.-H., Horn, B., & Huang, Z. (2018). Classification of Chinese wine varieties using 1H NMR spectroscopy combined with multivariate statistical analysis. *Food Control, 88*, 113–122. https://doi.org/10.1016/j.foodcont.2017.11.002.

Fernández de Simón, B., Cadahía, E., Sanz, M., Poveda, P., Perez-Magariño, S., Ortega-Heras, M., et al. (2008). Volatile compounds and sensorial characterization of wines from four Spanish denominations of origin, aged in Spanish Rebollo (Quercus pyrenaica Willd.) oak wood barrels. *Journal of Agricultural and Food Chemistry, 56*(19), 9046–9055. https://doi.org/10.1021/jf8014602.

Ferreira, V., & Lopez, R. (2019). The actual and potential aroma of winemaking grapes. *Biomolecules, 9*(12), 818. https://doi.org/10.3390/biom9120818.

Ferri, E., Galimberti, A., Casiraghi, M., Airoldi, C., Ciaramelli, C., Palmioli, A., et al. (2015). Towards a universal approach based on omics Technologies for the Quality Control of food. *BioMed Research International, 2015*, 365794. https://doi.org/10.1155/2015/365794.

Forleo, T., Zappi, A., Gottardi, F., & Melucci, D. (2020). Rapid discrimination of Italian prosecco wines by head-space gas-chromatography basing on the volatile profile as a chemometric fingerprint. *European Food Research and Technology, 246*(9), 1805–1816. https://doi.org/10.1007/s00217-020-03534-8.

Fotakis, C., & Zervou, M. (2016). NMR metabolic fingerprinting and chemometrics driven authentication of Greek grape marc spirits. *Food Chemistry*, *196*, 760–768. https://doi.org/10.1016/j.foodchem.2015.10.002.

Fraige, K., Pereira-Filho, E. R., & Carrilho, E. (2014). Fingerprinting of anthocyanins from grapes produced in Brazil using HPLC-DAD-MS and exploratory analysis by principal component analysis. *Food Chemistry*, *145*, 395–403. https://doi.org/10.1016/J.foodchem.2013.08.066.

Geana, E. I., Popescu, R., Costinel, D., Dinca, O. R., Ionete, R. E., Stefanescu, I., et al. (2016). Classification of red wines using suitable markers coupled with multivariate statistic analysis. *Food Chemistry*, *192*, 1015–1024. https://doi.org/10.1016/j.foodchem.2015.07.112.

Geană, E.-I., Ciucure, C. T., Apetrei, C., & Artem, V. (2019). Application of spectroscopic UV-vis and FT-IR screening techniques coupled with multivariate statistical analysis for red wine authentication: Varietal and vintage year discrimination. *Molecules (Basel, Switzerland)*, *24*(22). https://doi.org/10.3390/molecules24224166.

Gil, M., Reynes, C., Cazals, G., Enjalbal, C., Sabatier, R., & Saucier, C. (2020). Discrimination of rosé wines using shotgun metabolomics with a genetic algorithm and MS ion intensity ratios. *Scientific Reports*, *10*(1), 1170. https://doi.org/10.1038/s41598-020-58193-2.

Giraudeau, P., Silvestre, V., & Akoka, S. (2015). Optimizing water suppression for quantitative NMR-based metabolomics: A tutorial review. *Metabolomics: Official Journal of the Metabolomic Society*, *11*(5), 1041–1055. https://doi.org/10.1007/s11306-015-0794-7.

Godelmann, R., Fang, F., Humpfer, E., Schütz, B., Bansbach, M., Schäfer, H., et al. (2013). Targeted and nontargeted wine analysis by (1)h NMR spectroscopy combined with multivariate statistical analysis. Differentiation of important parameters: Grape variety, geographical origin, year of vintage. *Journal of Agricultural and Food Chemistry*, *61*(23), 5610–5619. https://doi.org/10.1021/jf400800d.

González-Barreiro, C., Rial-Otero, R., Cancho-Grande, B., & Simal-Gándara, J. (2015). Wine aroma compounds in grapes: A critical review. *Critical Reviews in Food Science and Nutrition*, *55*(2), 202–218. https://doi.org/10.1080/10408398.2011.650336.

Gougeon, L., da Costa, G., Guyon, F., & Richard, T. (2019). 1H NMR metabolomics applied to Bordeaux red wines. *Food Chemistry*, *301*, 125257. https://doi.org/10.1016/j.foodchem.2019.125257.

Gougeon, L., da Costa, G., Le Mao, I., Ma, W., Teissedre, P.-L., Guyon, F., et al. (2018). Wine analysis and authenticity using 1H-NMR metabolomics data: Application to Chinese wines. *Food Analytical Methods*, *11*(12), 3425–3434. https://doi.org/10.1007/s12161-018-1310-2.

Gougeon, L., da Costa, G., Richard, T., & Guyon, F. (2019). Wine authenticity by quantitative 1H NMR versus multitechnique analysis: A case study. *Food Analytical Methods*, *12*(4), 956–965. https://doi.org/10.1007/s12161-018-01425-z.

Green, J. A., Parr, W. V., Breitmeyer, J., Valentin, D., & Sherlock, R. (2011). Sensory and chemical characterisation of sauvignon blanc wine: Influence of source of origin. *Food Research International*, *44*(9), 2788–2797. https://doi.org/10.1016/j.foodres.2011.06.005.

Guerrero, R. F., Valls-Fonayet, J., Richard, T., & Cantos-Villar, E. (2020). A rapid quantification of stilbene content in wine by ultra-high pressure liquid chromatography – Mass spectrometry. *Food Control*, *108*, 106821. https://doi.org/10.1016/j.foodcont.2019.106821.

Haggerty, J., Bowyer, P. K., Jiranek, V., & Taylor, D. K. (2016). Optimisation and validation of a high-throughput semi-quantitative solid-phase microextraction method for analysis of fermentation aroma compounds in metabolomic screening studies of wines. *Australian Journal of Grape and Wine Research*, *22*(1), 3–10. https://doi.org/10.1111/ajgw.12167.

Herbert-Pucheta, J. E., Mejía-Fonseca, I., Zepeda-Vallejo, L. G., Milmo-Brittingham, D., & Maya, G. P. (2019). The "wine-T 1" NMR experiment for novel wine-metabolome fingerprinting with nuclear-spin relaxation. *BIO Web of Conferences*, *12*, 2029. https://doi.org/10.1051/bioconf/20191202029.

Herbert-Pucheta, J. E., Pino-Villar, C., Rodríguez-González, F., Padilla-Maya, G., Milmo-Brittingham, D., & Zepeda-Vallejo, L. G. (2019). "One-shot" analysis of wine parameters in non-saccharomyces large-scale alcohol reduction processes with one- and two-dimensional nuclear magnetic resonance. *BIO Web of Conferences, 15*, 2016. https://doi.org/10.1051/bioconf/20191502016.

Hernández-Orte, P., Franco, E., Huerta, C. G., García, J. M., Cabellos, M., Suberviola, J., et al. (2014). Criteria to discriminate between wines aged in oak barrels and macerated with oak fragments. *Food Research International, 57*, 234–241. https://doi.org/10.1016/j.foodres.2014.01.044.

Holmberg. (2010). Wine fraud. *International Journal of Wine Research, 2*, 105. https://doi.org/10.2147/IJWR.S14102.

Hu, B., Cao, Y., Zhu, J., Xu, W., & Wu, W. (2019). Analysis of metabolites in chardonnay dry white wine with various inactive yeasts by 1H NMR spectroscopy combined with pattern recognition analysis. *AMB Express, 9*(1), 140. https://doi.org/10.1186/s13568-019-0861-y.

Hu, B., Yue, Y., Zhu, Y., Wen, W., Zhang, F., & Hardie, J. W. (2015). Proton nuclear magnetic resonance-spectroscopic discrimination of wines reflects genetic homology of several different grape (V. Vinifera L.) cultivars. *PLoS One, 10*(12), e0142840. https://doi.org/10.1371/journal.pone.0142840.

International Organisation of Vine and Wine. (2019). *2019 statistical report of world viticulture*.

Isaac-Lam, M. F., & Zuiderweg, E. R. (2016). Determination of alcohol content in alcoholic beverages using 45 MHz benchtop NMR spectrometer. *International Journal of Spectroscopy, 2016*, 2526946. https://doi.org/10.1155/2016/2526946.

J. Baxter, M., M. Crews, H., John Dennis, M., Goodall, I., & Anderson, D. (1997). The determination of the authenticity of wine from its trace element composition. Food Chemistry, 60(3), 443–450. https://doi.org/10.1016/S0308-8146(96)00365-2.

Jiménez-Carvelo, A. M., González-Casado, A., Bagur-González, M. G., & Cuadros-Rodríguez, L. (2019). Alternative data mining/machine learning methods for the analytical evaluation of food quality and authenticity—A review. *Food Research International (Ottawa, Ont.), 122*, 25–39. https://doi.org/10.1016/j.foodres.2019.03.063.

Kamiloglu, S. (2019). Authenticity and traceability in beverages. *Food Chemistry, 277*, 12–24. https://doi.org/10.1016/j.foodchem.2018.10.091.

Kemsley, E. K., Defernez, M., & Marini, F. (2019). Multivariate statistics: Considerations and confidences in food authenticity problems. *Food Control, 105*, 102–112. https://doi.org/10.1016/j.foodcont.2019.05.021.

Kumšta, M., Pavloušek, P., & Kárník, P. (2014). Use of anthocyanin profiles when differentiating individual varietal wines and terroirs. *Food Technology and Biotechnology, 52*(4), 383–390. https://doi.org/10.17113/ftb.52.04.14.3650.

Kustos, M., Gambetta, J. M., Jeffery, D. W., Heymann, H., Goodman, S., & Bastian, S. E. P. (2020). A matter of place: Sensory and chemical characterisation of fine Australian chardonnay and shiraz wines of provenance. *Food Research International (Ottawa, Ont.), 130*, 108903. https://doi.org/10.1016/j.foodres.2019.108903.

Laghi, L., Versari, A., Marcolini, E., & Parpinello, G. P. (2014). Metabonomic investigation by 1H-NMR to discriminate between red wines from organic and biodynamic grapes. *Food and Nutrition Sciences, 05*(01), 52–59. https://doi.org/10.4236/fns.2014.51007.

Lambert, M., Meudec, E., Verbaere, A., Mazerolles, G., Wirth, J., Masson, G., et al. (2015). A high-throughput UHPLC-QqQ-MS method for polyphenol profiling in Rosé wines. *Molecules (Basel, Switzerland), 20*(5), 7890–7914. https://doi.org/10.3390/molecules20057890.

Lecat, B., Brouard, J., & Chapuis, C. (2017). Fraud and counterfeit wines in France: An overview and perspectives. *British Food Journal, 119*(1), 84–104. https://doi.org/10.1108/BFJ-09-2016-0398.

Liu, Z., Ezernieks, V., Reddy, P., Elkins, A., Krill, C., Murphy, K., et al. (2020). A simple GC-MS/MS method for determination of smoke taint-related volatile phenols in grapes. *Metabolites*, *10*(7). https://doi.org/10.3390/metabo10070294.

López-Rituerto, E., Savorani, F., Avenoza, A., Busto, J. H., Peregrina, J. M., & Engelsen, S. B. (2012). Investigations of La Rioja terroir for wine production using 1H NMR metabolomics. *Journal of Agricultural and Food Chemistry*, *60*(13), 3452–3461. https://doi.org/10.1021/jf204361d.

Lukić, I., Radeka, S., Budić-Leto, I., Bubola, M., & Vrhovsek, U. (2019). Targeted UPLC-QqQ-MS/MS profiling of phenolic compounds for differentiation of monovarietal wines and corroboration of particular varietal typicity concepts. *Food Chemistry*, *300*, 125251. https://doi.org/10.1016/j.foodchem.2019.125251.

Ma, Y., Tanaka, N., Vaniya, A., Kind, T., & Fiehn, O. (2016). Ultrafast polyphenol metabolomics of red wines using MicroLC-MS/MS. *Journal of Agricultural and Food Chemistry*, *64*(2), 505–512. https://doi.org/10.1021/acs.jafc.5b04890.

Magdas, D. A., Guyon, F., Feher, I., & Pinzaru, S. C. (2018). Wine discrimination based on chemometric analysis of untargeted markers using FT-Raman spectroscopy. *Food Control*, *85*, 385–391. https://doi.org/10.1016/j.foodcont.2017.10.024.

Magdas, D. A., Pirnau, A., Feher, I., Guyon, F., & Cozar, B. I. (2019). Alternative approach of applying 1H NMR in conjunction with chemometrics for wine classification. *LWT*, *109*, 422–428. https://doi.org/10.1016/j.lwt.2019.04.054.

Makris, D. P., Kallithraka, S., & Mamalos, A. (2006). Differentiation of young red wines based on cultivar and geographical origin with application of chemometrics of principal polyphenolic constituents. *Talanta*, *70*(5), 1143–1152. https://doi.org/10.1016/j.talanta.2006.03.024.

Mannu, A., Karabagias, I. K., Di Pietro, M. E., Baldino, S., Karabagias, V. K., & Badeka, A. V. (2020). 13C NMR-based chemical fingerprint for the varietal and geographical discrimination of wines. *Foods (Basel, Switzerland)*, *9*(8). https://doi.org/10.3390/foods9081040.

Mascellani, A., Hoca, G., Babisz, M., Krska, P., Kloucek, P., & Havlik, J. (2020). 1H NMR chemometric models for classification of Czech wine type and variety. *Food Chemistry*, *339*, 127852. https://doi.org/10.1016/j.foodchem.2020.127852.

Mayr, C., de Rosso, M., Dalla Vedova, A., & Flamini, R. (2018). High-resolution mass spectrometry identification of secondary metabolites in four red grape varieties potentially useful as traceability markers of wines. *Beverages*, *4*(4), 74. https://doi.org/10.3390/beverages4040074.

Mazzei, P., Spaccini, R., Francesca, N., Moschetti, G., & Piccolo, A. (2013). Metabolomic by 1H NMR spectroscopy differentiates "Fiano di Avellino" white wines obtained with different yeast strains. *Journal of Agricultural and Food Chemistry*, *61*(45), 10816–10822. https://doi.org/10.1021/jf403567x.

Medina, S., Pereira, J. A., Silva, P., Perestrelo, R., & Câmara, J. S. (2019). Food fingerprints - a valuable tool to monitor food authenticity and safety. *Food Chemistry*, *278*, 144–162. https://doi.org/10.1016/J.foodchem.2018.11.046.

Medina, S., Perestrelo, R., Silva, P., Pereira, J. A. M., & Câmara, J. S. (2019). Current trends and recent advances on food authenticity technologies and chemometric approaches. *Trends in Food Science & Technology*, *85*, 163–176. https://doi.org/10.1016/J.tifs.2019.01.017.

Monakhova, Y. B., Schäfer, H., Humpfer, E., Spraul, M., Kuballa, T., & Lachenmeier, D. W. (2011). Application of automated eightfold suppression of water and ethanol signals in 1H NMR to provide sensitivity for analyzing alcoholic beverages. *Magnetic Resonance in Chemistry: MRC*, *49*(11), 734–739. https://doi.org/10.1002/mrc.2823.

Moyano, L., Serratosa, M. P., Marquez, A., & Zea, L. (2019). Optimization and validation of a DHS-TD-GC-MS method to wineomics studies. *Talanta*, *192*, 301–307. https://doi.org/10.1016/j.talanta.2018.09.032.

Muhammad, A., & Countryman, A. M. (2019). In vino 'no' Veritas: Impacts of fraud on wine imports in China. *Australian Journal of Agricultural and Resource Economics*, *63*(4), 742–758. https://doi.org/10.1111/1467-8489.12333.

de Pascali, S. A., Coletta, A., Del Coco, L., Basile, T., Gambacorta, G., & Fanizzi, F. P. (2014). Viticultural practice and winemaking effects on metabolic profile of Negroamaro. *Food Chemistry*, *161*, 112–119. https://doi.org/10.1016/j.foodchem.2014.03.128.

Pasvaska, K., Tzachristas, A., & Proestos, C. (2019). Quality tools in wine traceability and authenticity. In *Quality control in the beverage industry* (pp. 289–334). Elsevier.

Perutka, Z., Voženílek, V., & Šebela, M. (2020). Wine contaminations and frauds from the bioanalytical and biochemical points of view. *Reference Module in Food Science*. Elsevier https://doi.org/10.1016/B978-0-08-100596-5.22835-7.

Picard, M., Franc, C., de Revel, G., & Marchand, S. (2018). Dual solid-phase and stir bar sorptive extraction combined with gas chromatography-mass spectrometry analysis provides a suitable tool for assaying limonene-derived mint aroma compounds in red wine. *Analytica Chimica Acta*, *1001*, 168–178. https://doi.org/10.1016/j.aca.2017.11.074.

Picone, G., Trimigno, A., Tessarin, P., Donnini, S., Rombolà, A. D., & Capozzi, F. (2016). (1)H NMR foodomics reveals that the biodynamic and the organic cultivation managements produce different grape berries (Vitis vinifera L. cv. Sangiovese). *Food Chemistry*, *213*, 187–195. https://doi.org/10.1016/j.foodchem.2016.06.077.

Pinto, J., Oliveira, A. S., Azevedo, J., Freitas, V. de, Lopes, P., Roseira, I., … Guedes de Pinho, P. (2018). Assessment of oxidation compounds in oaked chardonnay wines: A GC-MS and 1H NMR metabolomics approach. Food Chemistry, 257, 120–127. https://doi.org/10.1016/j.foodchem.2018.02.156.

Pinu, F. (2018). Grape and wine metabolomics to develop new insights using untargeted and targeted approaches. *Fermentation*, *4*(4), 92. https://doi.org/10.3390/fermentation4040092.

Pisano, P. L., Silva, M. F., & Olivieri, A. C. (2015). Anthocyanins as markers for the classification of Argentinean wines according to botanical and geographical origin. Chemometric modeling of liquid chromatography-mass spectrometry data. *Food Chemistry*, *175*, 174–180. https://doi.org/10.1016/j.foodchem.2014.11.124.

Poitou, X., Thibon, C., & Darriet, P. (2017). 1,8-cineole in French red wines: Evidence for a contribution related to its various origins. *Journal of Agricultural and Food Chemistry*, *65*(2), 383–393. https://doi.org/10.1021/acs.jafc.6b03042.

Pons, A., Lavigne, V., Eric, F., Darriet, P., & Dubourdieu, D. (2008). Identification of volatile compounds responsible for prune aroma in prematurely aged red wines. *Journal of Agricultural and Food Chemistry*, *56*(13), 5285–5290. https://doi.org/10.1021/jf073513z.

Pons, A., Lavigne, V., Landais, Y., Darriet, P., & Dubourdieu, D. (2008). Distribution and organoleptic impact of sotolon enantiomers in dry white wines. *Journal of Agricultural and Food Chemistry*, *56*(5), 1606–1610. https://doi.org/10.1021/jf072337r.

Pons, A., Lavigne, V., Landais, Y., Darriet, P., & Dubourdieu, D. (2010). Identification of a sotolon pathway in dry white wines. *Journal of Agricultural and Food Chemistry*, *58*(12), 7273–7279. https://doi.org/10.1021/jf100150q.

Ragone, R., Crupi, P., Piccinonna, S., Bergamini, C., Mazzone, F., Fanizzi, F. P., et al. (2015). Classification and chemometric study of southern Italy monovarietal wines based on NMR and HPLC-DAD-MS. *Food Science and Biotechnology*, *24*(3), 817–826. https://doi.org/10.1007/s10068-015-0106-z.

Robinson, A. L., Boss, P. K., Heymann, H., Solomon, P. S., & Trengove, R. D. (2011). Development of a sensitive non-targeted method for characterizing the wine volatile profile using headspace solid-phase microextraction comprehensive two-dimensional gas chromatography time-of-flight mass spectrometry. *Journal of Chromatography. A*, *1218*(3), 504–517. https://doi.org/10.1016/j.chroma.2010.11.008.

Robles, A., Fabjanowicz, M., Chmiel, T., & Płotka-Wasylka, J. (2019). Determination and identification of organic acids in wine samples. Problems and challenges. *TrAC Trends in Analytical Chemistry*, *120*, 115630. https://doi.org/10.1016/j.trac.2019.115630.

Rocchetti, G., Gatti, M., Bavaresco, L., & Lucini, L. (2018). Untargeted metabolomics to investigate the phenolic composition of chardonnay wines from different origins. *Journal of Food Composition and Analysis*, *71*, 87–93. https://doi.org/10.1016/j.jfca.2018.05.010.

Rochfort, S., Ezernieks, V., Bastian, S. E. P., & Downey, M. O. (2010). Sensory attributes of wine influenced by variety and berry shading discriminated by NMR metabolomics. *Food Chemistry*, *121*(4), 1296–1304. https://doi.org/10.1016/j.foodchem.2010.01.067.

Romanini, E., Colangelo, D., Lucini, L., & Lambri, M. (2019). Identifying chemical parameters and discriminant phenolic compounds from metabolomics to gain insight into the oxidation status of bottled white wines. *Food Chemistry*, *288*, 78–85. https://doi.org/10.1016/j.foodchem.2019.02.073.

Roullier-Gall, C., Witting, M., Tziotis, D., Ruf, A., Gougeon, R. D., & Schmitt-Kopplin, P. (2015). Integrating analytical resolutions in non-targeted wine metabolomics. *Tetrahedron*, *71*(20), 2983–2990. https://doi.org/10.1016/j.tet.2015.02.054.

Roullier-Gall, C., Hemmler, D., Gonsior, M., Li, Y., Nikolantonaki, M., Aron, A., et al. (2017). Sulfites and the wine metabolome. *Food Chemistry*, *237*, 106–113. https://doi.org/10.1016/j.foodchem.2017.05.039.

Roullier-Gall, C., Witting, M., Gougeon, R. D., & Schmitt-Kopplin, P. (2014). High precision mass measurements for wine metabolomics. *Frontiers in Chemistry*, *2*, 102. https://doi.org/10.3389/fchem.2014.00102.

Rubert, J., Zachariasova, M., & Hajslova, J. (2015). Advances in high-resolution mass spectrometry based on metabolomics studies for food—A review. *Food Additives & Contaminants. Part A, Chemistry, Analysis, Control, Exposure & Risk Assessment*, *32*(10), 1685–1708. https://doi.org/10.1080/19440049.2015.1084539.

Sáenz, C., Cedrón, T., & Cabredo, S. (2010). Classification of wines from five Spanish origin denominations by aromatic compound analysis. *Journal of AOAC International*, *93*(6), 1916–1922.

Schueuermann, C., Bremer, P., & Silcock, P. (2017). PTR-MS volatile profiling of pinot noir wines for the investigation of differences based on vineyard site. *Journal of Mass Spectrometry: JMS*, *52*(9), 625–631. https://doi.org/10.1002/jms.3956.

Schueuermann, C., Khakimov, B., Engelsen, S. B., Bremer, P., & Silcock, P. (2016). GC-MS metabolite profiling of extreme southern pinot noir wines: Effects of vintage, barrel maturation, and fermentation dominate over vineyard site and clone selection. *Journal of Agricultural and Food Chemistry*, *64*(11), 2342–2351. https://doi.org/10.1021/acs.jafc.5b05861.

Schwarzinger, S. (2018). Large-scale screening of food products for quality and authenticity. In G. A. Webb (Ed.), *Modern magnetic resonance* (pp. 1795–1817). Cham: Springer International Publishing. https://doi.org/10.1007/978-3-319-28388-3_91.

Sen, I., & Tokatli, F. (2014). Authenticity of wines made with economically important grape varieties grown in Anatolia by their phenolic profiles. *Food Control*, *46*, 446–454. https://doi.org/10.1016/j.foodcont.2014.06.015.

Sobolev, A. P., Circi, S., & Mannina, L. (2016). Advances in nuclear magnetic resonance spectroscopy for food authenticity testing. In *Advances in food authenticity testing* (pp. 147–170). Elsevier. https://doi.org/10.1016/B978-0-08-100220-9.00006-0.

Sobolev, A. P., Thomas, F., Donarski, J., Ingallina, C., Circi, S., Cesare Marincola, F., et al. (2019). Use of NMR applications to tackle future food fraud issues. *Trends in Food Science & Technology*, *91*, 347–353. https://doi.org/10.1016/j.tifs.2019.07.035.

Son, H.-S., Kim, K. M., van den Berg, F., Hwang, G.-S., Park, W.-M., Lee, C.-H., et al. (2008). 1H nuclear magnetic resonance-based metabolomic characterization of wines by grape varieties and production areas. *Journal of Agricultural and Food Chemistry*, *56*(17), 8007–8016. https://doi.org/10.1021/jf801424u.

Springer, A. E., Riedl, J., Esslinger, S., Roth, T., Glomb, M. A., & Fauhl-Hassek, C. (2014). Validated modeling for German white wine varietal authentication based on headspace solid-phase microextraction online coupled with gas chromatography mass spectrometry fingerprinting. *Journal of Agricultural and Food Chemistry, 62*(28), 6844–6851. https://doi.org/10.1021/jf502042c.

Stój, A., Czernecki, T., Domagała, D., & Targoński, Z. (2017). Application of volatile compounds analysis for distinguishing between red wines from Poland and from other European countries. *South African Journal of Enology & Viticulture, 38*(2). https://doi.org/10.21548/38-2-2079.

Stój, A., Czernecki, T., Domagała, D., & Targoński, Z. (2017b). Comparative characterization of volatile profiles of French, Italian, Spanish, and polish red wines using headspace solid-phase microextraction/gas chromatography-mass spectrometry. *International Journal of Food Properties, 20*(sup1), S830–S845. https://doi.org/10.1080/10942912.2017.1315590.

Stój, A., Kapusta, I., & Domagała, D. (2020). Classification of red wines produced from Zweigelt and rondo grape varieties based on the analysis of phenolic compounds by UPLC-PDA-MS/MS. *Molecules (Basel, Switzerland), 25*(6), 1342. https://doi.org/10.3390/molecules25061342.

Stupak, M., Goodall, I., Tomaniova, M., Pulkrabova, J., & Hajslova, J. (2018). A novel approach to assess the quality and authenticity of scotch whisky based on gas chromatography coupled to high resolution mass spectrometry. *Analytica Chimica Acta, 1042*, 60–70. https://doi.org/10.1016/j.aca.2018.09.017.

Thibon, C., Pons, A., Mouakka, N., Redon, P., Méreau, R., & Darriet, P. (2015). Comparison of electron and chemical ionization modes for the quantification of thiols and oxidative compounds in white wines by gas chromatography-tandem mass spectrometry. *Journal of Chromatography. A, 1415*, 123–133. https://doi.org/10.1016/j.chroma.2015.08.027.

Tominaga, T., Des Peyrot Gachons, C., & Dubourdieu, D. (1998). A new type of flavor precursors in Vitisvinifera L. cv. Sauvignon blanc: S-cysteine conjugates. *Journal of Agricultural and Food Chemistry, 46*(12), 5215–5219. https://doi.org/10.1021/jf980481u.

Tufariello, M., Chiriatti, M. A., Grieco, F., Perrotta, C., Capone, S., Rampino, P., et al. (2014). Influence of autochthonous Saccharomyces cerevisiae strains on volatile profile of Negroamaro wines. *LWT - Food Science and Technology, 58*(1), 35–48. https://doi.org/10.1016/J.lwt.2014.03.016.

Úbeda, C., del Barrio-Galán, R., Peña-Neira, Á., Medel-Maraboli, M., & Durán-Guerrero, E. (2017). Location effects on the aromatic composition of Monovarietal cv. Carignan wines. *American Journal of Enology and Viticulture, 68*(3), 390–399. https://doi.org/10.5344/ajev.2017.16086.

Uttl, L., Hurkova, K., Kocourek, V., Pulkrabova, J., Tomaniova, M., & Hajslova, J. (2019). Metabolomics-based authentication of wines according to grape variety. *Czech Journal of Food Sciences, 37*(4), 239–245. https://doi.org/10.17221/82/2019-CJFS.

Vaclavik, L., Lacina, O., Hajslova, J., & Zweigenbaum, J. (2011). The use of high performance liquid chromatography-quadrupole time-of-flight mass spectrometry coupled to advanced data mining and chemometric tools for discrimination and classification of red wines according to their variety. *Analytica Chimica Acta, 685*(1), 45–51. https://doi.org/10.1016/j.aca.2010.11.018.

Valls, J., Agnolet, S., Haas, F., Struffi, I., Ciesa, F., Robatscher, P., et al. (2017). Valorization of Lagrein grape pomace as a source of phenolic compounds: Analysis of the contents of anthocyanins, flavanols and antioxidant activity. *European Food Research and Technology, 243*(12), 2211–2224. https://doi.org/10.1007/s00217-017-2923-1.

Valls, J., Millán, S., Martí, M. P., Borràs, E., & Arola, L. (2009). Advanced separation methods of food anthocyanins, isoflavones and flavanols. *Journal of Chromatography. A, 1216*(43), 7143–7172. https://doi.org/10.1016/j.chroma.2009.07.030.

Versari, A., Laurie, V. F., Ricci, A., Laghi, L., & Parpinello, G. P. (2014). Progress in authentication, typification and traceability of grapes and wines by chemometric approaches. *Food Research International, 60*, 2–18. https://doi.org/10.1016/j.foodres.2014.02.007.

Villano, C., Lisanti, M. T., Gambuti, A., Vecchio, R., Moio, L., Frusciante, L., et al. (2017). Wine varietal authentication based on phenolics, volatiles and DNA markers: State of the art, perspectives and drawbacks. *Food Control, 80*, 1–10. https://doi.org/10.1016/J.foodcont.2017.04.020.

Wadood, S. A., Boli, G., Xiaowen, Z., Hussain, I., & Yimin, W. (2020). Recent development in the application of analytical techniques for the traceability and authenticity of food of plant origin. *Microchemical Journal, 152*, 104295. https://doi.org/10.1016/j.microc.2019.104295.

Weldegergis, B. T., de Villiers, A., & Crouch, A. M. (2011). Chemometric investigation of the volatile content of young south African wines. *Food Chemistry, 128*(4), 1100–1109. https://doi.org/10.1016/J.foodchem.2010.09.100.

Welke, J. E., Manfroi, V., Zanus, M., Lazzarotto, M., & Alcaraz Zini, C. (2013). Differentiation of wines according to grape variety using multivariate analysis of comprehensive two-dimensional gas chromatography with time-of-flight mass spectrometric detection data. *Food Chemistry, 141*(4), 3897–3905. https://doi.org/10.1016/j.foodchem.2013.06.100.

Wu, H., Tian, L., Chen, B., Jin, B., Tian, B., Xie, L., et al. (2019). Verification of imported red wine origin into China using multi isotope and elemental analyses. *Food Chemistry, 301*, 125137. https://doi.org/10.1016/j.foodchem.2019.125137.

Zanuttin, F., Gurian, E., Ignat, I., Fornasaro, S., Calabretti, A., Bigot, G., et al. (2019). Characterization of white wines from North-Eastern Italy with surface-enhanced Raman spectroscopy. *Talanta, 203*, 99–105. https://doi.org/10.1016/j.talanta.2019.05.024.

Zhang, J., Li, L., Gao, N., Wang, D., Gao, Q., & Jiang, S. (2010). Feature extraction and selection from volatile compounds for analytical classification of Chinese red wines from different varieties. *Analytica Chimica Acta, 662*(2), 137–142. https://doi.org/10.1016/j.aca.2009.12.043.

Zhang, X.-K., Li, S.-Y., Zhao, X., Pan, Q.-H., Shi, Y., & Duan, C.-Q. (2020). HPLC-MS/MS-based targeted metabolomic method for profiling of malvidin derivatives in dry red wines. *Food Research International (Ottawa, Ont.), 134*, 109226. https://doi.org/10.1016/j.foodres.2020.109226.

Ziółkowska, A., Wąsowicz, E., & Jeleń, H. H. (2016). Differentiation of wines according to grape variety and geographical origin based on volatiles profiling using SPME-MS and SPME-GC/MS methods. *Food Chemistry, 213*, 714–720. https://doi.org/10.1016/j.foodchem.2016.06.120.